The Plant Geography of Korea

Geobotany 19

Series Editor

M.J.A. WERGER

The titles published in this series are listed at the end of this volume.

The Plant Geography of Korea

with an emphasis on the Alpine Zones

by

WOO-SEOK KONG and DAVID WATTS

Springer Science+Business Media, B.V.

ISBN 978-94-010-4708-1 ISBN 978-94-011-1624-4 (eBook)
DOI 10.1007/978-94-011-1624-4

To Kong, Yon-Kyu and Kim, Ok-Soon

respected parents of Kong, Woo-Seok,
without whose understanding, support, and many kindnesses
to both authors, the research which forms the basis of this book
would not have been completed.

Table of contents

Foreword & acknowledgments ix

1. The environment 1

2. Biogeographic divisions 11

3. Vegetation history 46

4. Arctic-alpine and alpine floristic elements 83

5. Evergreen arboreal arctic-alpine and alpine plants and their environments 105

6. Alpine and subalpine vegetation on Mt. Sorak 141

7. Conclusions 173

Appendices 185

References 201

Index of Subject & Place Names 217

Index of Plant Names 221

Foreword & acknowledgments

This is the first book to consider as a whole the rich, diverse and complex vegetation systems of the Korean peninsula and it is perhaps long overdue. It also differs in its approach to previous analyses of Korean vegetation. Formerly, most studies were based on the use of a few key indicator species to define vegetation communities, a procedure which is not wholly satisfactory in a region in which so many plant species (4164 spp. for the peninsula) are present. In contrast, we have used comprehensive species data from a wide-ranging series of selected sites chosen from throughout the peninsula, north and south, concentrating in particular on evergreen species, including evergreen broadleafed species, conifers and bamboos. Evergreens are frequently highly significant in the biogeography of Korea in that their provenances are in part north Asian, some being at the southern limits of their world range; in part subtropical, some being at the northern limits of their world range; and in part endemic species. The distribution of these species has been analysed by numerical rapid-sorting computer techniques, to produce a new biogeographic map of Korea, which relates closely to details of the Korean environment; and the eight new biogeographic zones which this analysis produced were confirmed by a joint field visit to the southern part of the peninsula, which the authors undertook in 1987. In this book we also consider the history of Korean vegetation, and suggest how present vegetation can be seen to be attuned to present environments through the analysis of physiognomy, spatial and vertical range extents, and detailed field study. We have additionally paid particular attention to arctic-alpine vegetation communities, since all these are potentially endangered through global warming. The result, we trust, is a more complete account and analysis of Korean vegetation and environmental pattern than any presented heretofore, and one which will be of especial use to biogeographers, ecologists and environmental scientists, planners, conservationists, and all those interested in the preservation of Korea's magnificent natural heritage.

In a work of this nature, the authors are clearly indebted to many people for their support, their time and the interchange of ideas. Financial support is acknowledged from the Royal Society, and the Korean Science and Engineering Foundation. In Korea, the following individuals were particularly helpful: Professor Dr Won Hack-Lee of the Department of Geography, Kyung-Hee University, Chong-ju; Dr Ahn Wontae, President of the Korean Economics & Social Institute; Shuck Chang Lee of the Royal Horticultural Society of Korea, and Director, the Cheju Resort Botanic Garden; Drs Lee Ki-Suk and Hwang Manik of the Department of Geography, Seoul National University; Dr Kyun Hyung Han of the Department of Geography Education, Korean National University, Chung Buk; Dr Hae Un Ri, Department of Geography, Dongguk University, Seoul; Lee Seok Soo, Director, National Park Division of the National Physical Environmental Planning Bureau, Ministry of Construction, Seoul; Son Hak Lee, also of the National Park Division (above); Yong Jun Sung, National Park Authority, Sorak-San National Park, Sokcho; and the Directors of the Taecheongbong Shelter, Sorak San.

We are grateful to Dr. M. Šrûtek, Třeboň, for making so many instructive photographs from North Korea available. Furthermore, we thank in particular for their comments and advice, Professor H.C. Li, Director of the Morris Arboretum, University of Pennsylvania, U.S.A.; Professor J.R. Flenley, Department of Geography, Massey University, New Zealand; Dr R.G.H. Bunce, Institute of Terrestrial Ecology, Merlewood, Cumbria, U.K.; Dr S. Ellis and Dr. F. di Giovanni, School of Geography, University of Hull, U.K.; Dr Y.B. Kim, Department of Politics, University of Hull, U.K.; Dr. D. Beard, Director of the Computer Centre, University of Hull, U.K.; the Directors of the Natural History Museum, London, and the Royal Botanic Gardens, Kew and Edinburgh, U.K. and finally, but certainly not least Dr. M.J.A. Werger of the Department of Plant Ecology and Evolutionary Biology, University of Utrecht, The Netherlands.

Woo-Seok Kong & David Watts
Department of Geography, Chonju University, Chonju, Republic of Korea, and School of Geography and Earth Resources, University of Hull, Hull HU6 7RX, U.K.

CHAPTER 1

The environment

Location and geology

The Korean peninsula extends southward from eastern Siberia and Manchuria (northeastern China) to within 112.7 km (70 miles) of Japan at its southern tip. To the north, it is separated from Russia by the Tuman River, and from Manchuria by the Amnok (Yalu) River. The East Sea (Sea of Japan) and the Korean Straits separate Korea from Japan to the east, while on the west the West Sea (Yellow Sea) lies between Korea and China (Fig. 1.1). The peninsula has more than 9656 km (ca. 6000 miles) of coastline, dotted with 3418 islands. Approximately 966 km (ca. 600 miles) long, and 209 km (ca. 130 miles) to 322 km (ca. 200 miles) wide, it has an area of 222 740 km^2 (ca. 86 000 square miles), and lies between the parallels of 33° 06′ and 43° 01′ North and 124° 11′ and 130° 53′ East.

A mountainous country, Korea is composed geologically largely of Pre-Cambrian rocks such as granite, gneiss and other metamorphics. To the west of the drainage divide one finds geosynclinical deposits, ranging up to Triassic in age. Cretaceous and Tertiary sedimentary rocks interspersed with Late Tertiary granitic intrusions characterise the low plains of the southwestern part of the peninsula. Unlike nearby Japan, Korea has no active volcanoes, and only very rarely experiences earthquake shocks, although the islands of Ulreung in the east and Cheju in the south are volcanic in origin, and crater lakes are found capping a few mountain peaks, the most pronounced being those on Mt. Paikdoo in the north and Mt. Halla on Cheju Island (Fig. 1.2).

Korea has in essence been in a geologically stable condition since the end of the Mesozoic, but this quiescence has been broken by small-scale earth movements from time to time during the late Tertiary and Quaternary. Also, it was not greatly affected by ice accumulation during the Pleistocene. However, cirques which are found near to the summits of the northern high mountains, mainly at altitudes of 1700 m to 2000 m a.s.l. (e.g. Mts. Paikdoo [2744 m a.s.l.], Kwanmo [2540 m], Solryong [2442 m], Nampodae [2435 m] and Pukpodae [2289 m]; Kano, 1937; Sasa & Tanaka, 1938; Lautensach, 1945) infer the presence of some small-scale ice accumulation, and severe climatic conditions during glacial cold phases. It is estimated that the snow line at the time of their development was approximately 2000 m a.s.l. Recent discoveries of patterned soil and solifluction phenomena further indicate the existence of strong periglacial activity during such cold phases below this snow line (Kwon, 1987), and it is accordingly likely that all of Korea, except possibly for the southern coastlands and Cheju Island, was very cold during these phases.

Structure and geomorphology

In geomorphological terms, the western slopes of the Korean peninsula today are primarily gentle, with an interspersion of basins and plains and low intervening ridges following one upon the other. The drainage is dominated by westward-flowing rivers, and the main divide is near the East Sea. In contrast, the eastern areas are much more mountainous and steeply-sloped, particularly where they fall away to the East Sea. Within this general pattern, there are three major peninsular mountain range systems, each with a distinctive directional thrust. First is a system which runs the length of the peninsula from north to south close

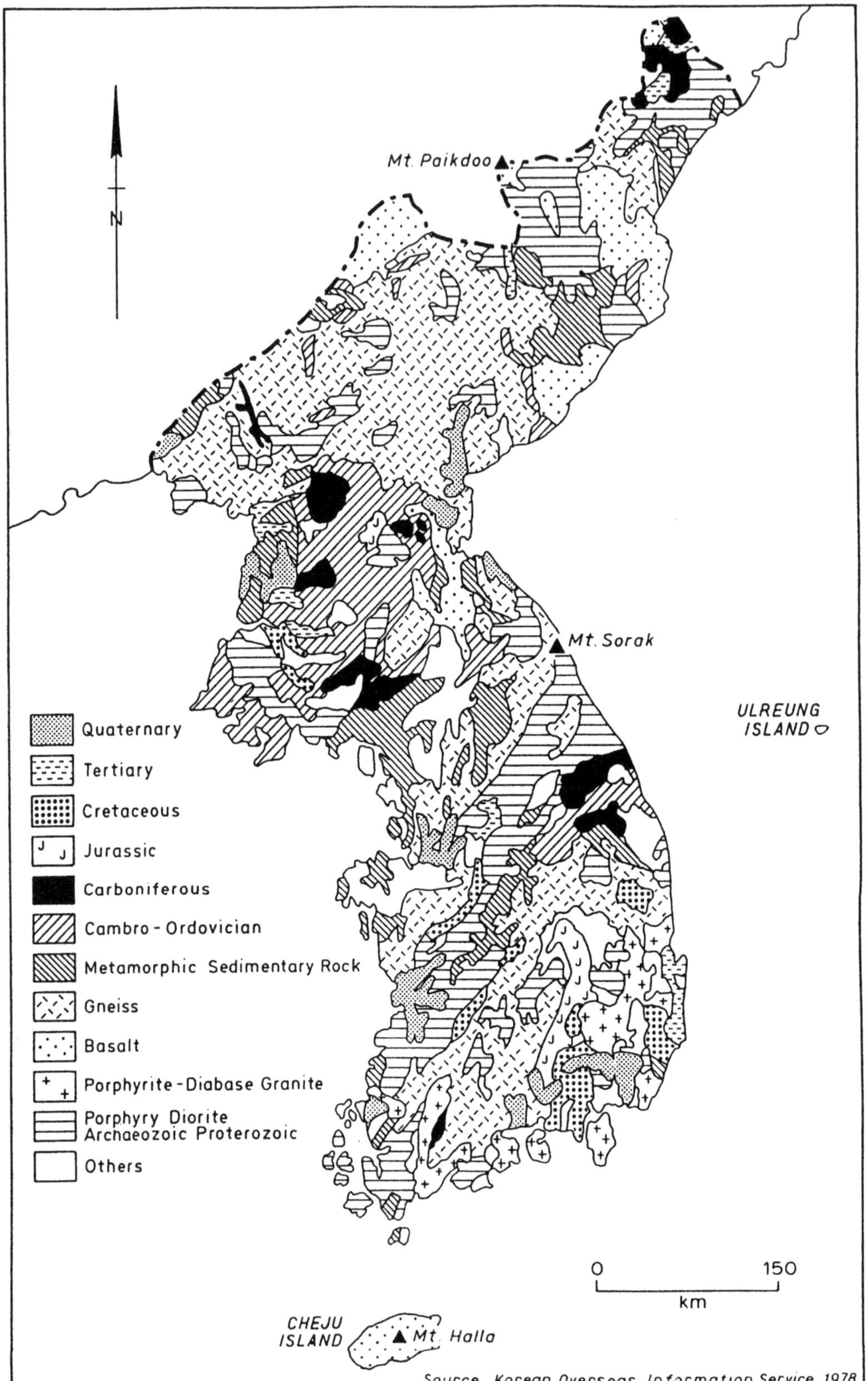

Fig. 1.1. Geology of Korea.

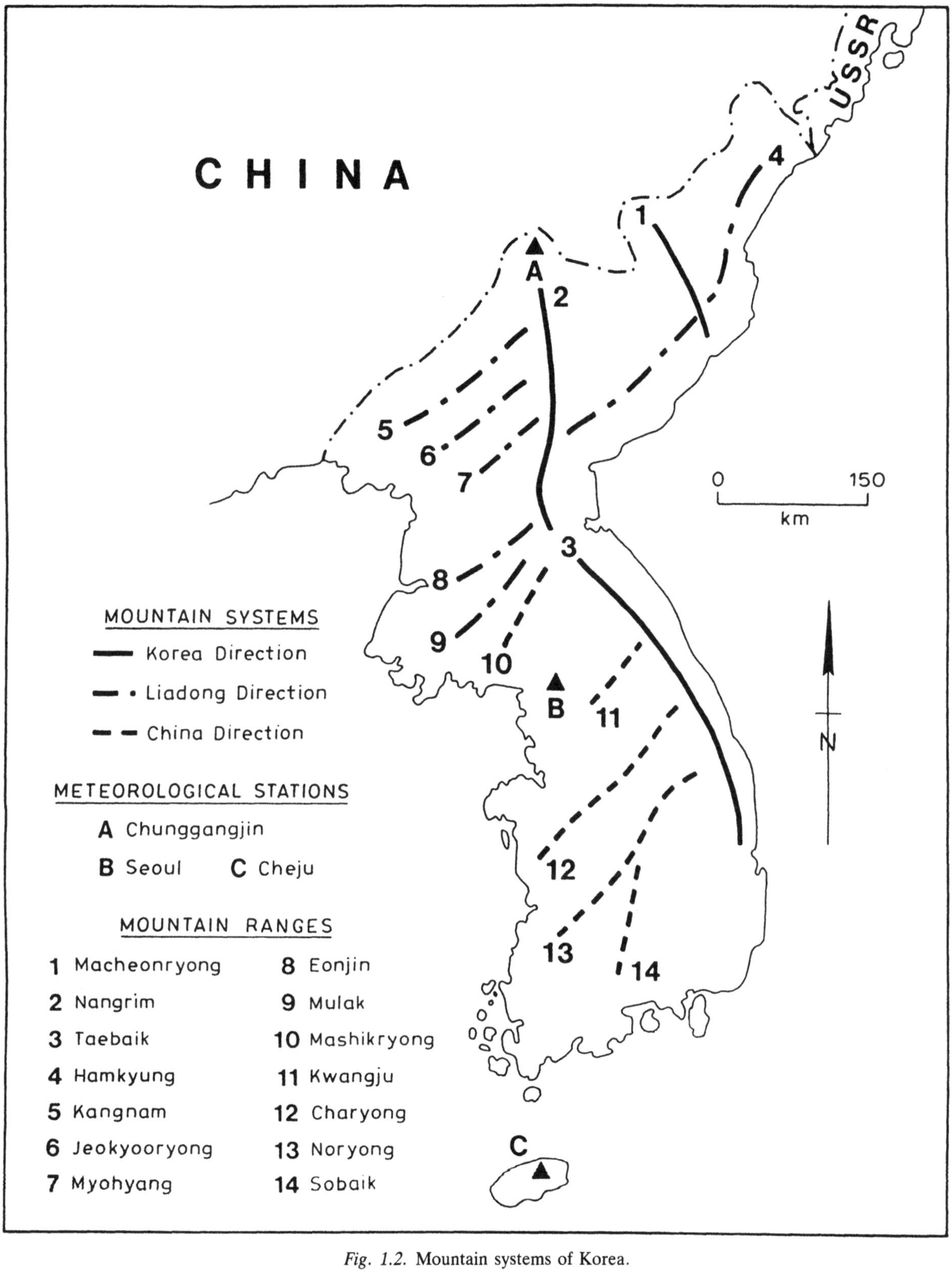

Fig. 1.2. Mountain systems of Korea.

Photo 1.1. The summit of Mt. Paikdoo (2744 m a.s.l.), the highest peak of the Korean Peninsula, which is situated on the border between North Korea and China, Kaema Upland. The old caldera is occupied by Sky Lake (Tianchi), 3–4 km in diameter and a maximum depth of 375 m. Mt. Paikdoo has erupted four times since the 15th century: in 1413, 1597, 1668 and 1702 BCE. The old surfaces in the Mt. Paikdoo area have been completely covered by masses of basalt of 50–200 m thick. The pumice layer covers the top of the basalt (the thickness of the pumice layer varies from 1.5 to 7 m). (Photograph by M. Šrůtek.)

to the eastern coast, roughly forming the drainage divide. This includes the predominantly parallel Nangrim and Taebaik and also Macheonryong mountain ranges. Secondly, the Liadong system, which includes the Hamkyung, Kangnam, Jeokyooryong, Myohyang, Eonjin and Mulak mountain ranges of northern Korea, runs from NE to SW toward the Liadong peninsula of northern China. Thirdly, a 'China' system, runs from NNE to SSW towards southern China, and includes the Mashikryong, Kwangju, Charyong, Sobaik and Noryong mountain ranges of central and southern Korea.

Overall, mountains and hills occupy nearly 70 percent of Korea's territory, and these contain 79 'high' mountains, which reach above 2000 m a.s.l., and are located entirely in northern Korea; and 264 mountains with heights of from 1500 m to 2000 m a.s.l., located again mainly in the north, but also including 23 mountains in the midlands and in the south, among which are Mts. Halla (1950 m), Chiri (1915 m) and Sorak (1708 m). There are also 788 mountains which reach to between 1000 m to 1500 m a.s.l. Overall, 60 percent of the peninsula lies between 200 m to 1000 m a.s.l., and the mean altitude of Korea is 482 m. The main areas of lowland are found in the western and southern parts of Korea, especially close to the outlets of the major rivers.

Adjacent oceanic waters

The Korean peninsula is open to the sea in three directions, namely to the deep East Sea (mean depth of ca. 1700 m), the South Sea (mean depth of ca. 101 m) and to the shallow West Sea (mean depth of 44 m). There are 3418 islands, which are to be found mainly in the south (ca. 2900 islands) off the southern and southwestern coasts, with the other 518 predominantly in the north and midlands. However, only five of these islands are larger than 100 km^2; and only 259 islands *in toto* are larger than 1 km^2.

Two major sea currents exist around Korea. One is diverted from the warm Kuroshio Current

Photo 1.2. Old crater lake at summit of Mt. Halla (1950 m), Cheju Island, Korea. In the grassland-shrub vegetation is *Empetrum nigrum* var. *asiaticum* and *Diapensia lapponica* var. *obovata*. (Photograph by D. Watts.)

of Japan, which originates far away to the south, off the east coast of the Philippines; and the other is a cold current, which splits away from the main Liman Current, coming from the Okhotsk Sea. The warm Tonghan (East Korean) Current, which flows northward along the east coast after passing the Korean Straits between the Korean peninsula and the Japanese Islands, is a quite strong current, with a high temperature gradient compared with adjacent waters. A tributary of this, the warm West Sea Current, flows northward along the west coast, and then back southward along the east coast of China; but this is a weak current due to the shallowness of the sea. The cold Pukhan (North Korean) Current flows southward along the northeastern coast. The strength of the cold Pukhan Current becomes greater in winter, and its influence then reaches further south, pressing the warm Tonghan Current to lower latitudes. The surface temperature of the West Sea shows a low temperature in winter, but conversely a high temperature in summer, as compared with that of the East Sea. The southern portion of the east coast shows a warmer temperature in winter than that of the west, due to the influence of the warm current there (Fig. 1.3), and this influences vegetation patterns to some extent.

Climate

In terms of atmospheric dynamics, the Korean peninsula is located at the meeting point of the polar and the tropical domains, but it also lies under the strong influence of continental climatic conditions. During the winter season, continental high-pressure air masses develop over Siberia,

Photo 1.3. Coniferous forest on Mt. Chiri (1915 m).

from which strong northerlies bring dry and cold air into Korea. The summer monsoon in contrast brings southerly winds, and abundant moisture from the ocean, producing heavy rainfall.

The annual range of mean monthly temperature between January and August for Chunggangjin in the north (41°47′ N, 126°53′ E; see Fig. 1.3) is 44°C, 30°C for Seoul in the midlands (37°34′ N, 126°58′ E), and 20°C for Cheju in the south (33°31′ N, 126°32′ E). These ranges of temperature are much greater in the north and inland than in the south and along the coast due to the greater continental influence in the former areas. The comparison of mean daily minimum temperatures for January in Chunggangjin (−37.6°C) and mean daily maximum temperatures for August in Seoul (34.9°C) shows an extremely large temperature difference and emphasises both the

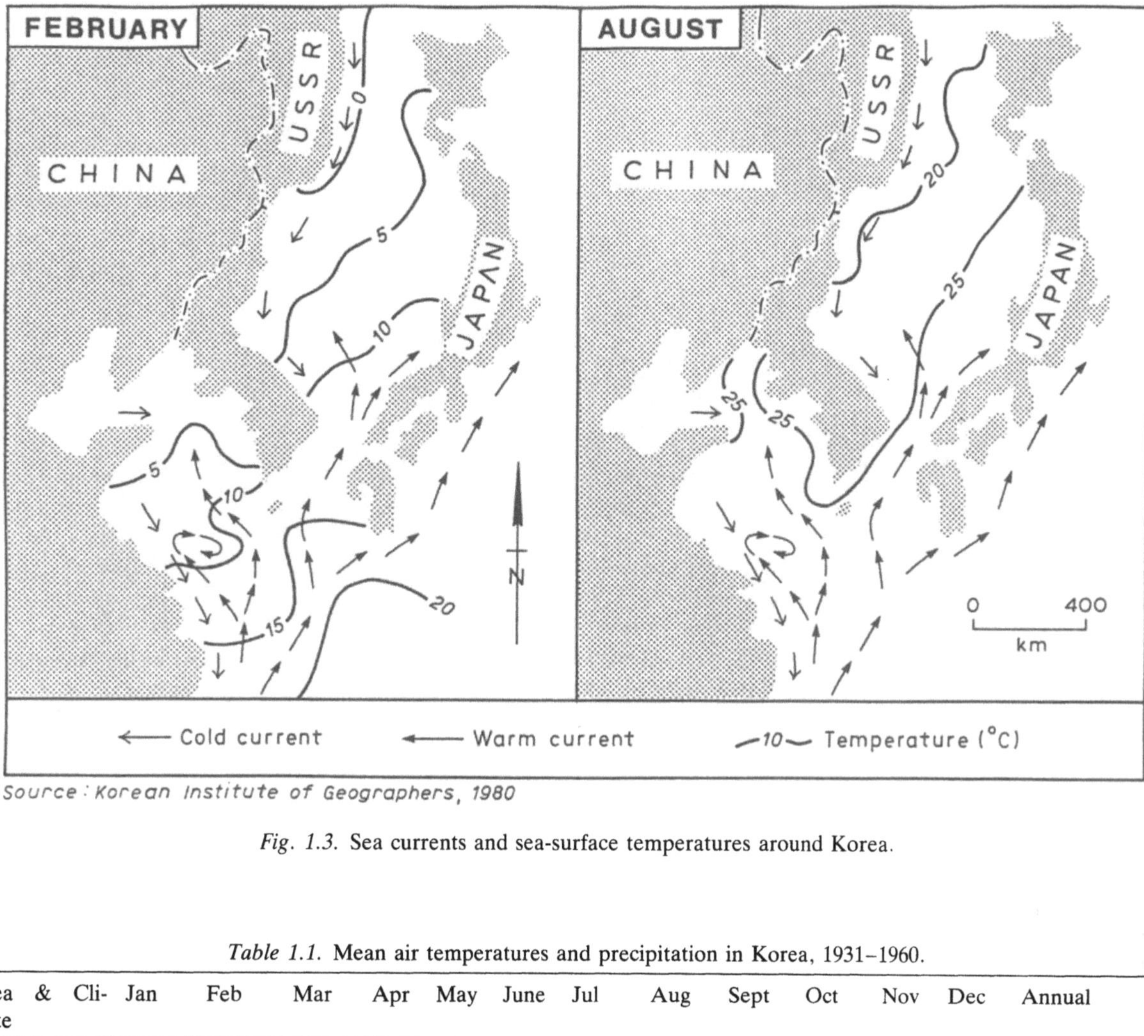

Source: Korean Institute of Geographers, 1980

Fig. 1.3. Sea currents and sea-surface temperatures around Korea.

Table 1.1. Mean air temperatures and precipitation in Korea, 1931–1960.

Area & Climate		Jan	Feb	Mar	Apr	May	June	Jul	Aug	Sept	Oct	Nov	Dec	Annual
Chunggangjin	(°C)	−20.8	−15.3	−4.0	6.5	13.5	18.7	22.7	21.7	14.5	6.9	−3.1	−15.1	3.8
	(mm)	11.0	8.6	24.8	39.9	82.2	117.1	184.3	171.2	81.1	43.5	28.9	19.9	812.5
Seoul	(°C)	−4.9	−1.9	3.6	10.5	16.3	20.8	24.5	25.4	20.3	13.4	6.3	−1.2	11.1
	(mm)	17.1	21.0	55.6	68.1	86.3	169.3	358.0	224.2	142.3	49.2	36.0	32.0	1259.2
Cheju	(°C)	4.8	5.2	8.0	12.3	16.2	20.4	25.1	25.8	21.7	16.8	12.1	7.6	14.7
	(mm)	59.2	75.6	73.1	82.3	88.8	158.1	209.8	226.6	249.5	87.5	69.2	60.2	1439.2

(Source: Central Meteorological Service, Korea, 1968).

distinctive seasonal, and the clear regional differences of temperature within the peninsula (Table 1.1). The number of days with minimum temperatures of less than 0°C (viz. 184 days in Chunggangjin, 114 days in Seoul and 17 days in Cheju) also suggest clear regional differences, and the presence of particularly harsh climatic conditions in northern Korea during the winter months.

Though Korea has a relatively large amount of rainfall (a mean annual figure of ca. 1190 mm for the peninsula as a whole), 45–60 percent of it comes in summer, and only 3–10 percent in winter (Central Meteorological Office, Korea, 1968). The annual rainfall totals increase from 500 to 600 mm in the northwestern highlands to 1800 mm in the southeastern coastal zone of Cheju (Korean Institute of Geography, 1980). Overall, the Korean climate can be described as

Photo 1.4. Pibong Fall in the Kumgangsan Mts., Okryu Valley, SE part of North Korea near the coast of the Japanese Sea. *Pinus densiflora* and *Quercus mongolica* are common in the forests. (Photograph by M. Šrûtek.)

Photo 1.5. Warm-temperate, species-rich broadleaved evergreen forest (Biogeographic Region VII), southwestern Korea. (Photograph by D. Watts.)

Table 1.2. Previous vegetation zonations of Korean forests.

Vegetation zone	Wilson (1919)	Honda (1922)	Zon & Sparhawk (1923)	Uyeki (1933)	Lautensach (1941, 1945)	Hyan (1956)
North	Conifers	Arctic forest	Cold zone	Cold zone Northern temperate zone	Areas above timberline Coniferous forest	Cold zone
Midland	Mixed forests of conifers & hardwood trees	Temperature forest	Temperate zone	Cold temperate zone	Summergreen mixed forest	Temperate zone
South	Deciduous leaf trees	Warm temperate forest	Subtropical zone	Southern temperate zone Warm zone	Evergreen broadleaved forest	Subtropical zone
Vegetation zone	**Wang (1961)**	**Chung & Lee (1965)**	**Park (1966)**	**Yim (1968)**	**Noda (1971)**	**Kim (1973)**
North	Northern high mountain area Northern mountain area	Frigid area	Forage, *Carex Sasamorpha* type	Subarctic zone Northern cool temperate forest zone	Alpine zone Coniferous forest zone	Evergreen needle-leaved forest zone
Midland	Midland area	Temperate zone	*Miscanthus*, *Pteridium*, *Festuca* type	Central cool temperature forest zone	Summergreen tree zone forest zone	Deciduous broadleaved
South	Southern area Southernmost area	Sub-tropical zone	*Zoysica*, *Imperata*, *Phragmites* type	Southern cool temperate forest zone Warm temperate forest zone		Evergreen broadleaved forest zone
Vegetation zone	**Hämet-Ahti *et al.* (1974)**	**Okumura (1974)**	**Lee (1976b)**	**F.A.O. (1976)**	**Hagman *et al.* (1978)**	**Eyre (1984)**
North	Hemiboreal subzone	Arctic forest	Cold temperate plant region	*Picea – Quercus* type	Conifer belt	Mixed boreal & deciduous forest
Midland	Temperate zone	Temperate forest	Temperate plant region	Mixed hardwood type Dry hardwood type	*Quercus – Abies* belt	Deciduous summer forest
South	Hemitemperate subzone Southern meridional zone	Warm temperate forest	Warm temperate plant region	*Pinus densiflora* type	*Pinus – Acer* belt *Camellia* belt	Broadleaved evergreen forest

having hot-humid summers, and cold-dry winters, as well as a large seasonal variation and distinctive regional differences of temperature and precipitation, both in terms of latitude and geographical location. This variety of climatic conditions within the Korean peninsula itself has helped to stimulate the development of the rich vascular flora found therein.

Previous biogeographic divisions in Korea

For Korea, the flora and vegetation remained little known to the West until Miquel published a report in 1865 (Chung, 1986). Floristic and vegetational works since that time can be categorised into three main branches, viz: first, the demarcation of forest vegetation zones; secondly, the

Table 1.3. Previous floristic divisions of Korean vegetation.

	Authors (Year)					
Area	Nakai (1919, 1935)	Good (1947)	K.C.C.N.N.R.* (1971)	Udvardy (1975)	Oh (1977)	Lee & Yim (1978)
North	Northern Region	Sino–Japanese Region of Boreal Kingdom	Northern Area	Manchuria – Japanese mixed forest region	Northern Region	Kapsan Kwanseobuk Region
Midland	Central Region		Central Area	Oriental deciduous forest region	Central & Southern Region	Middle Region
South	Southern Region		Southern Area	Japanese evergreen forest region	Southern	South Region
Islands	Cheju Region				Southern Insular Region	Southern Coastal Region
	Ulreung Region				Cheju Region Ulreung Region	Cheju Region Ulreung Region

* K.C.C.N.N.R.: The Korean Commission for Conservation of Nature and Natural Resources.

delimitation of evergreen broadleaved plant ranges; and, thirdly, subdivisions into floristic divisions. Thus, in 1872 Grisebach listed two *vegetation* zones, north and south, divided by the 40° N parallel; and Schimper adopted a similar approach in 1898, with the 38° N line as a divide. Following on from these two, Brockmann-Jerosch used the 37° N parallel as a divide in 1919; and Miyoshi in 1911 recognised three zones, north, midland and south, separated by the 34° N and 38° N lines of latitude. Also in 1911, Uyeki produced probably the first true regional zonation of Korean forest vegetation based on his own observations, and this was later revised in 1933. This zonation and other divisions of Korean forest vegetation types, normally comprising three zones, along with representative plants from each division, are summarised in Table 1.2.

A study of the northern distributional limit of *evergreen broadleaved trees* in Korea is of particular ecological value, and provides useful information for the production of a vegetation map. Uyeki (1941) has described the northern limit of Korean evergreen broadleaved trees as a line connecting the mid-western islands, the southernmost coast and Ulreung Island. Recently this has been shifted northwards into the inland area of southwestern Korea (Kong & Chung, 1984). At present, the northernmost limit of evergreen broadleaved plants and their typical communities have been assigned as protected areas by the Natural Monuments Commission for conservation and research purposes (Lee, 1969; Korean Ministry of Culture and Information, 1970; Korean Ministry of Home Affairs, 1972).

Very few divisions of floristics in Korea have been compiled, possibly because of the large numbers of species found there. Six previous studies which delineate floristic divisions in an unsophisticated way have been presented to date, as indicated in Table 1.3. In general, five or six floristic regimes have been recognised, namely the Northern, Central, Southern, Southern Insular, Cheju and Ulreung regions. The validity of these divisions, and those relating to vegetation zones, are examined further in a later stage of this book.

Despite the richness and the floristic diversity of East Asia as a whole and Korea in particular, very little is known about the overall biogeographical relationships in either area, or the categorisation of the component plant species into more detailed biogeographic regions. Most categorisations so far have been of a general nature, or limited to the examination of one particular, or a few common groups of plants. In contrast, we attempt here to develop the first broadly-based biogeographical analysis of the whole area of the Korean peninsula, including a full and detailed assessment of biogeographic regions and their environments.

CHAPTER 2

Biogeographic divisions

It is still not clear how many species are present in the rich flora and fauna of the Korean peninsula. Nor is much yet known of either in distributional terms. Concerning the flora, Nakai (1952) recorded 223 families, 968 genera and 4196 'species'* (3176 spp., 841 var. and 174 forms.), of which 1116 species and 11 genera were endemic to the region. Recently Lee (1982a) reported 4164 species from Korea (170 fam., 897 gen., 2899 spp., 7 subspp., 929 var. and 329 forms.). Floristic information from Lee (1982a) and Yim & Baik (1985) indicates that there are only 407 species (61 fam., 172 gen., 339 spp., 46 var. and 22 forms.) of Korean endemic plants in the peninsula; and five endemic genera, namely *Pentactina*, *Echinosophora*, *Abeliophyllum*, *Hanabusaya* and *Megaleranthis*, have been reported in Korea by the Korean Ministry of Home Affairs (1979). In this research, we have largely followed the figures of Lee (1982a,b) and of Yim & Baik (1985), particularly in respect of endemic species. Classification and nomenclature of the Korean flora has been employed from Lee & Ahn (1963).

Data sources

According to Balgooy (1971), three possible types of floristic source material exist for Korea, i.e. literature, herbaria material, and unpublished personal information. In this chapter, information from the literature has mainly been used because of the large number of sites and species listed within it. A full list of the floristic reports available is given in Table 2.1, and these have provided valuable information in respect both of the spatial and vertical distributions of plants throughout the Korean peninsula.

Because of the vast number of Korean plants available for study, this chapter has concentrated, as a pilot study, on the evergreen flora alone, since Korean evergreen plants constitute a wide range of species, ranging from cold-tolerant northern species, to temperate species, to warm-temperature southern species and bamboos; and geographically, they cover almost the entire peninsula. The use of a relatively small data set also makes spatial analysis easier. Accordingly, the distributions of 204 evergreen broadleaved, coniferous and bamboo species from 46 families (see Appendix 1) have been collated and analysed. These data have been collected from 146 sample sites in the peninsula, taken both from the mainland and on islands.

Analysis

First we present a background discussion on the various methods of site and species classification, and their implications for the study of the Korean peninsula, along with the selection of a satisfactory method for the quantitative treatment of the available data; and secondly we make a division of data which involves the selection of sites for biogeographic groupings.

Background

'Vegetation' may be taken to be the total plant cover of a region, area or place. On the other hand, the term 'flora' is used when speaking of individual plants, kinds of plants or lists of plants that occupy an area (Fosberg, 1961). 'Forest' is

* For this paragraph alone, and in the absence of any other suitable terminology, the term 'species' is used collectively, and is taken to include subspecies, varieties and forms.

Table 2.1. List of Korean floristic reports used.

Ann, H.S. & Lee, C.R., 1965
Ann, H.S. *et al.*, 1982
Cha, J.W. *et al.*, 1975
Chejudo, 1985
Chung, T.H., 1933, 1943, 1958, 1965
Chung, T.H. & Lee, W.C., 1961, 1962, 1963, 1965, 1966
Chung, T.H. & Paik, S.E., 1963
Chung, Y.H., 1986
Huh, K.S. & Lee, I.K., 1981
Hwang, K.S., 1979
Kim, C.H. *et al.*, 1982
Kim, D.S., 1972
Kim, I.T. & Lee, I.K., 1980
Kim, J.H. *et al.*, 1984
Kim, T.W. & Kim, S.S., 1978
Kim, W. & Yang, I.S., 1973
Kim, Y.S. *et al.*, 1981
Kong, W.S., 1985
Lee, B.H. *et al.*, 1957
Lee, D.B., 1955, 1957, 1959, 1968
Lee, D.B. & Joo, S.W., 1958
Lee, H.J., 1981
Lee, H.J. & Lee, W.H., 1962
Lee, H.J. *et al.*, 1982
Lee, I.K., 1964, 1980, 1981
Lee, I.K. & Lee,C.B., 1959
Lee, I.K. *et al.*, 1973, 1980, 1982
Lee, I.K. & Kim, I.T., 1980
Lee, J.S., 1962
Lee, K.N. & Chae, J.K., 1981
Lee, T.B., 1982a
Lee, W.C., 1976, 1977
Lee, W.C. & Chon, S.K., 1975
Lee, W.C. & Chung, H.B., 1976
Lee, W.C. *et al.*, 1980
Lee, W.C. & Paik, W.K., 1984
Lee, W.C. & Yim, Y.J., 1975
Lee, W.H., 1975
Lee, Y.N., 1957, 1976
Lee, Y.N. & Oh, Y.J., 1970, 1982
Lim, J.W. *et al.*, 1982
Nakai, T., 1915–1939, 1935, 1952
Oh, S.Y., 1971a,b, 1972, 1973, 1983
Park, B.K., 1971
Park, M.K. & Lee, E.B., 1970
Park, S.E. & Kim, H.E., 1972
Park, S.H., 1970
Yang, I.S., 1958, 1961, 1969
Yang, I.S. & Kim, W., 1970, 1971, 1972
Yoon, H.S., 1982

defined as the natural plant cover in both an untouched state, and also after it has been interfered with by man provided, in the latter case, it has been allowed to redevelop naturally without human interference (Whitmore, 1984).

On the basis of the above three definitions, we aim first, to delineate a biogeographic division of present-day Korean vegetation based on the distribution of evergreen woody forest species in the peninsula; and secondly, to analyse the species composition and structure (especially physiognomy and altitudinal range) of Korean biogeographic regions.

Quantitative biogeography, as is by now well known, is defined as the use of quantitative methods and/or computers in biogeography and can be used to relate distributional patterns to environmental factors (Proctor, 1967; Crovellor, 1981). Such numerical methods have three further advantages over other methods in descriptive phytogeography. First, the obtained results can be verified by others for better accuracy; secondly, the rapid handling of a large quantity of information is made relatively easy; and thirdly, the former tendency of selecting particular patterns of distribution as 'vegetation types' is avoided (Birks, 1976). However, one has to admit that all such numerical analyses until recently have been completed on a relatively minor spatial scale, because of computer-processing limitations.

We have used TWINSPAN as a method of classification in this study because of its simplicity of use in the analysis of the original vegetation data; its ability to categorise and classify with good statistical validity both sample sites and species; its production of a data matrix which shows increasing refinement at each successive level of analysis; its ordering of the sample sequence, which enables one to place similar samples together and results in clear hierarchical diagrams; and not least, the minimum requirements of computer time associated with it. One ordination method also was attempted, this being DECORANA (Detrended Correspondence Analysis) (Hill, 1979b; Hill & Gauch, 1980). The results obtained, however, were not entirely satlsfactory, the ordination sequences being more clumped than anticipated, perhaps because of the relatively large quantity of data outliers recorded. These latter, however, were picked out by the

TWINSPAN programme as well, and since TWINSPAN ultimately provided much better results overall it has formed the predominant basis for the presentation of biogeographic divisions in the Korean peninsula, as indicated herein.

Division of data: the selection of site groupings

The processes regarding the selection of site grouping are based on Birks & Deacon (1973), who have suggested four stages for the compilation of lists of vascular plant taxa. Translated into a Korean context, these are: first, the region studied is confined to the Korean peninsula alone; secondly 204 evergreen broadleaved plants, evergreen coniferous plants and evergreen bamboo species were selected for study; thirdly, these species, and the floristic data relating to them are extracted from the detailed floral reports compiled from 1915 to 1986, as noted in Table 1.1; and fourthly, a tabulation of occurrences of the taxa concerned in 146 selected sites was made. This was followed by error-free coding of the 204 species and 146 sites, on a presence-absence base, onto the computer, prior to its input into the TWINSPAN programme.

This programme appears to work well in a Korean context. In particular the hierarchical division of site groupings which ensued from the TWINSPAN analysis of data (Fig. 2.1), although imbalanced at levels 1 and 2 (this was primarily due to the presence of particular endemics on a limited range of sites, and no attempt has been made to exclude any of these from the data input), thereafter showed a good balance, so that even at level 7 of the division, 16 new groups of sites are distinguishable. In fact, 22 groups of sites are recognisable down to level 7, including one group each at levels 1, 2, 4 and 5; two groups at level 6 and the remainder at level 7. Each of these groups has both biogeographical and ecological relevance.

Although a preliminary analysis of the results obtained has recently been presented by Watts & Kong (1987) and Kong & Watts (1987), a more detailed review is presented here. First, the relative proportion of endemics among Korean evergreen plants is presented in Table 2.2.

The relatively high degree of endemism among evergreen coniferous plants (ECP) is an indicator of the biogeographical and ecological uniqueness of the Korean environment; however, other endemics are also to be found among evergreen broadleaved plants (EBP) and among the evergreen bamboos (EB). The endemic evergreen broadleaved species include *Distylium racemosum* var. *latifolium*, *Buxus koreana*, *B. koreana* var. *elongata*, *B. koreana* var. *insularis*, *Eurya japonica* var. *integra*, *Textoria morbifera*, *Ledum palustre* var. *maximum* and *Ligustrum quihoui* var. *latifolium*. The three endemic bamboos are *Sasa coreana*, *S. quelpaertensis* and *Sasamorpha chiisanensis*; and there are 12 endemic conifers, namely *Abies koreana*, *Picea intercedens*, *P. koraiensis*, *P. pungsanensis*, *P. tonaiensis*, *Thuja koraiensis*, *Juniperus coreana*, *J. coreana* var. *rigida*, *J. sibirica*, *J. utilis* var. *longicarpa*, *J. utilis* var. *modesta* and *Sabina chinensis* var. *horizontalis*. The biogeographical significance of these endemics will be accorded a further discussion later (pp. 22–25). Including all species, endemics and non-endemics, within the data set, eight major biogeographic regions have been established from the analysis, arising from the disposition of site groupings. These are listed in Table 2.3.

We have employed both dot and linear methods first to plot the location of sites in Korea, and then to delineate the biogeographic divisions of present-day vegetation in the peninsula. Three stages have been involved in constructing the biogeographic regions from the hierarchical division of site groups produced by the TWINSPAN programme. The first was the plotting of individual site points on the map on the basis of species similarity; then, the connecting of dots was used to build up a site-distribution area; and finally, areas of each group were agglomerated to form the biogeographic divisions designated in Table 2.4. The spatial patterning of these eight biogeographic regions is presented in Fig. 2.2.

Altogether 88 inland and 58 island sites have been employed in the construction of these biogeographic regions. The site composition of each region is as follows (site numbers from the original data set are in parentheses; for details of site names see Table 2.6, pp. 32–34)

I The Northern Alpine Region (11 sites): Moosan (3), Mt. Paikdoo (5), Mt. Kwanmo (6), Mt. Mantap (7), Keelju (8), Kapsan (10), Mt. Chilbo (11), Mt. Hoochi (13), Mt. Ro (15), Mt. Keumpae (17), Dancheon (18).

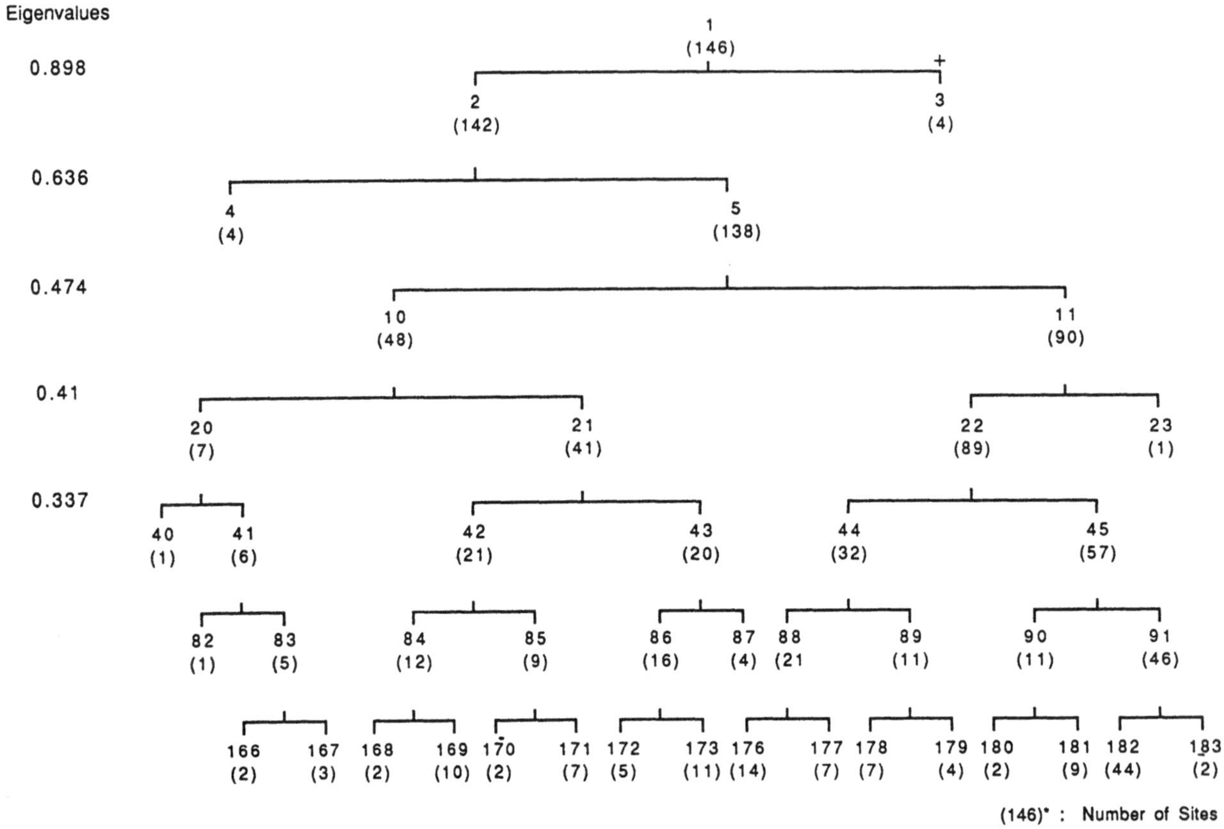

Fig. 2.1. Classification of sample sites in Korea from the TWINSPAN analysis.

Table 2.2. Proportion of evergreen plants and endemism in Korea.

	EBP	EB	ECP	Total
No. of Species	146	15	43	204
No. of Endemics	8	3	12	23
% of Endemics	5.5%	20.0%	27.9%	11.3%

Note: the terms EBP, ECP and EB in this and following Tables and Figures refer respectively to evergreen broad-leaved plant, evergreen coniferous plant, and evergreen bamboo.

II The North-South Subalpine Region (21 sites):
Mt. Jung (1), Mt. Songjin (2), Mt. Chayoo (4), Mt. Seungjuk (14), Mt. Pirae (16), Mt. Peenanduk (19), Mt. Nangrim (20), Mt. Myohyang (21), Mt. Sasoo (22), Mt. Haram (23), Mt. Chuae (24), Mt. Keumkang (25), Mt. Sorak (33), Mt. Myungji (36), Mt. Odae (40), Mt. Kaebang (43), Mt. Taeji (44), Mt. Chiak (49), Mt. Daesung (50), Mt. Taebaik (51), Mt. Dukyoo (70).

Table 2.3. Biogeographic divisions and their component site groups in Korea.

Regions	Names of Divisions	Site Groups*
I	Northern Alpine Region	4, 40, 82, 166, 167
II	North-South Subalpine Region	168, 169, 170, 171
III	Midland Mountain Region	172, 173
IV	Southern Mountain Region	87, 176, 177
V	Mid-Western Insular Region	178, 179
VI	Southern Insular Region	180, 181
VII	West-South-Eastern Insular and Associated Inland Region	23, 182, 183
VIII	North-South Disjunctive Regions	3

* Taken from Fig. 2.1.

III The Midland Mountaln Region (16 sites):
Mt. Koowol (26), Mt. Myulak (28), Mt. Changsoo (29), Mt. Sooyang (32), Mt. Whaak (34), Mt. Kari (37), Mt. Samak (38),

Table 2.4. Proportion of species composition in the biogeographic regions in Korea.

Region	Total No. Species	No. ECP	No. EB	No. EBP
I	22	4 (18.0%)	0	18 (82.0%)
II	28	15 (53.6%)	3 (10.7%)	10 (35.7%)
III	16	12 (75.0%)	0	4 (25.0%)
IV	33	15 (45.5%)	6 (18.1%)	12 (36.4%)
V	15	5 (33.3%)	0	10 (66.7%)
VI	77	8 (10.4%)	4 (5.2%)	65 (84.4%)
VII	41	8 (19.5%)	5 (12.2%)	28 (68.3%)
VIII	5	3 (60.0%)	2 (40.0%)	0

Mt. Daeryong (39), Kwangnung (41), Mt. Sori (42), Mt. Yongmoon (45), Mt. Kwanak (47), Mt. Mantap (48), Mt. Wolak (54), Mt. Ilwol (55), Mt. Sokri (58).

IV The Southern Mountain Region (25 sites): Mt. Kachil (35), Mt. Pukhan (46), Dodamsambong (52), Mt. Sobaik (53), Mt. Chuheul (56), Mt. Kunja (57), Mt. Chuwang (59), Mt. Kaeryong (60), Mt. Hyangro (61), Mt. Booso (62), Mt. Pohyun (63), Mt. Keumo (64), Mt. Doduk (65), Mt. Palgong (66), Mt. Kai (67), Mt. Kaya (68), Mt. Choijung (71), Mt. Biseul (72), Mt. Kaji (74), Mt. Woonmoon (75), Mt. Naejang (77), Mt. Ipam (78), Mt. Keumjung (82), Mt. Moodeung (83), Mt. Chogae (84).

V The Mid-Western Insular Region (11 sites): Songjiho (30), Cape Changsan (31), Is. Kwangwha (91), Is. Yungjong (92), Is. Mooie (93), Is. Dukjeok (94), Is. Seungbong (95), Is. Poong (96), Is. Ong (97), Is. Juk (144), Is. Dok (146).

VI The Southern Insular Region (11 sites): Mt. Daedun (88), Is. Hong (111), Is. Hucksan (112), Is. Chin (115), Is. Cheju (119), Is. Bogil (120), Is. Wan (123), Is. Keomoon (130), Is. Sonjuk (131), Is. Pijin (140), Is. Keojae (141).

VII The West-South-Eastern Insular and Associated Inland Region (47 sites): Kimjae (73), Pyunsan (76), Mt. Paikyang (79), Mt. Chiri (80), Youngkwang (81), Mokpo (85), Mt. Wolchul (86), Mt. Manduk (87), Is. Paikryung (89), Is. Daechung (90), Is. Anmyon (98), Is. Oeyeon (99), Is. Eochung (100), Is. Mal (101), Is. Bangchuk (102), Is. Seonyoo (103), Is. Yami (104), Is. Shinsi (105), Is. Pian (106), Is. Wi (107), Is. Anma (108), Is. Imja (109), Is. Pikeum (110), Is. Docho (113), Is. Wooie (114), Is. Hatae (116), Is. Sohucksan (117), Is. Chuja (118), Is. Nowha (121), Is. Soan (122), Is. Daemo (124), Is. Shinji (125), Is. Kokeum (126), Is. Chungsan (127), Is. Yeoseo (128), Is. Sangil (129), Is. Pyung (132), Is. Kwang (133), Is. Oenaro (134), Is. Dolsan (135), Is. Namhae (136), Is. Kal (137), Is. Yeonwha (138), Is. Kuk (139), Is. Chishim (142), Is. Chinwoo (143), Is. Ulreung (145).

VIII The North-South Disjunctive Region (4 sites): Myungcheon (9), Pungsan (12), Pyungkang (27), Wanu (69).

The major areas of endemism are dominated by particular genera as follows: area 1, *Hanabusaya* and *Echinosophora*; area 2, *Hanabusaya*, *Pentactina*, *Echinosophora* and *Megaleranthis*; area 3, *Hanabusaya* and *Megaleranthis*; area 4, *Abeliophyllum*; and area 5, *Megaleranthis*.

Biogeographic regions

In general, floristic composition and physiognomy are regarded as the two most essential features of vegetation analysis, since they give information regarding the response of a community to particular environmental factors (Küchler 1962; Mueller-Dombois & Ellenberg, 1974). Floristic analyses often give the impression of being more sophisticated in their approach than those relating to physiognomy. However, the latter have several practical advantages over the former, for they are normally less laborious to compile, do not require detailed floristic knowledge and are often capable of being widely applied in related (e.g. environmental) analyses. A comparison of the overall results of floristic and physiognomic classification of communities shows that the similarity between the two is often high (Werger & Springer, 1982). Both types of analysis are employed here, in each biogeographic region of Korea.

Species composition

The relative numerical species composition of the

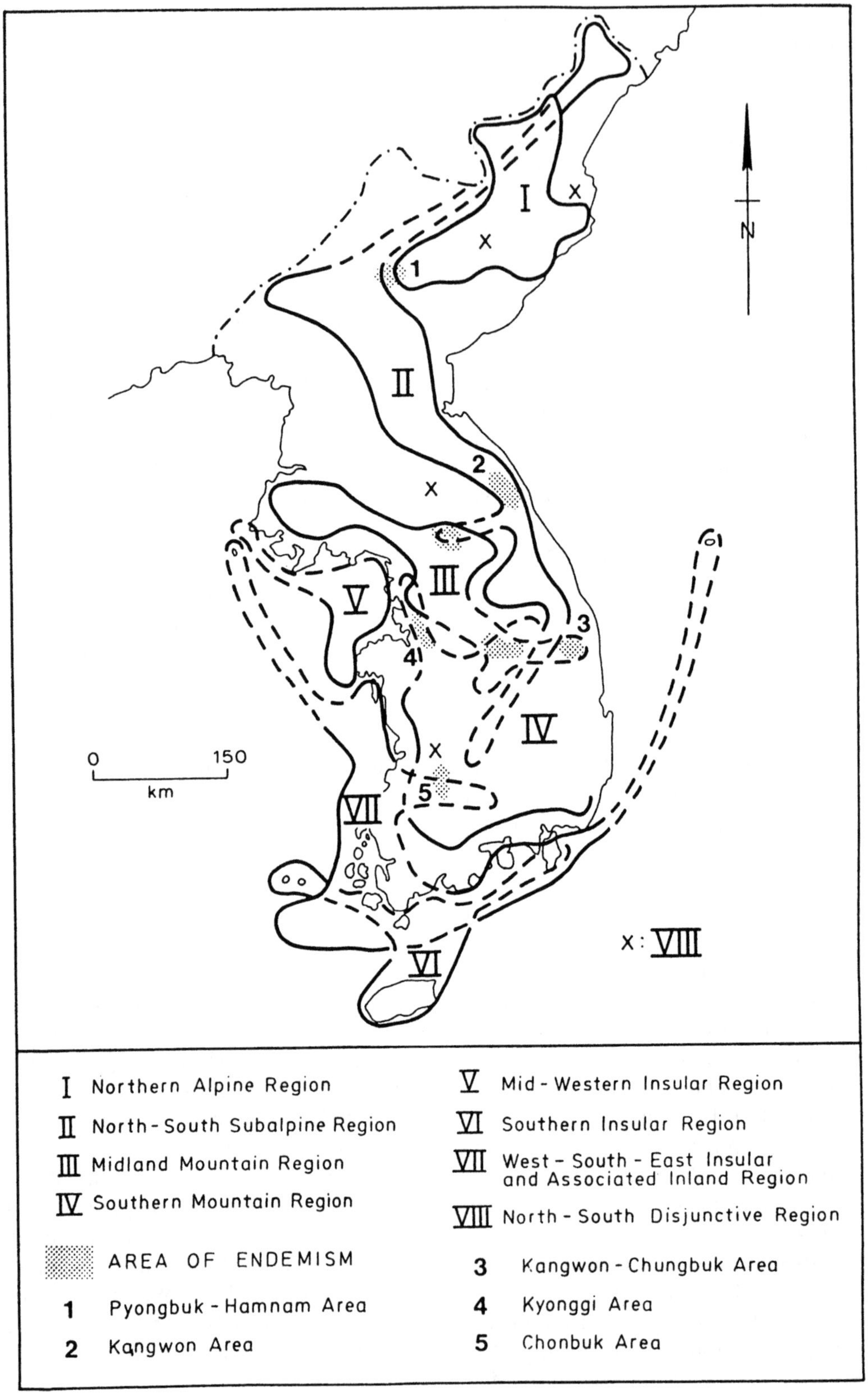

Fig. 2.2. The biogeographic regions of Korea.

Photo 2.1. Landscape near the top of Mt. Paikdoo (2744 m a.s.l.), the highest peak of the Korean Peninsula, which is situated on the border between North Korea and China, Kaema Upland. *Larix olgensis* is the dominant tree at the timber-line as well as the forest limit in the Korean part of Mt. Paikdoo. It grows as trees and as krummholz at the timber-line. The latitudinal range of the timber-line is approximately 1950 to 2100 m a.s.l. (Photograph by M. Šrûtek.)

Korean biogeographic regions designated above is displayed in Fig. 2.3.

In general, total species numbers tend to increase from north to south within the Korean peninsula, most being located in the Southern Insular Region (VI). Evergreen coniferous plants (ECP) are rich in the inland regions, especially the North-South Subalpine Region (II), the Midland Mountain Region (III) and the Southern Mountain Region (IV). In contrast, evergreen broadleaved plants (EBP) are the dominant element in the warmer island regions, such as the Mid-Western Insular Region (V), the Southern Insular Region (VI) and the West-South-Eastern Insular and Associated Inland Region (VII). However, the dominance of evergreen broadleaved plants in the Northern Alpine Region (I) is due to a very different environmental situation, in which large numbers of cold-tolerant plants are present in this category. The sharp decrease in total species numbers in the Midland Mountain Region (III) and the Mid-Western Insular Region (V) is possibly related to the fact that these two regions form transitional zones of vegetation between the better-defined northern and southern floristic elements. Evergreen bamboos (EB) are common in the inland region, such as the North-South Subalpine Region (II), the Southern Mountain Region (IV) and the North–South Disjunctive Region (VIII), but also can be found in island regions.

The total number of species in Table 2.4 is larger than the original 204 species listed in Appendix 1, simply because some species occur in more than one region. The detailed species composition of each region, as produced by the TWINSPAN programme, is as follows (Groups are those designated by the programme).

I. The Northern Alpine Region (22 species):
   Group 4: *Andromeda polifolia*, *A. polifolia* form. *acerosa*, *Chamaedaphne calyculata*, *Ledum palustre* var. *latifolium*, *Oxycoccus quadripetalus*, *Vaccinium vitis-idaea* var. *genuinum*.
   Group 40: *Ledum palustre* var. *yesoense*, *Phyllodoce caerulea*, *Linnaea borealis*, *Taxus cuspidata* var. *latifolia*.
   Group 82: *Empetrum nigrum* var. *japon-*

Photo 2.2. The old surfaces in the Mt. Paikdoo area have been completely covered by masses of basalt 50–200 m thick. A pumice layer covers the basalt (the thickness of the pumice layer varies from 1.5 to 7 m). The pumice surface is strongly eroded, mainly by water from melting snow in spring. Trunks of burned trees which were destroyed at the last eruption (1702 BCE) are common in the alpine zone of Mt. Paikdoo, Kaema Upland, North Korea. (Photograph by M. Šrůtek.)

icum, *Ledum palustre* subsp. *decumbens*, **L. palustre* var. *maximum*, *L. palustre* var. *minus*, *L. palustre* var. *procumbens*, *Oxycoccus microcarpus*, *Rhododendron aureum*, *R. parvifolium*, **Juniperus sibirica*.

Group 166: *Pinus pumila*, *Juniperus utilis*.

Group 167: *Rhododendron fauriae* form. *rufescens*.

II. The North-South Subalpine Region (28 species):

Group 168: *Rhododendron dauricum*.

Group 169: *Viscum coloratum* var. *lutescens*, **Buxus koreana*, *Rhododendron aureum*, *R. fauriae* form. *rufescens*, *Vaccinium vitis-idaea* var. *minus*, *Sasamorpha purpurascens* var. *borealis*, *Taxus cuspidata*, *T. cuspidata* var. *latifolia*, *Abies nephrolepis*, *Picea jezoensis*, *Pinus koraiensis*, *P. pumila*, **Thuja koraiensis*, *Sabina sargentii*.

Group 170: *Hypear tanakae*, *Rhododendron micranthum*, *Bladhia crenata* var. *taquetii*, *Psudosasa japonica*, *Sinoarundinaria reticulata*, *Pinus densiflora* var. *globosa*, *P. parvifolia*, *P. thunbergii*, *Biota orientalis*, *Juniperus utilis*, *Sabina chinensis*, **S. chinensis* var. *horizontalis*.

Group 171: *Rhododendron fauriae* var. *roseum*.

III The Midland Mountain Region (16 species)

Group 172: *Hypear tanakae*, *Bifaria japonica*, *Masakia japonica*, *M. japonica* var. *latifolia*, *Taxus cuspidata*, *Cephalotaxus nana*, *Abies nephrolepis*, *Pinus bungeana*, *Biota orientalis*, *B. orientalis* var. *stricta*, **Thuja koraiensis*, **Juniperus utilis* var. *longicarpa*, *J. virginia*, *Sabina pacifera*.

Group 173: *Pinus rigida*, **Juniperus coreana* var. *rigida*.

IV The Southern Mountain Region (33 species)

Group 87: *Buxus microphylla*, *Masakia japonica* var. *latifolia*, *M. radicans*, *Eurya japonica* var. *montana*, *Thea sinensis* var. *bohea*, *Elaeagnus macrophylla*, *Bladhia japonica* var. *typica*, *Trachelospermum asiaticum* var. *intermedium*, *Pseudosasa japonica*, *Sasamorpha purpurascens* var. *borealis*, *Sinoarundinaria nigra* var. *henonis*, *S. reticulata*, *Cephalotaxus koreana*, **Abies koreana*, *Pinus globosa*, *P. thunbergii*, *Biota orientalis*, *B. orientalis* var. *stricta*, **Juniperus coreana* var. *rigida*, *Sabina sargentii*.

Group 176: *Viscum coloratum* var. *lutes-*

* Endemic Species.

Photo 2.3. ***Rhododendron*** **community on the peak of Mt. Sobaik (1439 m).**

cens, *Stauntonia hexaphylla*, *Pinus rigida*, *Sabina chinensis*.

Group 177: *Zanthoxylum planispinum*, *Hedera tobleri*, *Sinoarundinaria pubescens*, *S. reticulata*, *Torreya nucifera*, *Abies holophylla*, *Pinus koraiensis*, *Chamaecyparis obtusa* var. *nepalensis*, *Cryptomeria japonica*.

V The Mid-Western Insular Region (15 species):

Group 178: *Masakia radicans*, *Camellia japonica*, *Vitex rotundifolia*, *Pinus densiflora*, **Juniperus coreana* var. *rigida*, *Sabina chinensis*, *S. pacifera*.

Group 179: *Machilus japonica*, *Pittosporum tobira*, *Elaeagnus glabra* var. *eug-*

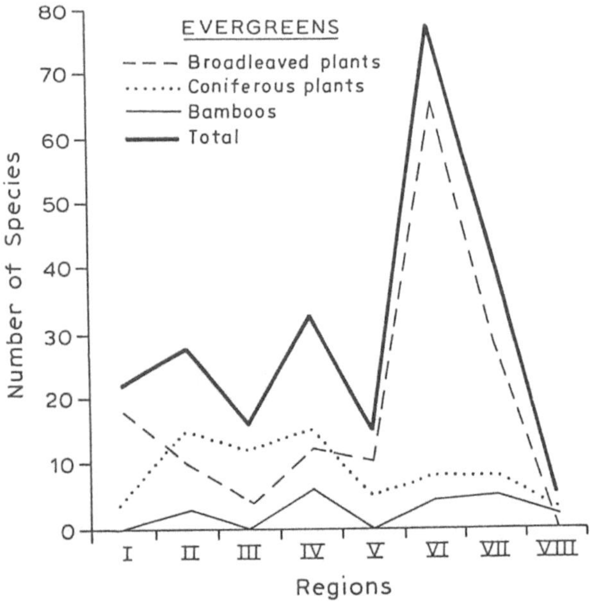

Fig. 2.3. Species composition of biogeographic regions in Korea.

laba, E. macrophylla, Hedera tobleri, Bladhia japonica var. *typica, Damnacanthus major, *Thuja koraiensis.*

VI The Southern Insular Region (77 species):
Group 180: *Piper kadzura, Sarcandra glabra, Myrica rubra, Castanopsis latifolia, Cyclobalanopsis gilva, C. stenophylla, C. stenophylla* var. *latifolia, Ficus erecta* var. *longependulata, Scurrula yadoriki, Viscum coloratum* var. *lutescens, V. coloratum* var. *rubroaurantiacum, Stephania japonica, Cinnamomum camphora, C. loureiri, *Distylium racemosum* var. *latifolium, Citrus kinokuni, C. media* var. *sarcodactylus, C. natsudaidai, C. nobilis, C. sinensis, C. unshiu, Daphniphyllum glaucescens, *Buxus koreana* var. *insularis, B. microphylla, Ilex cornata* var. *typica, Elaeocarpus sylvestrus* var. *ellipticus, Eurya emarginata, *E. japonica* var. *integra, Sasakia ochnacea, Ternstroemia mokof, Xylosma apathis, Elaeagnus maritima, Diapensia lapponica* subsp. *obovata, Empetrum nigrum* var. *japonicum, Rhododendron dauricum, R. fauriae* var. *roseum, R. fauriae* form. *rufescens, Vaccinium bracteatum, V. vitis-idaea* var. *minus, Bladhia crenata, B. crenata* var. *taquetii, B. crispa* var. *typica, B. villosa* var. *typica, Dicalix lucida, D. prunifolia, Ligustrum lucidum, *Textoria morbifera, Trachelospermum jasminoides, Marsdenia tomentosa, Damnacanthus indicus, D. major, Viburnum awabuki, Pseudosasa japonica* var. *purpurascens, *Sasa quelpaertensis, Taxus cuspidata, Abies holophylla, *A. koreana, Sabina sargentii.*

Group 181: *Machilus thunbergii* form. *obovata, Euonymus chibai, Eurya japon-*

Photo 2.4. Taxus cuspidata.

* Endemic Species.

Photo 2.5. Natural montane forest with *Abies nephrolepis* as the dominant tree. Surroundings of the Naegog Spa, Kaema Upland, northern part of North Korea. (Photograph by M. Šrûtek.)

Photo 2.6. Abies holophylla.

ica var. *aurescens*, *E. japonica* var. *montana*, *Fatsia japonica*, **Ligustrum quihoui* var. *latifolium*, *Osmanthus zantaranus*, *Trachelospermum asiaticum var. oblanceolatum*, *T. jasminoides* var. *pubescens*, *Gardenia jasminoides* form. *glandifolia*, *Serissa japonica*, *Nandina domestica*, *Sinoarundinaria nigra*, *S. nigra* var. *henonis*, *Pinus rigida*, *Biota orientalis*, *Chamaecyparis obtusa* var. *nepalensis*, * *Juniperus coreana* var. *rigida*, *Cryptomeria japonica*.

VII The West-South-Eastern Insular and Associated Inland Region (41 species).

Group 23: *Pinus densiflora* var. *globosa*.

Group 182: *Castanopsis cuspidata* var. *sieboldii*, *C. cuspidata* var. *thunbergii*, *Cyclobalanopsis acuta*, *Ficus erecta*, *F. nipponica*, *Stauntonia hexaphylla*, *Machilus japonica*, *M. thunbergii*, *Neolitsea sericea*, *Pittosporum tobira*, *Raphiolepis umbellata*, *Ilex integra* var. *typica*, *Masakia japonica*, *M. radicans*, *Eurya japonica* var. *montana*, *Elaeagnus glabra* var. *euglaba*, *E. macrophylla*, *Hedera tobleri*, *Aucuba japonica* var. *typica*, *Bladhia japonica* var. *typica*, *Vitex rotundifolia*, *Pinus thunbergii*.

Group 183: *Cyclobalanopsis myrsinaefolia*, **Buxus koreana*, *Ilex cornuta* var. *typica*, *I. crenata* var. *microphylla*, *Thea sinensis* var. *bohea*, *Rhododendron fauriae* form. *rufescens*, *Ligustrum ovalifolium*, *Pleistoblastus simoni*, **Sasamorpha chiisanensis*, *S. purpurascens* var. *borealis*, *Sinoarundinaria nigra* var. *henonis*, *S. reticulata*, *Taxus cuspidata*, *Cephalotaxus koreana*, **Abies koreana*, *Picea jezoensis*, *Pinus koraiensis*, *Sabina sargentii*.

VIII The North-South Disjunctive Region (5 species):

Group 3: *Sinoarundinaria compressa*, **Sasa coreana*, **Picea pungsanensis*, *Pinus tabulaeformis*, **Juniperus utilis* var. *modesta*.

A comparative analysis of these biogeographic divisions and their component groups (Table 2.3), the broad numerical species composition of regions (Fig. 2.3) and the proportion of species composition in each region (Table 2.4) provides further information important to understanding the nature and history of the biogeography of Korea. Supplementary information also comes from the distribution of endemic plants within these divisions (Fig. 2.4), as discussed further below.

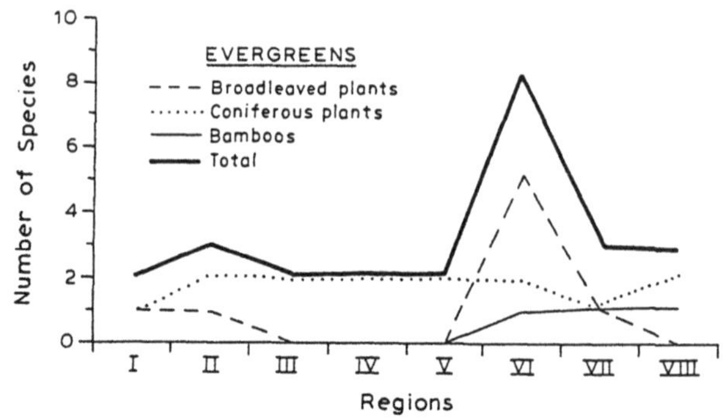

Fig. 2.4. Distribution of endemic plants in the biogeographic regions of Korea.

The Northern Alpine Region (I) is composed of 18 arctic-alpine and alpine evergreen broadleaved plants, and four circumpolar evergreen coniferous plants; among these are one endemic evergreen broadleaved plant, *Ledum palustre* var. *maximum*, and three endemic evergreen coniferous plants, namely *Picea intercedens*, *P. tonaiensis* and *Juniperus sibirica*, the first two of which are not represented in the computer source data. Two endemic evergreen coniferous plants, *Picea intercedens* and *P. tonaiensis*, at Moosan (site 3) and one evergreen broadleaved plant, *Dryas octopetala* var. *asiatica* at Mt. Paikdoo (site 8) and Mt. Ro (site 15), though present, also are not represented in the computer data for this site. The region is characterised by dominance of arctic-alpine and alpine evergreen broadleaved plants (18 spp.), along with four circumpolar evergreen coniferous plants, reflecting the mountainous and often harsh environments.

The North-South Subalpine Region (II) contains 15 evergreen coniferous plants, 10 evergreen broadleaved plants and three evergreen bamboos, including two endemic evergreen coniferous plants, *Thuja koraiensis* and *Sabina chinensis* var. *horizontalis*, and one endemic

* Endemic Species.

evergreen broadleaved plant, *Buxus koreana*. The region is dominated by the arctic-alpine and alpine evergreen broadleaved plants and evergreen coniferous plants, along with (in the south) the southern evergreen broadleaved plants and bamboos.

The Midland Mountain Region (III) includes 12 evergreen coniferous plants and four evergreen broadleaved plants, including three endemic evergreen coniferous plants, *Thuja koraiensis*, *Juniperus utilis* var. *longicarpa* and *J. coreana* var. *rigida*. Though the region is mainly comprised of northern evergreen coniferous plants it can be regarded as an inland transitional zone of vegetation from northern to southern elements, because of the sharp decrease in the total species number, and as a result of the appearances of four southern evergreen broadleaved plants, namely *Hypear tanakae*, *Bifaria japonica*, *Masakia japonica* and *M. japonica* var. *latifolia*.

The Southern Mountain Region (IV) consists of 15 evergreen coniferous plants, 12 evergreen broadleaved plants and six evergreen bamboos, including two endemic evergreen coniferous plants, *Abies koreana* and *Juniperus coreana* var. *rigida*. The dominance of southern evergreen broadleaved plants, evergreen bamboos and even evergreen coniferous plants, such as *Cephalotaxus koreana*, *Abies koreana*, *Pinus thunbergii*, *Biota orientalis*, *B. orientalis* var. *stricta* and *Torreya nucifera*, designates the region as a southern floristic and vegetational area.

The Mid-Western Insular Region (V) contains 10 evergreen broadleaved plants and five evergreen coniferous plants, including two endemic evergreen coniferous plants, *Juniperus coreana* var. *rigida* and *Thuja koraiensis*. The region can be regarded as being an insular transitional zone of vegetation from northern to southern elements, due to the appearances of characteristic southern evergreen broadleaved plants, such as *Camellia japonica*, *Machilus japonica*, *Pittosporum tobira*, *Elaeagnus glabra* var. *euglaba*, *E. macrophylla*, *Hedera tobleri*, *Bladhia japonica* var. *typica*, *Masakia radicans* and *Damnacanthus major*, along with many northern species.

The Southern Insular Region (VI) is composed of 65 evergreen broadleaved plants, eight evergreen coniferous plants and four evergreen bamboos, including five endemic evergreen broadleaved plants *viz*., *Distylium racemosum* var. *latifolium*, *Buxus koreana* var. *insularis*, *Eurya japonica* var. *integra*, *Ligustrum quihoui* var. *latifolium* and *Textoria morbifera*, two endemic evergreen coniferous plants, *Abies koreana* and *Juniperus coreana* var. *rigida*, and also one endemic evergreen bamboo, *Sasa quelpaertensis*. The region has the largest number of southern evergreen broadleaved plants of all the eight biogeographic regions, and this reflects the relatively warm and humid climatic conditions. The region also has a large number of southern elements which are also present in southern China and Japan. In contrast, it also contains several alpine or arctic-alpine species, such as *Diapensia lapponica* subsp. *obovata*, *Empetrum nigrum* var. *japonicum*, *Rhododendron dauricum*, *R. fauriae* var. *roseum*, *R. fauriae* form. *rufescens*, *Vaccinium bracteatum*, *V. vitis-idaea* var. *minus*, *Taxus cuspidata*, *Abies holophylla*, *A. koreana* and *Sabina sargentii* in its mountain region; and the importance of these to an understanding of the biogeographical and environmental history of the peninsula is to be discussed later (pp. 41–42).

The West-South-Eastern Insular and Associated Inland Region (VII) contains 28 evergreen broadleaved plants, eight evergreen coniferous plants and five evergreen bamboos, including one endemic evergreen broadleaved plant, *Buxus koreana*, one endemic evergreen coniferous plant, *Abies koreana*, and one endemic evergreen bamboo *Sasamorpha chiisanensis*. Southern evergreen broadleaved plants and evergreen bamboos are common in the insular and coastal areas, but northern evergreen coniferous plants are richer in the inland areas. The region possesses the second largest number of southern evergreen broadleaved plants, arising out of the relatively mild climatic conditions.

The North-South Disjunctive Region (VIII) consists of three evergreen coniferous plants and two evergreen bamboos, including two endemic evergreen coniferous plants, *Picea pungsanensis* and *Juniperus utilis* var. *modesta*, and one endemic evergreen bamboo, *Sasa coreana*. The region is characterised by a high number of endemics (60.0%) proportional to the total number of species in the region.

Overall, the percentage of endemic plants is also relatively high in the two transitional areas

Photo 2.7. Abies koreana and *Acer* spp. in autumn on Mt. Halla (1950 m).

of vegetation, namely, the Mid-Western Insular Region (V) (13.3%) and the Midland Mountain Region (III) (12.5%), and then tails away to 10.7% in the North-South Subalpine Region (V), 10.4% in the Southern Insular Region (VI), 9.0% in the Northern Alpine Region (I), 7.3% in the West-South-Eastern Insular and Associated Inland Region (VII) and 6.0% in the Southern Mountain Region (IV). This possibly suggests that more active evolution has taken place along the fringe of the main distribution zone of the northern and southern floristic elements due to increased competition and/or hybridisation in these marginal areas, and also frequent fluctuations of environmental conditions in the past. Such mechanisms might also explain the occurrences of five Korean endemic genera at the marginal areas of the designated biogeographic re-

Photo 2.8. Echinosophora koraiensis.

gions (Fig. 2.2), namely, first, *Hanabusaya latisepala* and *Echinosophora koraiensis* in area 1; secondly, *Hanabusaya asiatica*, *Pentactina rupicola*, *Echinosophora koraiensis* and *Megaleranthis saniculifolia* in area 2; thirdly, *Hanabusaya asiatica* and *Megaleranthis saniculifolia* in area 3; fourthly, *Abeliophyllum distichum* in area 4; and fifthly, *Megaleranthis saniculifolia* in area 5.

Similar patterns have been reported from southwest Asia, western Africa and western North America (Morton, 1972; Stebbins, 1972). Watts (1984) reviewed the status of endemics and suggests that the vast majority have evolved recently. For the explanation of their origins, one therefore needs to look at the degree of isolation and the characteristics of environment within these timescales; and Flenley (1984) has noted that the period from 10 000 to 100 000 years B.P. is particularly important in this respect. The significance of these endemics to an understanding of Korean biogeographical history, and in particular the relevance of the high number of endemics of northern evergreen coniferous plants and evergreen broadleaved plants, both in northern and southern Korea is discussed further in Chapter 4.

In Korea as a whole, the Northern Alpine Region (I), the North-South Subalpine Region (II) and the Southern Insular Region (VI) include arctic-alpine and alpine evergreen coniferous plants and evergreen broadleaved plants. Northern evergreen coniferous plants are rich in all inland areas, namely the North-South Subalpine Region (II), the Midland Mountain Region (III) and the Southern Mountain Region (IV). The southern evergreen broadleaved plants are most frequent in island areas, namely the Southern Insular Region (VI) and the West-South-Eastern Insular and Associated Inland Region (VII). The southern evergreen bamboos are most common in the Southern Mountain Region (IV), the West-South-Eastern Insular and Associated Inland Region (VII) and the Southern Insular Region (VI). These features form the basis for further examination below.

Physiognomy

The physiognomic approach to vegetation analysis is based on the study of community structure, or the morphological features of plant communities. As a basic proposition, one may agree with

Beard (1980) that the physiognomic treatment of vegetation, combined with floristic study, is indispensable for the broader biogeographical interpretation of the significance of plant communities, their origins and histories.

A system of physiognomic types of plants and plant communities was first worked out as early as ca. 300 B.C. by Theophrastos. But in 1806 it was Humboldt who first drew the attention of natural historians to the further examination of physiognomy. Since then, many people have worked with and modified Humboldt's ideas, including nineteenth-century scholars, such as De Candolle, Grisebach, Darwin, Kerner, Warming and Schimper. In 1908, Raunkiaer proposed a functional classification of plants as representing certain 'life-forms', on the basis of the height above the ground of their perennating organs, so differing from all the physiognomic studies of earlier works. A modification to Raunkiaer's life-form was presented by Braun-Blanquet in 1928, who made only slight alterations to the major units of Raunkiaer; and new terms were created a short time later for many of the subordinate units (Du Rietz, 1931).

Du Rietz (1931) subdivided the main life-form groups primarily according to the actual stratification of vegetation, and insisted that life-forms delimited independently of any adaptation theory cannot be of any use at all as units for the inductive study of the adaptation problem, and of the actual correlation between life-form and environment. Since then, Küchler (1949) has classified vegetation in terms of its height, density and spatial features. More recently, Ellenberg and Mueller-Dombois (1965–66, 1967) and Mueller-Dombois and Ellenberg (1974) have revised Raunkiaer's plant life-forms, giving rise to a tentative physiognomic-ecological classification of plant formations; and they have noted that a physiognomic analysis often gives information about the response of a community to particular environmental factors, on the utilisation of space and on the probable competitive relations within a community. Following this, Box (1978) has classified world terrestrial vegetation into 59 ecophysiognomic plant life-forms, all of which are presumed to be physiognomic adaptations to particular environmental situations. Box's concepts developed further in 1981 (Box, 1981a, b), when he suggested that each of his life-forms could be defined physiognomically by reference to six essential characteristics: the general structural form (tree, shrub etc.); the general plant size (tall, small, etc.); the leaf type (broad, linear, etc.); the relative leaf size (large, normal, small, etc.); and the seasonal habit (evergreen, summergreen, etc.). We have used a modified version of Box's classification of physiognomic features in this study, along with the use of some ecological characteristics.

This physiognomic analysis of present-day Korean vegetation accordingly concentrates on vegetational structure and the morphological adaptations of plants to environmental conditions, along with relevant ecological features. The physiognomic characteristics to be studied include life-form, height of trunk, length of leaves, width of leaves, shape of leaves, flower colour, flowering period, dates of fruit formation, and the ripening and shape of fruit (Table 2.5). The physiognomic characteristics of species are based on information from the regional flora reports (Table 2.1), and our own field studies.

Table 2.5 shows that the Northern Alpine Region (I) contains 22 evergreen plants, including 18 evergreen broadleaved plants and four evergreen coniferous plants. Shrubs, and small shrubs with relatively small leaves, are those most commonly represented. The large number of decumbent small shrubs (*Andromeda polifolia*, *Phyllodoce caerulea*, *Empetrum nigrum* var. *japonicum*, *Ledum palustre* var. *procumbens*, *Juniperus sibirica*, *Pinus pumila*), procumbent small shrubs (*Oxyoccus quadripetalus*, *Linnaea borealis*), species with small-sized coriaceous (leathery) leaves (*Andromeda polifolia*, *A. polifolia* form *acerosa*, *Chamaedaphne calyculata*, *Rhododendron parvifolium*, *R. aureum*, *R. fauriae* form. *rufescens*), those with revolute leaves (*Andromeda polifolia*, *A. polifolia* form. *acerosa*, *Oxycoccus quadripetalus*, *O. microcarpus*, *Vaccinium vitis-idaea* var. *genuinum*, *Ledum palustre* var. *yesoense*, *Phyllodoce caerulea*, *Empetrum nigrum* var. *japonicum*, *Rhododendron aureum*, *R. fauriae* form. *rufescens*), and those with leaf trichomes (hairy leaves) (*Chamaedaphe calyculata*, *Ledum palustre* var. *latifolium*, *L. palustre* var. *yesoense*, *L. palustre* var. *anguistrum*, *L. palustre* var. *maximum*, *L. palustre* var. *minus*, *L. palustre* var. *procumbens*,

Table 2.5. Physiognomy of species in the biogeographic regions of Korea*.

Physiognomy	Region	I	II	III	IV	V	VI	VII	VIII
Number of Species		22	28	16	33	15	77	41	5
Life Form	EBP	18	10	4	12	10	65	28	0
	EB	0	3	0	6	0	4	5	2
	ECP	4	15	12	15	5	8	8	3
Height of Trunk	Tree (15 m+)	2	11	6	12	3	19	13	2
	Small Tree (5–15 m)	1	2	1	0	2	9	1	0
	Shrub (1–5 m)	11	12	9	14	5	31	14	1
	Small Shrub (-1 m)	8	1	0	1	1	7	1	0
	Vine	0	0	0	5	4	6	6	0
	Bamboo	0	3	0	5	0	4	6	2
Length of Leaves	-5 cm	8	12	8	9	4	19	7	2
	5–10 cm	3	6	3	8	4	22	14	0
	10 cm+	1	7	0	9	4	21	18	1
Width of Leaves	-1 cm	9	7	2	5	0	16	6	0
	1–5 cm	3	9	5	13	7	20	20	1
	5 cm+	0	0	0	1	1	4	3	0
Texture of Leaves	Coriaceous	6	7	2	10	3	29	14	0
	Revolute	10	5	0	0	0	5	1	0
Hairs on Leaves	Hairs	4	2	1	1	1	4	3	0
	Hairs on one side	7	2	0	3	1	6	3	0
	Hairs only young stage	1	0	0	1	3	6	3	0
Flower Colour	Yellow	1	8	2	5	2	8	3	0
	Red & Pink	5	3	1	0	1	4	3	0
	White	11	4	2	6	4	33	11	0
	Purple	2	4	1	4	1	4	4	1
	Green	0	2	1	4	3	8	4	0
Flowering Period	Mar.–May	13	11	12	24	9	32	24	2
	Jun.–Aug.	8	8	4	4	3	34	13	1
	Sept.–Nov.	0	0	0	4	3	9	7	0
	Dec.–Feb.	0	0	0	0	1	3	2	0
Ripening Season	Jun.–Aug.	1	3	0	3	0	6	4	0
	Sep.–Nov.	17	16	10	15	8	48	34	1
	Before Jul. following year	0	8	0	2	3	4	3	0
	After Aug. following year	3	8	4	9	4	3	10	2
Shape of Fruit	Berry	4	3	2	4	4	23	8	8
	Drupe	1	2	1	2	3	12	7	0
	Nut	0	1	0	0	0	4	4	0
	Synconium	0	0	0	0	0	1	2	0
	Capsule	13	6	2	6	3	22	6	0
	Cone	3	13	10	14	5	8	6	3
	Caryopsis	0	3	0	5	0	2	5	2
	Pome	0	0	0	0	0	0	1	0

* Details of life-form modified and expanded from Box (1981a,b).

Linnaea borealis, *Empetrum nigrum* var. *japonicum*, *Oxycoccus microcarpus*, *Rhododendron parvifolium*, *R. fauriae* form. *rufescens*) is a good indication that there is a general morphological adaptation to cold climate conditions. Flowers bloom mainly in the spring and summer, and fruits ripen in the autumn, and are mainly of the capsule type.

The North-South Subalpine Region (II) includes 28 evergreen plants, including 15 evergreen coniferous plants, 10 evergreen broad-leaved plants and three evergreen bamboos. Trees and shrubs with small leaves are common. There exist diverse morphological adaptations to cold climatic conditions, such as small-sized coriaceous leaves (*Rhododendron dauricum*, *R. fauriae* form. *rufescens*, *R. fauriae* var. *roseum*, *R. aureum*, *Viscum coloratum* var. *lutescens*, *Buxus koreana*, *Vaccinium vitis-idaea* var. *minus*), revolute leaves (*Rhododendron aureum*, *R. fauriae* var. *roseum*, *R. fauriae* var. *rufescens*, *Vaccinium vitis-idaea* var. *minus*, *Bladhia crenata* var. *taquetii*), and leaf trichomes (*Rhododendron dauricum*, *R. fauriae* form. *rufescens*, *R. micranthum*, *Buxus koreana*). Yellow flowers are common, and plants bloom mainly in the spring and summer, with fruits maturing in the autumn. The cone type of fruit is the most dominant.

The Midland Mountain Region (III) consists of 12 evergreen coniferous plants and four evergreen broadleaved plants. Shrubs and trees with small leaves are the most common. The low proportion of species with coriaceous or hairy leaves as compared to regions I and II is due to the dominance of coniferous plants here. Flowers bloom mainly in the spring, and bear fruit in the autumn. Due to the high proportion of evergreen coniferous plants, the cone type of fruit is the most abundant.

The Southern Mountain Region (IV) contains 15 evergreen coniferous plants, 12 evergreen broadleaved plants and six evergreen bamboos. Shrubs and trees with small to large leaves are common. There are many large-sized coriaceous-leaved plants (*Buxus microphylla*, *Masakia japonica* var. *latifolia*, *M. radicans*, *Eurya japonica* var. *montana*, *Thea sinensis* var. *bohea*, *Trachelospermum asiaticum* var. *intermedium*, *Viscum coloratum* var. *lutescens*, *Stauntonia hexaphylla*, *Hedera tobleri*) and species with hairy leaves (*Buxus microphylla*, *Elaeagnus macrophylla*, *Trachelospermum asiaticum* var. *intermedium*, *Sinoarundinaria nigra* var. *henonis*, *Hedera tobleri*). White and yellow flowers are common. Flowers bloom in the spring and bear fruit both in the autumn and in the following year's autumn, and these latter are mainly of the cone type.

The Mid-Western Insular Region (V) contains 10 evergreen broadleaved plants and five evergreen coniferous plants. The admixture of northern floristic characteristics, such as the procumbent shrub (*Sabina pacifera*), and the small tree or shrub (*Thuia koraiensis*) with southern floristic features, such as vines (*Masakia radicans*, *Elaeagnus glabra* var. *euglabra*, *E. macrophylla*, *Hedera tableri*) again suggests that this region is an insular transitional one of vegetation lying between the main zones of the northern and southern floristic elements. Most species have middle-sized leaves. They bloom mainly in the spring, and bear fruit in the autumn, with various types of fruit present.

The Southern Insular Region (VI) includes 65 evergreen broadleaved plants, eight evergreen coniferous plants and four evergreen bamboos. The life-form is mainly of shrubs and trees, with small to large leaves. Most species have large-sized coriaceous leaves (*Sarcandra glabra*, *Myrica rubra*, *Cyclobalanopsis gilva*, *Scurrula yadoriki*, *Viscum coloratum* var. *lutescens*, *V. coloratum* form. *rubroaurantiacum*, *Distylium racemosum* var. *latifolium*, *Buxus koreana* var. *insularis*, *B. microphylla*, *Ilex integra* var. *typica*, *Elaeocarpus sylvestris* var. *ellipticus*, *Eurya emarginata*, *E. japonica* var. *integra*, *Sasakia ochnacea*, *Ternstromia mokof*, *Xylosma apathis*, *Ligustrum lucidum*, *Trachelospermum jasminoides*, *Damnacanthus indicus*, *Eurya japonica* var. *aurescens*, *E. japonica* var. *montana*, *Fatsia japonica*, *Trachelospermum oblanceolatum*, *T. jasminoides* var. *pubescens* and *Nandina domestica*), characteristic of a relatively warm climate. However, the occurrence of small-sized coriaceous-leaved species (*Diapensia lapponica* subsp. *obovata*, *Rhododendron dauricum*, *R. fauriae* var. *roseum*, *R. fauriae* form. *rufescens*, *Vaccinium vitis-idaea* var. *minus*) and the presence of northern evergreen coniferous plants (*Taxus cuspidata*, *Abies holophylla*, *Sabina sargentii*) is a reflection of the existence of both past and present cryo-climatic conditions, mainly at high altitudes on Mt. Halla (1950 m a.s.l.), Cheju Island. The dominant flower colour is white, and most flowers bloom in the spring and the summer, and bear fruit in the autumn. Berries and capsules are the two main fruit types.

The West-South-Eastern Insular and Associated Inland Region (VII) contains 28 evergreen broadleaved plants, coniferous plants and five evergreen bamboos. Shrubs and trees are common, as well as bamboos and vines, all with relatively large leaves. There are many large-leaved coriaceous species (*Cyclobalanopsis acuta*, *Ficus nipponica*, *Stauntonia hexaphylla*, *Ilex integra* var. *typica*, *I. cornata* var. *typica*, *Masakia japonica*, *M. radicans*, *Eurya japonica* var. *montana*, *Hedera tobleri*, *Aucuba japonica* var. *typica*, *Vitex rotundifolia*, *Buxus koreana*, *Thea sinensis* var. *bohea*), which indicate a relatively mild climate.

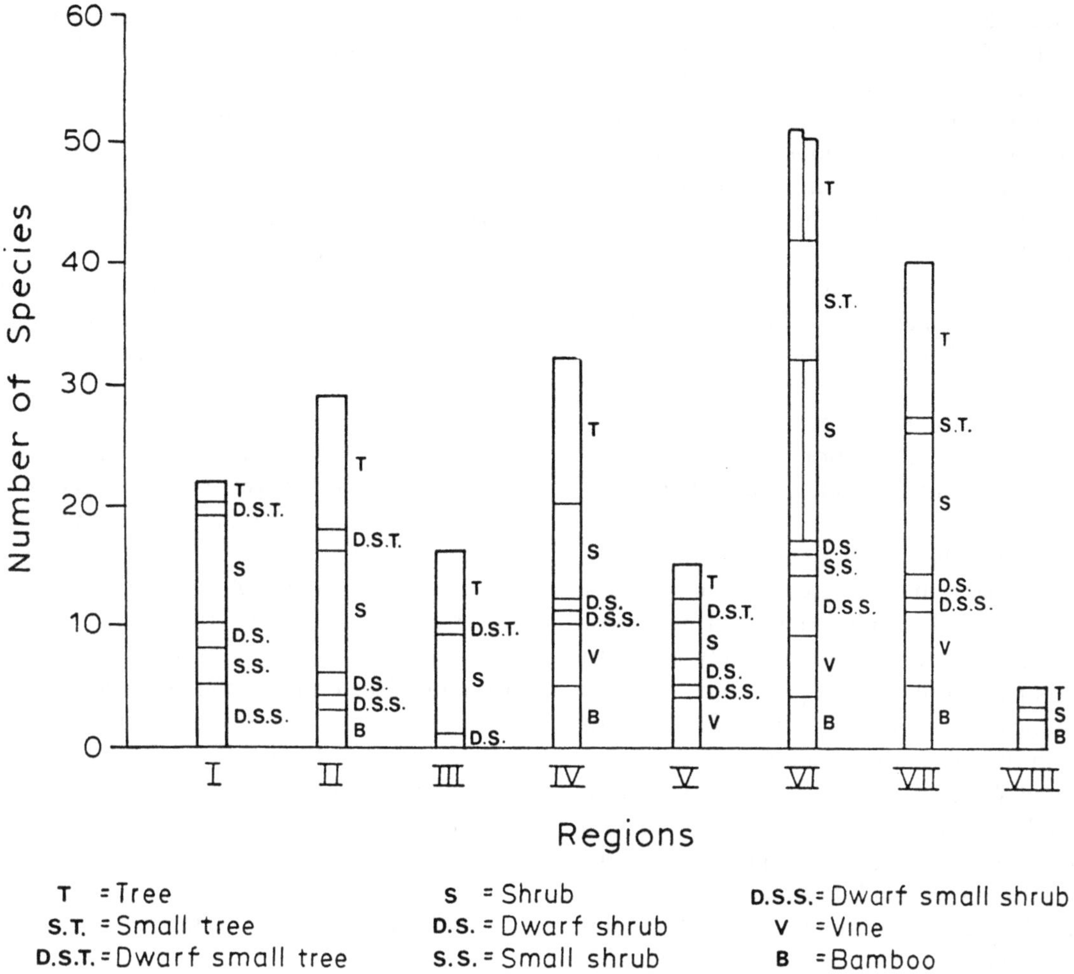

Fig. 2.5. Life forms of species in the biogeographic regions of Korea.

However, the occurrence of some coriaceous and revolute-leaved species (*Rhododendron fauriae* form. *rufescens*), and northern evergreen coniferous plants (*Taxus cuspidata*, *Picea jezoensis*, *Pinus koraiensis*, *Sabina sargentii*) suggests also the existence of severe climatic conditions, mainly in the subalpine areas of Mt. Chiri (1915 m a.s.l.). Most flowers are white, bloom in the spring-to-summer period, and bear fruit in the autumn.

The North-South Disjunctive Region (VIII) includes three evergreen coniferous plants and two evergreen bamboos. Trees and bamboos accordingly are the most abundant form. The evergreen coniferous plants bear cones in the following year's autumn, and caryopsis is a typical shape of the bamboo fruits.

The details of species' life-forms in each region is summarised in Fig. 2.5. Such data may help to explain the ecological status of regions particularly in connection with past and present environmental conditions, a topic which is discussed further below (pp. 65–68).

In brief, the Northern Alpine Region (I) shows six life forms, namely trees, dwarf small trees, shrubs, small shrubs, dwarf shrubs and dwarf small shrubs. The large number of decumbent, procumbent, dwarf shrub and small shrub forms reflects a dominance of arctic-alpine species, which may in turn be linked to the existence of refugia during past glacial periods. The morphological adaptations of arctic-alpine plants against severe environmental conditions, such as the existence of leaf trichomes, revolute leaves and coriaceous leaves, all of which are common in this region, help to ensure the survival of these species in the present cryo-climatic conditions.

The North-South Subalpine Region (II) contains six life forms, namely trees, dwarf small trees, shrubs, dwarf shrubs, dwarf small shrubs and bamboos. Bamboos here grow to the height of a small shrub. The various shrub and tree forms, caused by an admixture of arctic-alpine, northern and southern elements, infer a complex vegetation history.

The Midland Mountain Region (III) includes four life forms, namely trees, dwarf small trees, shrubs and dwarf shrubs. There are a relatively small number of life forms, mainly trees and shrubs, and this supports the idea of an inland transitional zone of vegetation, intermediate between the northern and southern floristic elements.

The Southern Mountain Region (IV) consists of six life forms, namely trees, shrubs, dwarf shrubs, dwarf small shrubs, vines and bamboos. The number of trees is approximately equal to the number of shrubs. The appearance of vines and bamboos indicates the presence of a southern element in the presently-moderate climate.

The Mid-Western Insular Region (V) has a relatively low number of species which fall into the categories of trees, dwarf small trees, shrubs, dwarf shrubs, dwarf small shrubs and vines, albeit in relatively even proportions. This latter point perhaps results from the fact that this region is transitional from northern to southern floristic elements. It is similar in this respect to region III, but in this case is placed in an island setting.

The Southern Insular Region (VI) contains the largest number of species of all Korean biogeographic regions, and includes eight life forms, namely trees, small trees, shrubs, dwarf shrubs, small shrubs, dwarf small shrubs, vines and bamboos. In the order of their abundance, shrubs, trees, vines and bamboos are the most important. The highly complicated range of life forms, and the floristic diversity, both are due to the presence of arctic-alpine, temperate and subtropical plants together. This reflects the diverse present-day environmental and habitat conditions, and also post-glacial environmental stability.

The West-South-Eastern Insular and Associated Inland region (VII) include seven life forms, namely trees, small trees, shrubs, dwarf shrubs, dwarf small shrubs, vines and bamboos. The proportion of shrubs and trees is nearly equal, and that of vines and bamboos also is nearly equal. Southern floristic elements are abundant in both island and coastal areas, but the inland districts, mainly at higher altitudes, have some northern floristic elements as well. The region is closely related biogeographically to region VI.

The North-South Disjunctive Region (VIII) is composed of three life forms, namely trees, shrubs and bamboos. There is a high degree of physiognomic diversity for the small number of species present, which is so far unexplained.

Overall the Northern Alpine Region (I) and the Southern Insular Reglon (VI) show highly adaptive leaf trichome forms. The high proportion of revolute leaves in the Northern Alpine Region (I), the North-South Subalpine Region (II) and the Southern Insular Region (VI) relates to the existence there of above-average numbers of arctic-alpine plants. In contrast, the large number of coriaceous leaves in the Southern Mountain Region (IV), the Southern Insular Region (VI) and the West-South-Eastern Insular and Associated Inland Region (VII) indicates a dominance there of both warm temperate and subtropical plants.

Latitudinal and altitudinal ranges

The increase in numbers of plant species per unit area as one moves from high latitudes into the tropics is a well-known fact, and five hypotheses have been suggested to explain this (Whittaker, 1965; Pianka, 1966). These are the time theory (high latitudinal areas have had less time, since the post-glacial phase, to develop large numbers of species than have the tropics); the theory of spatial heterogeneity (tropical environments are more heterogeneous than are those of high altitudes, and accordingly can support more species); the competition hypothesis (greater levels of competition in the tropics have produced more species than elsewhere); the theory of climatic stability (tropical climates are *relatively* more stable than those of high latitudes, and so support more species); and the productivity hypothesis (energy levels are higher in the tropics than elsewhere and so support more species).

It has already been made clear in Table 2.4 that, in general, plant species within Korea also increase per unit area from north to south. The reasons for this are likely to be complicated and do not fit clearly into any single one of the above hypotheses. Some idea of the scale of the complications involved may be seen when one considers the distribution of evergreen plants in Korea in a three-dimensional manner. For this, information relating to the latitudinal and altitudinal ranges of site and species distribution is required. The altitudinal limits for all of Korea's 79 alpine areas which reach over 2000 m, 264 subalpine areas which lie between 2000 to 1500 m, and 788 mountains which reach 1500 to 1000 m in height are known, and although many of these possess adequate floristic reports, much of the best data comes from the lower mountains; however, all mountains whose altitudes run from 409 m a.s.l. to 2744 m a.s.l. have been incorporated into this research, if floristic reports were available (see Table 2.1). The analysis of spatial distribution is based upon data collected not only from the mountain sites, but also from a few selected island sites in the south, west and east of Korea, which also have good floristic data. The sites selected from each region are summarized in Table 2.6.

The vertical distribution and latitudinal variation of species from Korean mountain sites have been compiled mainly from Chung & Lee's (1965) data, and then incorporated into Fig. 2.6. Some considerable variation in altitudinal ranges on an inter-regional basis is evident.

Although the Northern Alpine Region (I) has a species data base of 22 species, altitudinal ranges are only available for 11 of these. These are presented in Table 2.7.

There are four distinctive altitudinal range divisions in this region, one of relatively low elevation, from 450 to 550 m, as exemplified by the evergreen coniferous plant *Juniperus utilis*; a second, from 800 to 1500 m, in which the northern evergreen broadleaved species, *Rhododendron fauriae* form. *rufescens* and *R. parvifolium* are found; a third, characterised by the northern evergreen coniferous plants, *Pinus pumila* and *Juniperus sibirica*, from 2000 to 2205 m; and a fourth in which the northern evergreen broadleaved plants, *Ledum* spp., *Phyllodoce caerulea* and *Rhododendron aureum* are located, from 1500 to 2540 m. Overall, the northern evergreen broadleaved plants and evergreen coniferous plants are dominant from mid- to high-altitudes in the Northern Alpine Region. The endemic species *Ledum palustre* var. *maximum* occurs in site 5 up to 1600 m, and *Juniperus sibirica* is found in sites 5 and 7 up to 2205 m.

Also relevant here is the matter of timberline. Timberline, at 2000 m on Mt. Paikdoo (42°N, 2744 m a.s.l), the highest mountain in Korea, is relatively low compared with other peaks in East Asia. For example, it stands at 2300 m in the Kentai-shan (49°N, 2800 m) of Mongolia, at 2500 m in Hsiaoutai-shan (40°N, 2500 m) in North China and at 2750 m in Mt. Fui (35°N, 3789 m) in Japan (Wang, 1961). However Mt. Paikdoo displays an abundant number of species existing above the timberline as compared to these other peaks, among which are *Phyllodoce caerulea* and decumbent *Juniperus sibirica*. This is probably due to the above-average precipitation and greater relative humidity of the air at this site, along with the relative absence of impact by the Pleistocene glacial phases in Korea (Takahasi, 1944; Good, 1947).

The North-South Subalpine Region (II) contains 28 species, from which the range of 21 species is given in Table 2.8.

There are seven divisions of altitudinal species range within this group, and some of these ranges are very large, reflecting latitudinal variations from north to south. These include one from 100 to 700 m (the parasitic evergreen broadleaved plants *Viscum* and *Hypear*, the southern evergreen bamboo *Pseudosasa*, and the southern evergreen broadleaved endemic *Buxus koreana*); a second with a smaller altitudinal range characterised by two other evergreen bamboos, *Sasamorpha* and *Sinoarundinaria*, from 400 to 900 m; a third, involving the northern evergreen broadleaved plants *Rhododenron dauricum* and *R. fauriae* var. *roseum*, from 300 to 1100 m; a fourth, indicated by the northern evergreen coniferous plants *Pinus koraiensis* and the endemic *Thuja koraiensis*, from 300 to 2300 m; a fifth, in which the northern evergreen coniferous plants *Taxus cuspidata*, *Abies nephrolepis* and *Picea jezoensis* range from 500 to 1900 m; a sixth, in which the northern evergreen coniferous plants *Pinus pumila* and *Sabina sargentii* are found from 700 to 2300 m; and finally, the northern evergreen broadleaved plants *Rhododendron fauriae* form.

Table 2.6. Site composition in the biogeographic regions of Korea.

Region	Group	Site name & number	Height of mountain (m)* or size of island (km^2)*
The Northern Alpine Region I	4	Moosan (3)	606 m
		Keelju (8)	–
		Kapsan (10)	–
		Dancheon (18)	–
	40	Mt. Kwanmo (6)	2541 m
	82	Mt. Paikdoo (5)	2744 m
	166	Mt. Mantap (7)	2205 m
		Mt. Ro (15)	2210 m
	167	Mt. Chilbo (11)	906 m
		Mt. Hoochi (13)	1335 m
		Mt. Keumpae (17)	1676 m
The North-South Subalpine Region II	168	Mt. Jung (1)	1003 m
		Mt. Songjin (2)	1146 m
	169	Mt. Chayoo (4)	1500 m
		Mt. Seungjuk (14)	1970 m
		Mt. Pirae (16)	1470 m
		Mt. Peenanduk (19)	1316 m
		Mt. Nangrim (20)	2002 m
		Myohyang (21)	1909 m
		Mt. Sasoo (22)	1747 m
		Mt. Haram (23)	1486 m
		Mt. Chuae (24)	1530 m
		Mt. Myungi (36)	1250 m
	170	Mt. Keumkang (25)	1638 m
		Mt. Sorak (33)	1708 m
	171	Mt. Odae (40)	1563 m
		Mt. Kaebang (43)	1577 m
		Mt. Taeji (44)	1300 m
		Mt. Chiak (49)	1288 m
		Mt. Daesung (50)	1307 m
		Mt. Taebaik (51)	1561 m
		Mt. Dukyoo (70)	1508 m
The Midland Mountain Region III	172	Mt. Myulak (28)	816 m
		Mt. Whaak (34)	1468 m
		Mt. Sori (42)	472 m
		Mt. Yongmoon (45)	1157 m
		Mt. Kwanap (47)	629 m
		Mt. Sokri (58)	1057 m
	173	Mt. Koowol (26)	954 m
		Mt. Changsoo (29)	711 m
		Mt. Sooyang (32)	945 m
		Mt. Kari (37)	1051 m
		Mt. Samak (38)	654 m
		Mt. Daeryong (39)	899 m
		Kwangnung (41)	–
		Mt. Mantap (48)	578 m
		Mt. Wolak (54)	1093 m
		Mt. Ilwol (55)	1219 m
The Southern Mountain Region IV	87	Mt. Kachil (35)	1242 m
		Mt. Palgong (66)	1192 m
		Mt. Kaya (68)	1420 m
		Mt. Kaji (74)	1240 m
	176	Mt. Pukhan (46)	836 m
		Mt. Sobaik (53)	1421 m
		Mt. Chuheul (56)	1105 m
		Mt. Kunja (57)	948 m

Table 2.6. Continued.

Region	Group	Site name & number	Height of mountain (m)* or size of island (km^2)*
		Mt. Chuwang (59)	720 m
		Mt. Hyangro (61)	930 m
		Mt. Booso (62)	–
		Mt. Pohyun (63)	1124 m
		Mt. Keumo (64)	977 m
		Mt. Doduk (65)	703 m
		Mt. Ka (67)	902 m
		Mt. Choijung (71)	915 m
		Mt. Biseul (72)	1084 m
		Mt. Woonmoon (75)	1190 m
	177	Dodamsambong (52)	–
		Mt. Kaeryong (60)	845 m
		Mt. Naejang (77)	763 m
		Mt. Ipam (76)	655 m
		Mt. Keumjung (82)	801 m
		Mt. Moodeung (83)	1187 m
		Mt. Chogae (84)	887 m
The Mid-western Insular Region V	178	Songjiho (30)	–
		Cape Changsan (31)	–
		Is. Kangwha (91)	319.8 km^2
		Is. Mooie (93)	–
		Is. Dukjeok (94)	219 km^2
		Is. Poong (95)	1.5 km^2
		Is. Ong (97)	1.1 km^2
	179	Is. Yungjong (92)	39.9 km^2
		Is. Seungbong (95)	4.3 km^2
		Is. Juk (144)	2.1 km
		Is. Dok (146)	1.3 km_1
The Southern Insular Region VI	180	Is. Hucksan (112)	18.1 km^2
	181	Is. Cheju (= Mt. Halla) (119)	1950 m
		Mt. Daedun (88)	700 m
		Is. Hong (111)	6.1 km^2
		Is. Chin (115)	319.6 km^2
		Is. Bogil (120)	32.8 km^2
		Is. Wan (123)	87.0 km^2
		Is. Keomoon (130)	1.1 km^2
		Is. Sonjuk (131)	3.1 km^2
		Is. Pijin (140)	2.4 km^2
		Is. Keojae (141)	384.4 km^2
The West-South-Eastern Insular and Associated Inland Region VII	23	Youngkwang (81)	–
	182	Kimjae (73)	–
		Mt. Paikyang (76)	750 m
		Mokpo (85)	–
		Mt. Wolchul (86)	809 m
		Mt. Manduk (87)	409 m^2
		Is. Paikryung (89)	44.1 km^2
		Is. Daecheong (90)	12.6 km^2
		Is. Anmyon (98)	87.7 km^2
		Is. Oeyeon (99)	2.1 km^2
		Is. Eocheong (100)	2.0 km^2
		Is. Mal (101)	–
		Is. Bangchuk (102)	1.3 km^2
		Is. Seonyoo (103)	2.0 km^2
		Is. Yami (104)	–

Table 2.6. Continued.

Region	Group	Site name & number	Height of mountain (m)* or size of island (km^2)*
		Is. Shinsi (105)	4.4 km^2
		Is. Pian (106)	–
		Is. Wi (107)	11.1 km^2
		Is. Anma (108)	6.7 km^2
		Is. Imja (109)	43.2 km^2
		Is. Pikeum (110)	43.1 km^2
		Is. Docho (113)	–
		Is. Wooie (114)	9.3 km^2
		Is. Hatae (116)	2.0 km^2
		Is. Soheuksan (117)	8.6 km^2
		Is. Chuja (118)	4.0 km^2
		Is. Nowha (121)	25.0 km^2
		Is. Soan (122)	–
		Is. Daemo (124)	–
		Is. Shinji (125)	25.1 km^2
		Is. Kokeum (126)	43.4 km^2
		Is. Cheongsan (127)	38.4 km^2
		Is. Yeoseo (128)	1.2 km^2
		Is. Sangil (129)	11.3 km^2
		Is. Pyung (132)	–
		Is. Kwang (133)	1.2 km^2
		Is. Oenaro (134)	26.5 km^2
		Is. Dolsan (135)	66.6 km^2
		Is. Namhae (136)	297.0 km^2
		Is. Kal (137)	–
		Is. Yeonwha (138)	1.5 km^2
		Is. Kuk (139)	–
		Is. Chishim (142)	–
		Is. Chinwoo (143)	–
		Is. Ulreung (145)	66.0 km^2 984 m
	183	Pyunsan (76)	–
		Mt. Chiri (80)	1915 m
The North–South Disjunctive Region VIII	3	Myungcheon (9)	–
		Pungsan (12)	–
		Pyungkang (27)	–
		Wanju (69)	–

(Compiled from various sources)
* Where applicable.
– No data available.

rufescens, *R. aureum* and *Vaccinium vitis-idaea* var. *minus* grow from 800 to 2300 m.

Overall, the southern evergreen broadleaved plants and evergreen bamboo are widespread from low- to mid-altitudes, and the northern evergreen broadleaved plant and evergreen coniferous plants are dominant from mid- to high-altitudes. The endemic species *Buxus koreana* grows in sites 21, 23 and 40 from 300 to 500 m, and *Thuja koraiensis* is found in sites 14, 19, 20, 21, 22, 23, 24, 25, 33, 34 and 51 from 400 to 2300 m.

Though the Midland Mountain Region (III) comprises 16 species, only five species provide data which show their altitudinal ranges (Table 2.9).

Two distinctive vertical belts are noticeable: first, the southern evergreen broadleaved plant *Masakia japonica* and the evergreen coniferous plant *Biota orientalis* exist from 200 to 400 m;

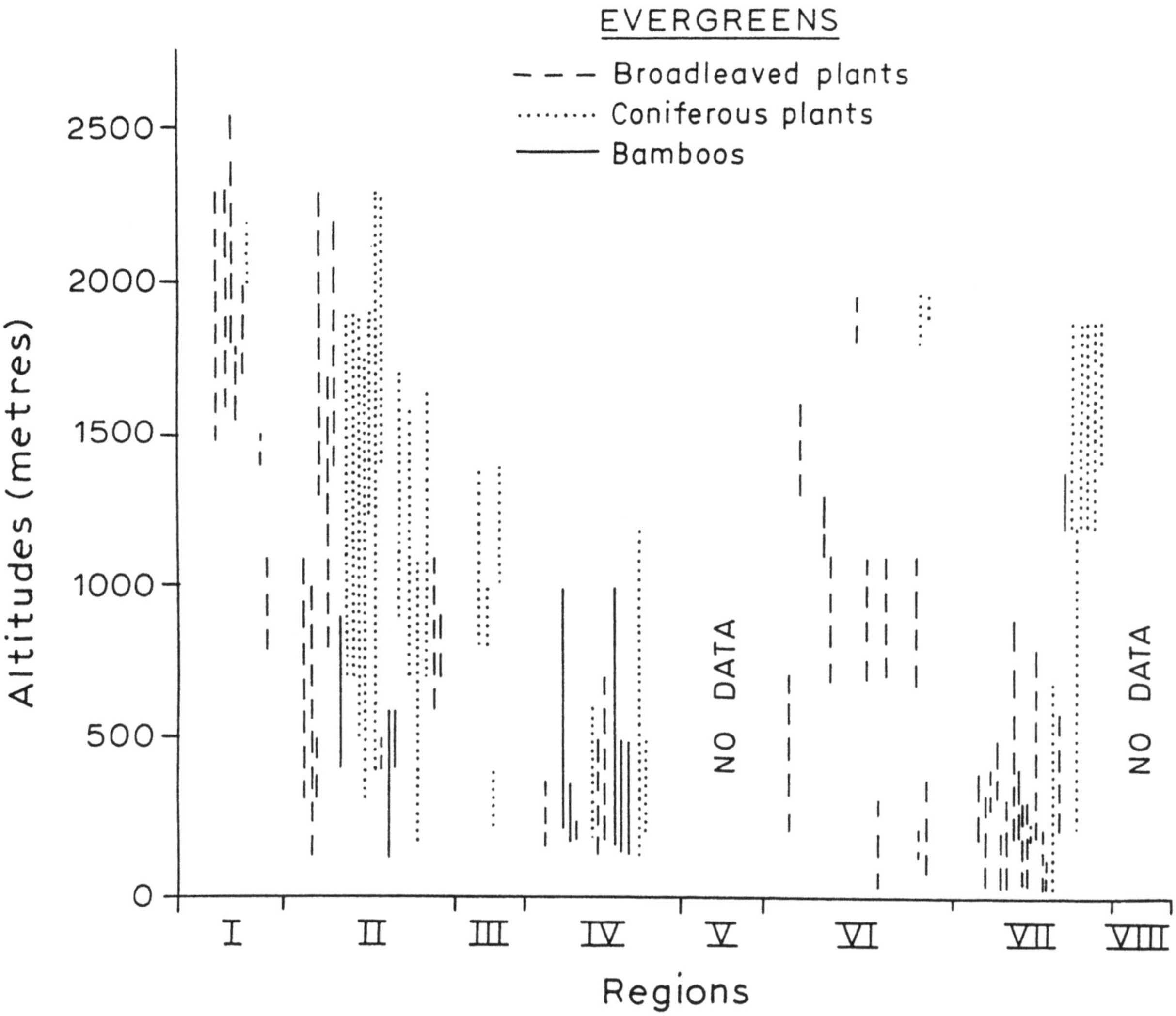

Fig. 2.6. Altitudinal ranges of species in the biogeographic regions of Korea.

and secondly, the northern evergreen coniferous representatives of *Taxus*, *Abies* and *Thuja* inhabit a zone of from 800 to 1400 m. Overall, the southern evergreen broadleaved plants and the southern evergreen coniferous plants are more common in the lower sites, while the northern evergreen coniferous plants are only found in high altitudes. The endemic species *Thuja koraiensis* grows in site 34 from 1100 to 1400 m. The demarcation of these distinctive altitudinal ranges reflects the transitional nature of this zone of vegetation, which contains both northern and southern floristic elements.

For the Southern Mountain Region (IV), which contains 33 species, information is available for the altitudinal ranges of 18 of these (Table 2.10).

Four vertical belts can be defined, namely one consisting of representatives of the southern evergreen broadleaved genera, *Buxus*, *Masakia*, *Thea*, *Trachelospermum*, *Zanthoxylum* and *Hedera* of from sea level up to 700 m; another recognised by the presence of the southern evergreen bamboos, *Sasamorpha*, *Sinoarundinaria* and *Pseudosasa*, from 100 to 1000 m; one in which the southern evergreen coniferous plants, *Pinus thunbergii*, *Biota orientalis* and *Torreya nucifera* are located, from sea level to 600 m; and finally one of the northern evergreen coniferous plants,

Table 2.7. Altitudinal ranges of species in the Northern Alpine Region of Korea (site number, and range in metres in parentheses).

Species Name	Altitudinal Range
Ledum palustre subsp. *decumbens*	(5, 1500–2300)
**L. palustre* var. *maximum*	(5, -1600)
L. palustre var. *minus*	(5, 1600–2300)
L. palustre var. *procumbens*	(5, 1800–2540)
Phyllodoce caerulea	(5, 1540–1800)
Rhododendron aureum	(7, 1800–2000) (15, 1700–2000)
Pinus pumila	(7, 2000–2200) (15, -2000)
**Juniperus sibirica*	(7, -2205)
J. utilis	(7, 450–550)
Rhododendron fauriae form. *rufescens*	(17, 1400–1500)
R. parvifolium	(13, 800–1100)

* Endemic species.

Abies holophylla and *Pinus koraiensis*, and one of the southern evergreen coniferous plants, *Abies koreana*, ranging up to 1200 m. Overall, the southern evergreen broadleaved plants and evergreen coniferous plants are more common at lower altitudes, and the southern evergreen bamboo and northern evergreen coniferous plants reach higher altitudes than other species. The endemic species *Abies koreana* occurs on sites 68, 70, 74, 80 and 83, and grows up to 1000 m a.s.l.

The Mid-Western Insular Region (V) includes 15 species, but no vertical range data is available for any of them.

For the Southern Insular Region (VI), which contains 77 species, the vertical ranges of 31 species have been determined in Table 2.11.

There are four distinctive altitudinal range groups here. First, the southern evergreen broadleaved plants, *Myrica*, *Machilus*, *Elaeocarpus* and *Bladhia crenata* var. *taquetii*, occur up to 350 m.

Photo 2.9. Timber-line at 2000 m a.s.l. on Mt. Paikdoo. *Betula ermanii* and *Rhododendron aureum*. (Photograph by Y. N. Lee.)

Photo 2.10. Some species of the herb layer at the timber-line of Mt. Paikdoo: *Androsace capitata* (white), *Potentilla nivea*, *Saussurea alpicola* (silver leaves). (Photograph by M. Šrůtek.)

Secondly, additional southern evergreen broadleaved plants range up to 700 m. Thirdly, other southern evergreen broadleaved plants, e.g. *Viscum*, *Buxus*, *Viburnum* and *Eurya emarginata*, grow from 700 to 1600 m; and finally, the northern evergreen broadleaved plants, *Diapensia*, *Empetrum* and *Rhododendron*, and the northern evergreen coniferous plants, representatives of *Taxus* and *Sabina*, are found from 1800 to 1950 m. Overall, most southern evergreen broadleaved plants are located in the lowlands, but some grow from mid- to high-altitudes as well; in contrast the northern evergreen broadleaved plants and evergreen coniferous plants are confined to the higher altitudes. The endemic species *Abies koreana* in site 119 grows from 1200 to 1900 m.

The West-South-Eastern Insular and Associated Inland Region (VII) has 41 species, and 21 altitudinal ranges are obtainable from this group (Table 2.12).

Two clear altitudinal range divisions are present. First, southern evergreen broadleaved plants and evergreen coniferous plants grow up to 800 m; and secondly, northern evergreen coniferous plants are present from 1200 to 1900 m. Overall, the southern evergreen broadleaved plants and evergreen coniferous plants grow best in the lowlands, but the northern evergreen coniferous plants are confined to high altitudes. The endemic species *Abies koreana* occurs in site 80, from 1200 to 1900 m. The limits of this region nearly coincides with the northern distributional limits of many southern evergreen broadleaved trees, as designated by Uyeki (1933, 1941).

The North-South Disjunctive Region (VIII) includes five species, but no altitudinal range data are available for these. However, most are local endemics (pp. 24–25).

In brief, the number of arctic-alpine evergreen broadleaved plants and evergreen coniferous plants decreases sharply as one moves from the northern alpine region (11 spp.) to the southern insular region (5 spp.), and they are confined in their range to high altitudes in the south. The northern evergreen coniferous plants, which flourish both at mid- and high-altitudes in the north, also are limited in their ranges to high altitudes in the south. The southern evergreen broadleaved plants, which grow only in the lowlands of northern Korea, expand their ranges to low- and mid-altitudes in the south.

Table 2.8. Altitudinal ranges of species in the North-South Subalpine Region of Korea (site number, and range in metres in parentheses).

Species Name	Altitudinal Range
Rhododendron dauricum	(1, 300–1000), (2, 600–1100)
Viscum coloratum var. *lutescens*	(14, −700), (16, −700), (22, 100–200), (24, 900–1000), (51, 900–1000), (70, 600–800)
**Buxus koreana*	(21, -500), (23, 300–500)
Rhododendron aureum	(16, -1300), (20, 1700–2300), (21, 1600–1900), (22, 1600–1800), (24, 1400–1500)
R. fauriae form. *rufescens*	(22, 800–1700), (23, 1100–1200)
Vaccinium vitis-idaea var. *minus*	(4, -1400), (20, 1900–1200), (22, 1800–1860)
Sasamorpha purpurascens var. *borealis*	(21, 800–900), (22, 400–600), (24, 700–900)
Taxus cuspidata	(14, 700–1500), (16, 750–1400), (21, 1000–1900), (23, 900–1500), (24, 1300–1550)
Abies nephrolepis	(4, 800–1400), (14, 700–1600), (16, 700–1400), (19, 1000–1250), (20, 1300–1800), (21, 700–1900), (22, 900–1750), (23, 1000–1400), (24, 900–1450)
Picea jezoensis	(4, 500–1400), (14, 700–1600), (16, 800–1450), (19, 1000–1250), (20, 1300–1800), (21, 800–1900), (22, 1200–1750), (23, 1000–1400)
Pinus koraiensis	(4, 800–1200), (14, 600–1500), (16, 300–1450), (19, 700–1200), (20, 1300–1700), (21, 300–1500), (22, 400–1750), (23, 700–1300), (24, 700–1450)
Pinus pumila	(14, 1500–1600), (16, -1350), (21, 1600–1900), (22, 1400–1750),(23, -1450), (25, 900–1680), (33, 1500–1700)
**Thuja koraiensis*	(14, 1000–1600), (19, 700–900), (20, 1900–2300), (21, 1000–1800), (22, 400–1750), (23, 1300–1400), (24, 1400–1500), (25, 800–1600),(33, 700–1600)
Sabina sargentii	(16, -1600), (20, -2300), (21, 1600–1900), (22, 1600–1750), (24, -1400), (25, 1000–1650), (23, 700–800)
Hypear tanakae	(33, 400–500)
Pseudosasa japonica	(25, 100–500), (33, 400–600)
Sinoarundinaria reticulata	(25, 400–600)
Biota orientalis	(33, -300)
Juniperus utilis	(25, 800–1100), (33, 150–550)
Sabina chinensis	(25, -800)
Rhododendron fauriae var. *roseum*	(44, 700–900)

* Endemic species.

Results and discussion

From the basis of the distribution of 204 evergreen broadleaved plants, evergreen coniferous plants and evergreen bamboos on 146 sample sites, eight biogeographic regions of the Korean peninsula have been established by the use of the TWINSPAN programme.

The Northern Alpine Region (I) consists of 18 arctic-alpine and alpine evergreen broadleaved plants, along with four circumpolar and alpine evergreen coniferous plants and includes 11 sample sites. Six life forms are present, namely trees, dwarf small trees, shrubs, small shrubs, dwarf shrubs and dwarf small shrubs. Shrubs and small shrubs with relatively small leaves are the most common. The large number of decumbent procumbent, dwarf shrub and small shrub forms reflects an abundance of arctic-alpine species, which may in turn be linked to the existence of refugia during past glacial periods. The morphological adaptations of these arctic-alpine and alpine species against severe environmental conditions is thought by us to have encouraged the continued survival of these species in such a cryo-climatic area. There are four distinctive altitudinal range divisions; namely one at 450 to 550 m, exemplified by northern evergreen coniferous plants, such as *Juniperus utilis*; another at 800 to 1500 m, in which the northern evergreen broadleaved plants, *Rhododendron fauriae* form. *rufescens* and *R. parvifolium*, are present; one at 2000 to 2205 m in which the northern evergreen coniferous plants *Pinus pumila* and *Juniperus sibirica* occur; and a fourth at 1500 to 2540 m, indicated by the northern evergreen broadleaved plants,

Photo 2.11. Falls near Samjiyon Lakes, about 900 m. a.s.l., Kaema Upland, northern part of North Korea. *Larix olgensis* and *Abies nephrolepis* are dominant trees in the forest. *Betula platyphylla*, *Picea koraiensis* and *P. jezoensis* are frequent. (Photograph by M. Šrůtek.)

Table 2.9. Altitudinal ranges of species in the Midland Mountain Region of Korea (site number, and range in metres in parenthesis).

Species Name	Altitudinal Range
Masakia japonica var. *latifolia*	(58, -400)
Taxus cuspidata	(34, 1100–1400), (45, 800–1150)
Abies nephrolepis	(45, 800–1000)
Biota orientalis	(24, 200–400)
**Thuja koraiensis*	(34, 1100–1400)

* Endemic species.

Ledum palustre var. *maximum*, *L. palustre* var. *minus*, *L. palustre* var. *procumbens*, *L. palustre* subsp. *decumbens*, *Phyllodoce caerulea* and *Rhododendron aureum*. The segregation of arctic-alpine and alpine species toward the mountain tops, above 1500 m a.s.l. in height, might be due to the upslope retreat of these species from a former wider range since the last Pleistocene glaciation, mainly due to subsequent climatic amelioration.

The North-South Subalpine Region (II) comprises 15 northern evergreen coniferous plants, 10 northern and southern evergreen broadleaved plants and three evergreen bamboos, and covers 21 sites. Trees and shrubs with small leaves are the typical life forms, together with dwarf small trees, dwarf shrubs, dwarf small shrubs and bamboos. A wide range of shrub and tree forms, caused by an admixture of arctic-alpine northern and southern species elements, infers a complex vegetation history, along with diverse present-day habitat conditions. Seven patterns of altitudinal range may be demarcated; namely one at 100 to 700 m, indicated by the southern evergreen broadleaved plants *Viscum coloratum* var. *lutescens*, *Hypear tanakae* and the endemic *Buxus koreana* and the southern bamboo *Pseudosasa japonica*; another at 300 to 1100 m, characterised by the northern evergreen broadleaved plants *Rhododendron dauricum* and *R. fauriae* var. *roseum*; a third from 300 to 2300 m, in which are the northern evergreen coniferous plants, *Pinus koraiensis* and the endemic *Thuja koraiensis*; a fourth from 400 to 900 m containing the southern evergreen bamboos *Sasamorpha purpurascens* var. *borealis* and *Sinoarundinaria reticulata*; a fifth from 500 to 1900 m, with the northern ever-

Table 2.10. Altitudinal range of species in the Southern Mountain Region of Korea (site number and range in metres in parentheses).

Species Name	Altitudinal Range
Buxus microphylla	(74, -200)
Masakia japonica var. *latifolia*	(74, 150–350)
M. radicans	(66, -770)
Thea sinensis var. *bohea*	(74,-220)
Trachelospermum asiaticum var. *intermedium*	(74, -200), (77, -300)
Sasamorpha purpurascens var. *borealis*	(66, 200–800), (74, 200–1000)
Sinoarundinaria nigra var. *henonis*	(74, 150–350), (83, 100–500)
S. reticulata	(60, -200), (66, -150), (74, -200), (77, -250), (82, 100–500)
**Abies koreana*	(74, -1000)
Pinus thunbergii	(74, -150)
Biota orientalis	(66, -600), (74, 150–350)
Zanthoxylum planispinum	(77, 200–300), (83, 100–500)
Hedera tobleri	(77, -300), (83, 150–700)
Pseudosasa japonica	(60, -150), (83, 150–1000)
Torreya nucifera	(77, -400)
Cephalotaxus koreana	(60, 100–700), (77, 100–300), (83, 150–1200)
Abies holophylla	(60, -200), (77, -400), (83, -800)
Pinus koraiensis	(60, -1200)

* Endemic species.

Table 2.11. Altitudinal ranges of species in the Southern Insular Region of Korea (site number and range in metres in parentheses).

Species Name	Altitudinal Range
Piper kadzura	(119, -700)
Sarcandra glabra	(119, -700)
Myrica rubra	(119, -250)
Cyclobalanopsis stenophylla	(119, -250)
Scurrula yadoriki	(119, 200–700)
Viscum coloratum var. *lutescens*	(119, 1300–1600)
Stephania japonica	(119, -700)
Cinnamomum camphora	(119, -700)
Daphniphyllum glaucescens	(119, -700)
Buxus microphylla	(119, 1100–1300)
Elaeocarpus sylvestrus var. *ellipticus*	(119, -300)
Eurya emarginata	(119, 700–1100)
Sasakia ochnacea	(119, -700)
Ternstroemia mokof	(119, -700)
Xylosma apathis	(119, -700)
Diapensia lapponica subsp. *obovata*	(119, 1800–1950)
Empetrum nigrum var. *japonicum*	(119, 1800–1950)
Rhododendron dauricum	(119, -1950)
Bladhia crenata	(119, -700)
B. crenata var. *taquetii*	(119, 0–300)
B. villosa var. *typica*	(119, 700–1100)
Dicalix prunifolia	(119, -700)
Ligustrum lucidum	(119, -700)
Damnacanthus indicus	(119, -700)
D. major	(119, -700)
Viburnum awabuki	(119, 700–1100)
Taxus cuspidata	(119, 1800–1950)
**Abies koreana*	(119, 1200–1900)
Sabina sargentii	(119, -1950)
Machilus thunbergii form. *obovata*	(88, 100–200)
Eurya japonica var. *montana*	(88, 50–350)

* Endemic species.

green coniferous plants *Taxus cuspidata*, *Abies nephrolepis* and *Picea jezoensis*; a sixth from 800 to 2300 m with the northern evergreen broadleaved plants *Rhododendron aureum*, *R. fauriae* form. *rufescens* and *Vaccinium vitis-idaea* var. *minus*; and finally, one from 1350 to 2300 m, in which are the northern evergreen coniferous plants *Pinus pumila* and *Sabina sargentii*. The isolation of arctic-alpine and alpine species in subalpine belts of above ca. 1500 m in altitude can be explained by the same mechanisms suggested in the case of region I.

The Midland Mountain Region (III) contains 12 northern evergreen coniferous plants and four evergreen broadleaved plants, and three evergreen bamboos, and possesses 16 sites. Four life forms are found, namely trees, dwarf small trees, shrubs and dwarf shrubs; shrubs and trees with small leaves are characteristic. The sharp decrease in the total number of species (16 spp.) as compared with region II (above) (2 spp.), and the adjacent southern mountain region (33 pp.), and the appearance of four southern evergreen broadleaved plants, viz. *Hypear tanakae*, *Bifaria japonica*, *Masakia japonica* and *M. japonica* var. *latifolia*, and three southern evergreen coniferous plants, *Cephalotaxus nana*, *Biota orientalis* and *B. orientalis* var. *stricta*, indicate that this region is probably a major inland transitional zone of vegetation, intermediate between the northern

Table 2.12. Altitudinal ranges of species in the West-South-Eastern Insular and Associated Inland Region of Korea (site number and range in metres in parentheses).

Species Name	Altitudinal Range
Cyclobalanopsis acuta	(86, 170–400), (87, -300)
Machilus thunbergii	(87, -200), (145, 0–300)
Neolitsea sercea	(86, 270–400)
Pittosporum tobira	(145, 300–500)
Ilex integra var. *typica*	(145, 0–200)
Masakia japonica	(86, -300), (145, 0–300)
M. radicans	(86, 170–300), (145, 300–900)
Eurya japonica var. *montana*	(86, 170–400), (87, 100–200)
Elaeagnus glabra var. *euglabra*	(145, 0–300)
E. macrophylla	(87, -256),(145, 0–300)
Hedera tobleri	(86, -170), (87, -200), (145, 200–800)
Aucuba japonica var. *typica*	(145, 0–200)
Vitex rotundifolia	(145, 0–100)
Pinus thunbergii	(86, 200–400), (87, 0–400), (145, 0–700)
Thea sinensis var. *bohea*	(80, 200–600)
Sasamorpha purpurascens var. *borealis*	(80,1200–1400)
Taxus cuspidata	(80, 1200–1900)
Cephalotaxus koreana	(80, 200–1200)
**Abies koreana*	(80,. 1200–1900)
Pinus koraiensis	(80, 1200–1900)
Sabina sargentii	(80, 1400–1900)

* Endemic species.

and southern floristic elements. There are two clear vertical range divisions, namely one from 200 to 400 m characterised by the southern evergreen broadleaved plant *Masakia japonica* var. *latifolia* and the southern evergreen coniferous plant *Biota orientalis*, and another from 800 to 1400 m, dominated by the northern evergreen coniferous plants *Taxus cuspidata*, *Abies nephrolepis* and the endemic *Thuja koraiensis*.

The Southern Mountain Region (IV) includes 15 northern and southern evergreen coniferous plants, 12 southern evergreen broadleaved plants and six southern evergreen bamboos, and covers 25 sites. Six life forms are present, namely trees, shrubs, dwarf shrubs, dwarf small shrubs, vines and bamboos; shrubs and trees having small to large leaves are most common. Four vertical range divisions are determined, namely one up to 600 m, in which the southern evergreen coniferous plants *Pinus thunbergii*, *Biota orientalis* and *Torreya nucifera* grow; another of up to 700 m, characterised by the southern evergreen broadleaved plants *Buxus microphylla*, *Masakia japonica* var. *latifolia*, *M. radicans*, *Thea sinensis* var. *bohea*, *Trachelospermum asiaticum* var. *intermedium*, *Zanthoxylum planispinum* and *Hedera tobleri*; a third, from 100 to 1000 m, exemplified by the southern evergreen bamboos *Sasamorpha purpurascens* var. *borealis*, *Sinoarundinaria nigra* var. *henonis*, *S. reticulata* and *Pseudosasa japonica*; and a fourth, up to 1200 m, in which the northern evergreen coniferous plants *Abies holophylla* and *Pinus koraiensis* are found. These suggest an admixture of species, some of which favour the presently-moderate climates of the lowlands, and others being tolerant of the harsh climatic conditions of the highlands. It is also worth noting that many species in this region have a very large range of amplitudes. Overall, the dominance of southern evergreen broadleaved plants, evergreen bamboo and evergreen coniferous plants, and the appearances of southern vines and bamboos, makes the region 'southern' in character.

The Mid-Western Insular Region (V) contains 10 southern evergreen broadleaved plants and five northern evergreen coniferous plants, and includes 11 sites. Shrubs and vines with middle-sized leaves are the most common of the six life forms, namely trees, dwarf small trees, shrubs, dwarf shrubs, dwarf small shrubs and vines. Even though there are no altitudinal range data available, the admixture of northern and southern species elements found within this region suggests that it is an insular transitional zone of vegetation between northern to southern floristic elements.

The Southern Insular Region (VI) comprises 65 northern and southern evergreen broadleaved plants, with eight northern and southern evergreen coniferous plants and four bamboos; it has 11 sites. Eight life forms are present, namely trees, small trees, shrubs, dwarf shrubs, small shrubs, dwarf small shrubs, vines and bamboos; shrubs and trees with small to large leaves are the most common. A large number of southern evergreen broadleaved plants, and several northern evergreen broadleaved plants and evergreen coniferous plants, occur in four different altitudinal-range groupings. These are, first, one found up to 350 m, in which southern evergreen broadleaved plants, such as *Myrica rubra*, *Machilus thunbergii* form. *obovata*, *Elaeocarpus sylvestrus* var. *ellipticus* and *Bladhia crenata* var. *taquetii* are found; second, one up to 700 m, for the southern evergreen broadleaved plants *Piper kadzura*, *Sarcandra glabra*, *Scurrula yadoriki*, *Stephania japonica*, *Cinnamomum camphora*, *Daphniphyllum glaucescens*, *Sasakia ochnacea*, *Ternstroemia mokof*, *Xylosma apathis*, *Bladhia crenata*, *Dicalix prunifolia*, *Ligustrum lucidum*, *Damnacanthus indicus* and *D. major*; thirdly, one from 700 to 1600 m, exemplified by the southern evergreen broadleaved plants *Viscum coloratum* var. *lutescens*, *Buxus microphylla*, *Eurya emarginata*, *Bladhia villosa* var. *typica* and *Viburnum awabuki*; and finally, one from 1800 to 1950 m, characterised by the northern evergreen broadleaved plants *Diapensia japponica* subsp. *obovata*, *Empetrum nigrum* var. *japonicum* and *Rhododendron dauricum*, and the northern evergreen coniferous plants *Taxus cuspidata*, the endemic *Abies koreana* and *Sabina sargentii*. It should be noted that the region has the largest number of southern evergreen broadleaved plants of all the eight biogeographic regions, which reflects the relatively warm and humid climatic conditions of the present-day. However, a disjunctive distribution of several arctic-alpine and alpine species, present near the peak of Mt. Halla (above ca. 1800 m to the summit at 1950 m), relevant to other regions of Korea, seems to point to a much wider distribution for these species during the Pleistocene glaciations (see p. 139). In short, the highly complicated range of life forms, and the high floristic diversity are due to the survival of arctic-alpine, alpine, temperate, and subtropical plants together, which reflect the diverse present-day environment, habitat conditions and vegetation history.

The West-South-Eastern Insular and Associated Inland Region (VII), which includes 28 southern evergreen broadleaved plants, eight northern and southern evergreen coniferous plants and five evergreen bamboos, has 47 sites. Large-leaved shrubs, trees, bamboos and vines are the most frequent of the seven life forms found; the others are small trees, dwarf shrubs and dwarf small shrubs. Southern evergreen broadleaved plants and evergreen bamboos are common in the insular and coastal areas, but northern evergreen coniferous plants occur more frequently in the inland area. There are two clear altitudinal range divisions first, one up to 800 m, characterised by southern evergreen broadleaved plants like *Cyclobalanopsis acuta*, *Machilus thunbergii*, *Neolitsea sericea*, *Pittosporum tobira*, *Ilex integra* var. *typica*, *Masakia japonica*, *M. radicans*, *Eurya japonica* var. *montana*, *Elaeagnus glabra* var. *euglabra*, *E. macrophylla*, *Hedera tobleri*, *Aucuba japonica* var. *typica*, *Vitex rotundifolia* and *Thea sinensis* var. *bohea*, and one southern evergreen coniferous plant, *Pinus thunbergii*; and second, one from 1200 to 1900 m, in which northern evergreen coniferous plants, including *Taxus cuspidata*, *Pinus koraiensis* and *Sabina sargentii* grow. The second largest number of southern evergreen broadleaved plant species present here, as compared to all other regions, reflects the existence of relatively mild climatic conditions today. The proportion of shrubs and trees is nearly equal, and so too in the case of vines and bamboos. Overall, this region is biogeographically closely related to the southern insular region (VI).

The North-South Disjunctive Region (VIII), containing three northern evergreen coniferous plan and two southern evergreen bamboos, comprises four sites. There are three life forms, namely trees, shrubs and bamboos. The region is characterised by a high proportion of endemics (60.0%) for the total number of species in the region, and a high degree of physiognomic diversity for the small number of species present, which is so far unexplained.

Taking the Korean peninsula as a whole, one may speculate that the arctic-alpine and alpine evergreen broadleaved and evergreen coniferous plants are found mainly in regions dominated by higher altitudes, viz. the Northern Alpine Region (I), the North-South Subalpine Region (II) and in the mountains of the Southern Insular Region (VI). Northern evergreen coniferous plants are particularly common in inland areas, such as the North-South Subalpine Region (II), the Midland Mountain Region (III) and the Southern Mountain Region (IV), while southern evergreen broadleaved plants are frequent in island areas, such as the Southern Insular Region (VI) and the West-South-Eastern Insular and Associated

Inland Region (VII). Southern evergreen bamboos occur frequently in the Southern Mountain Region (IV), the West-South-Eastern Insular and Associated Inland Region (VII) and the Southern Insular Region (VI).

The percentage of endemic plants is relatively high in the two transitional areas of vegetation, namely the Mid-Western Insular Region (13.3%) and the Midland Mountain Region (12.5%), as compared to values of 10.7% in the North-South Subalpine Region, 10.4% in the Southern Insular Region, 9.0% in the Northern Alpine Region, 7.3% in the West-South-Eastern Insular and Associated Inland Region and 6.0% in the Southern Mountain Region. The high degree of endemism in the North-South Disjunctive Region (60.0%) reflects mainly the small number of species present overall in this region.

Shrub forms of plants, namely shrubs, dwarf shrubs, small shrubs and dwarf small shrubs, are predominant in the Northern Alpine Region, are abundant in the North-South Subalpine Region and in the southern Insular Region, but equal in number to the number of tree forms in other regions. Tree forms are rare in the Northern Alpine Region, but are common in the Midland and Southern Mountain regions. Bamboos begin to appear in inland areas from the North-South Subalpine Region southwards, and grow well in the Southern Insular Region, the West-South-Eastern Insular and Associated Inland Area and in the North-South Disjunctive Region. Vines are confined to the south.

Highly adaptive leaf trichome forms are found in the Northern Alpine Region and the Southern Insular Region, mainly due to the occurrences of arctic-alpine plants in these regions. The high proportion of revolute leaves in the Northern Alpine Region, the North-South Subalpine Region and the Southern Insular Region also relates to the existence of arctic-alpine plants, such as *Andromeda polifolia*, *A. polifolia* form. *acerosa*, *Oxycoccus quadripetalus*, *O. microcarpus*, *Vaccinium vitis-idaea* var. *genuinum*, *Ledum palustre* var. *yesoense*, *Phyllodoce caerulea*, *Empetrum nigrum* var. *japonicum*, *Rhododendron aureum* and *R. fauriae* var. *rufescens* in the Northern Alpine Region, *Rhododendron aureum*, *R. fauriae* form. *rufescens*, and *Vaccinium vitis-idaea* var. *minus* in the Southern Insular Region. The large number of coriaceous leaves in the Southern Mountain Region, the Southern Insular Region and the West-South-Eastern Insular and Associated Inland Region indicates the dominance there of both warm-temperate and subtropical plants, as well as the presence of several arctic-alpine and alpine species such as *Diapensia lapponica* subsp. *obovata*, *Rhododendron dauricum*, *R. fauriae* var. *roseum*, *R. fauriae* form. *rufescens* and *Vaccinium vitis-idaea* var. *minus*.

Arctic-alpine and alpine evergreen broadleaved plants and evergreen coniferous plants are the dominant group in the north, but are confined to high altitudes in the south. Northern evergreen coniferous plants are common in mid- to high-altitudes in the north, but also limit their range to high altitudes in the south. Generally, southern evergreen broadleaved plants are common both in low- to high-altitudes but grow only in lowland districts in the north.

Overall, the rich species composition, various physiognomic types and the distinctive horizontal and vertical distribution patterns of species growth in Korea reflect the existence of not only a wide variety of habitats, but also the diverse vegetation history of the peninsula. This makes the Korean flora, vegetation and biogeography quite unique.

It is probably a fair reflection of previous work that earlier classifications of Korean vegetation have failed to account for these features satisfactorily. Thus, in 1933, Uyeki (Table 1.2) categorised the Korean flora into only five vegetation zones, namely a cold zone, in which representatives of *Abies*, *Picea*, *Pinus* and *Thuja* were present; secondly, a northern temperate zone, with *Acer*, *Populus*, *Juglans*, *Prunus*, *Tilia* and *Betula* species; thirdly, a central temperate zone, with *Acer*, *Prunus*, *Quercus* and *Maackia* species; fourthly, a southern temperate zone, with *Meliosma*, *Acer*, *Ulmus* and *Cephalotaxus* species; and finally, a warm temperate zone, with *Textoria* and *Ilex* representatives. This division undervalues the arctic-alpine plant element, and the southern evergreen elements, which are especially important for an understanding of both the past vegetation history and the present-day environmental conditions of the peninsula. In 1961, Hyun identified three vegetation zones: first, a cool zone, with *Picea*, *Abies*, *Pinus* and

Larix representatives; secondly, a temperate zone, with *Quercus*, *Betula*, *Acer*, *Tilia*, *Styrax*, *Platycarya*, *Euonymus*, *Zanthoxylum*, *Abies*, *Pinus*, *Larix*, *Juniperus* and *Cephalotaxus* species; and thirdly, a subtropical zone, with *Quercus*, *Castanopsis*, *Cinnamomum*, *Machilus* and *Vaccinium* species. This grouping reflects the general floristic elements reasonably well, but is too simplified for both the inland and island areas. In the same year, Wang demarcated five vegetation zones: first, a northern high mountain area, with *Picea* and *Abies* species; secondly, a northern mountain area, with *Acer*, *Tilia* and *Betula* species; thirdly, a midland area, with *Quercus* species; fourthly, a southern mountain area, with mixed mesophytic forest; and finally, a southernmost area, with *Castanopsis* and *Cinnamomum* species. However, he failed to represent truly the overall pattern of species composition in the peninsula, and ignored island areas entirely. In 1968, Yim also determined five vegetation zones for Korea; first, a subarctic zone, with *Picea* and *Abies* species; secondly, a northern cool temperate forest zone, with *Betula*, *Larix* and *Pinus* species; thirdly, a central cool temperate zone with *Quercus* and *Abies* representatives; fourthly, a southern cool temperate forest zone, with *Carpinus*, *Acer* and *Pinus* species; and finally, a warm temperate forest zone in the south. This classification is similar to Wang's (1961) division, but again ignores island areas.

Several other classifications followed during the 1970s. Hamet-Ahti *et al.* (1974) subdivided the peninsula into four vegetation zones, namely first, a hemiboreal subzone, with *Pinus* and *Abies* species; secondly, a temperate zone, with deciduous forest; thirdly, a hemitemperate subzone; and finally, a southern meridional subzone, with *Castanopsis*, *Shilla*, *Machilus* and *Cyclobalanopsis* species. However, the northern floristic element, and the influence of vertical ranges on vegetation pattern are poorly understood in his presentation. Lee (1976) produced three plant regions, viz. first, a cold temperate plant region, with *Abies*, *Picea*, *Pinus*, *Juniperus*, *Taxus*, *Thuja*, *Larix*, *Vaccinium*, *Quercus*, *Chosenia*, *Betula*, *Salix* and *Syringa* species; secondly. a temperate plant region, with *Pinus*, *Abies*, *Quercus*, *Carpinus*, *Betula*, *Acer*, *Fraxinus*, *Salix*, *Tilia* and *Rhododendron* species; and thirdly, a warm temperate plant region, with *Camellia*, *Cinnamomum*, *Machilus*, *Neolitsea*, *Ilex*, *Castanopsis*, *Citrus* and *Pittosporum* species. But the work is too simple to reflect environmental realities and also does not pay attention to the influence of vertical ranges on distribution patterns. In 1977, Yim & Kira categorised five vegetation zones; a subarctic zone, with *Abies* and *Picea* species; secondly, a northern zone, with *Ulmus*; thirdly, a middle zone, with *Quercus*; fourthly, a southern zone, with *Carpinus*; and finally, a warm-temperate zone, with evergreen broadleaved species. Their aim to select characteristic plants for comparison with the observed thermal climate was, in the event, not a wholly satisfactory means of categorisation. Finally, Hagman *et al.* (1978) produced five vegetation belts: first, a conifer belt; secondly, an *Abies–Betula* belt; thirdly, a *Quercus–Abies* belt; fourthly, a *Pinus–Acer* belt; and finally, a *Camellia* belt in the south. This scheme is also based on indicator species, and floristics overall are not considered.

Previous floristic divisions of Korea are either based on regional species composition or on species diversity. Nakai (1919, 1935) reported five floristic zones: first, a northern zone; secondly, a central zone; thirdly, a southern zone; fourthly, Cheju Island; and finally, Ulreung Island. However, he failed to give criteria for the demarcation of these zones. Oh (1977) suggested a slightly modified form of Nakai's division, with five floristic regions, namely a northern region; a central and southern region; a southern insular region; Cheju Island; and Ulreung Island. The inland areas are slightly undervalued in his scheme of zonation. Lee & Yim (1978) recognised six floral regions, viz. a Gabsan, Gwanbug and Gwanseo region in the north; a middle region; a southern region; a southern coastal region; Cheju Island; and Ulreung Island. The demarcation of the northernmost floral region and a southern coastal floral region shows some similarity with the present work.

Overall, however, there are several major differences within the analysis presented in this study and those of previous workers. This is due to the following reasons. First although there are a large number of plant species (ca. 4164 species (Lee, 1982a)) present in Korea, we have mainly concentrated on an understanding of the distribution of 204 evergreen broadleaved plants, evergreen coniferous plants and evergreen bamboos at the species level, which are together widely distributed within the peninsula. Therefore, we have put together a broad and manageable floristic base relevant for the whole peninsula, which not only provides the delineation of biogeographic regions, but also emphasises habitat differences. Secondly, we have employed a quantitative, non-subjective method of data analysis, notably by the use of the TWINSPAN programme. Thirdly, we have incorporated analyses of floristic, physiognomic and bioclimatic characteristics of species, which are especially useful in interpreting vegetation pattern and structure. Fourthly, the three-dimensional presentation of vegetation data includes the analyses of horizontal and vertical distribution patterns of plant species, which makes a major contribution not only to the understanding of present vegetation but also to past vegetation history and present-day ecological amplitudes. Finally, we have produced a more sophisticated biogeographic division of present-day vegetation in the Korean peninsula than any presented heretofore, paying considerable attention to floristics, physiognomy and range (spatial patterning). The eight biogeographic regions delineated have relevance not only to present-day, but also to past habitat situations, and this relevance is further stressed and elaborated upon in the next chapter.

CHAPTER 3

Vegetation history

This chapter reviews the vegetation history of Korea and also, through this, suggests origins and formation processes for the present-day Korean biogeographic divisions. Since there is as yet no information relevant to the vegetation history of the peninsula prior to the Carboniferous period, this study is restricted solely to those periods between the Carboniferous and the present day.

Data sources

Collection of data

Data required for the interpretation of early prehistoric vegetational patterns comes from plant fossils, both macroscopic and microscopic, including the remains of wood, bark, leaves, budscales, fruits, seeds, pollen, spores and diatoms. For more recent vegetation history, dendrochronology or tree-ring analysis, may also be utilised; and written records of past vegetation, phenological phenomena, ancient inscriptions and landscape paintings are helpful particularly in the reconstruction of the latest historic vegetational changes (Lamb, 1970, 1982; Ingram *et al.*, 1981a, b; Farmer & Ford, 1984; Bradley, 1985).

We have employed most of these now-traditional data sources here, along with some further indirect evidence of environmental change, such as glacial and periglacial history, recent palaeoclimates, sea level fluctuations and archaeological artefacts, to assess and evaluate both the long-term and more recent vegetation changes of the Korean peninsula, which have particular relevance for the understanding of the origins of the existing plant communities within it.

Analysis of data

The collected data are presented chronologically, using as a base the recent schema produced by Lee (1987), as displayed in Table 3.1. However, on the basis of the fossil flora data, a more complete picture is provided than that presented in Lee's work. For the sake of convenience this review of the vegetation history of Korea is further divided into three: that of the pre-Pleistocene, the Pleistocene and the Holocene. The boundary of each geological phase is based on the time scales adopted by Holmes (1978).

Vegetation of the pre-Pleistocene Period

Three geologic eras are represented within this discussion, namely the Palaeozoic, the Mesozoic and the early Cenozoic (Tertiary). According to Cox (1974), Korea has always been part of the Asian continent and was located at roughly the same latitudinal position as at present between the Silurian period of the Palaeozoic and the Upper Cretaceous (Fig. 3.1.)

Most of the evidence for the existence of particular vegetation types in the pre-Pleistocene of the peninsula is of a macrofossil nature, coming in particular from preserved fossil leaves. The geographic locations of the known pre-Pleistocene sites, which are to be discussed, are presented in Fig. 3.2; and the main indications arising out of the data they contain are as follows.

The Palaeozoic Era (ca. 600–270 my B.P.)

As indicated above, the available evidence is restricted to data from the Carboniferous and Permian periods.

Table 3.1. Chronology of Korea.

Geologic Period			Stratigraphy
Era	Period		Layer
		Holocene	Lake Youngland
Cenozoic	Quaternary	Pleistocene	Lake Younglang, Hwadae, Sokchang-ni Turubong, Chummal Yonggul Hesang, Cheonchong-am
			–
	Tertiary	Pliocene	Seoguipo
		Miocene	Yeonil, Hamjindong
			Kogeonweon, Janggungri, Tongchon
			Yongdong, Changii
		Oligocene	Pongsan
		Eocene	–
		Palaeocene	–
Mesozoic	Cretaceous		Sub-Nagdong
	Jurassic		Myogog Seonyeon
	Triassic		Kobangsan
Palaeozoic	Permian		Sadong
	Carboniferous		Hongjum

(Compiled from Huzioka, 1951, 1972; Yoo, 1970; Kim, 1980; Yasuda *et al.*, 1980; Choi, 1983, 1986; Park, 1983; Sohn, 1984; Lee, 1987.)

Carboniferous Period (ca. 350–270 my B.P.)
Plant macrofossils of four genera are present in the Hongjum layer of the Korean Carboniferous, viz. *Lepidodendron*, *Neuropteris*, *Linopteris* and *Odontopteris*. The occurrence of such hydrophilous plants suggests a warm and humid climate. In this period, indeed, climates were relatively warm all over the world, though they became somewhat colder in the Upper Carboniferous period (Garrels & Mackenzie, 1971; Chaloner & Meyen, 1973; Dickins, 1984).

Permian Period (ca. 270–225 my B.P.)
About 118 species of plant macrofossils have been described from the Sadong layer of the Korean Permian, among which are representatives of the following families: Equisetales (10 spp.), Sphenophyllales (15 spp.), Filicales and Pteridosperms (58 spp.), Cordaitales (12 spp.), Cycadophyta (6 spp.), Ginkgoales (5 spp.), Semina Gymnosperms (9 spp.) and Incertae Sedis (6 spp.). The dominance of ferns and gymnospermae, and the presence of a relatively rich flora, suggest that the local climates had started to warm up again after the relatively cool phase of the Upper Carboniferous. For the first time in geological history, clear evidence of the presence of distinctive world floristic provinces exists from the Permian (Mclean & Ivimey-Cook, 1967), though little in the way of local floristic differentiation within this period has yet been evidenced in Korea.

The Mesozoaic Era (ca. 225–70 my B.P.)

Triassic Period (ca. 225–180 my B.P.)
The Kobangsan layer of the Upper Permian and Lower Triassic is dominated by Filicales and Pteridospermae (45 spp.), along with Equisetales (11 spp.), Sphenophyllales (3 spp.), Psilophytales (3 spp.), Cordaitales (2 spp.), Coniferales (1 sp.), Semina Gymnosperms (4 spp.) and Incertae Sedis

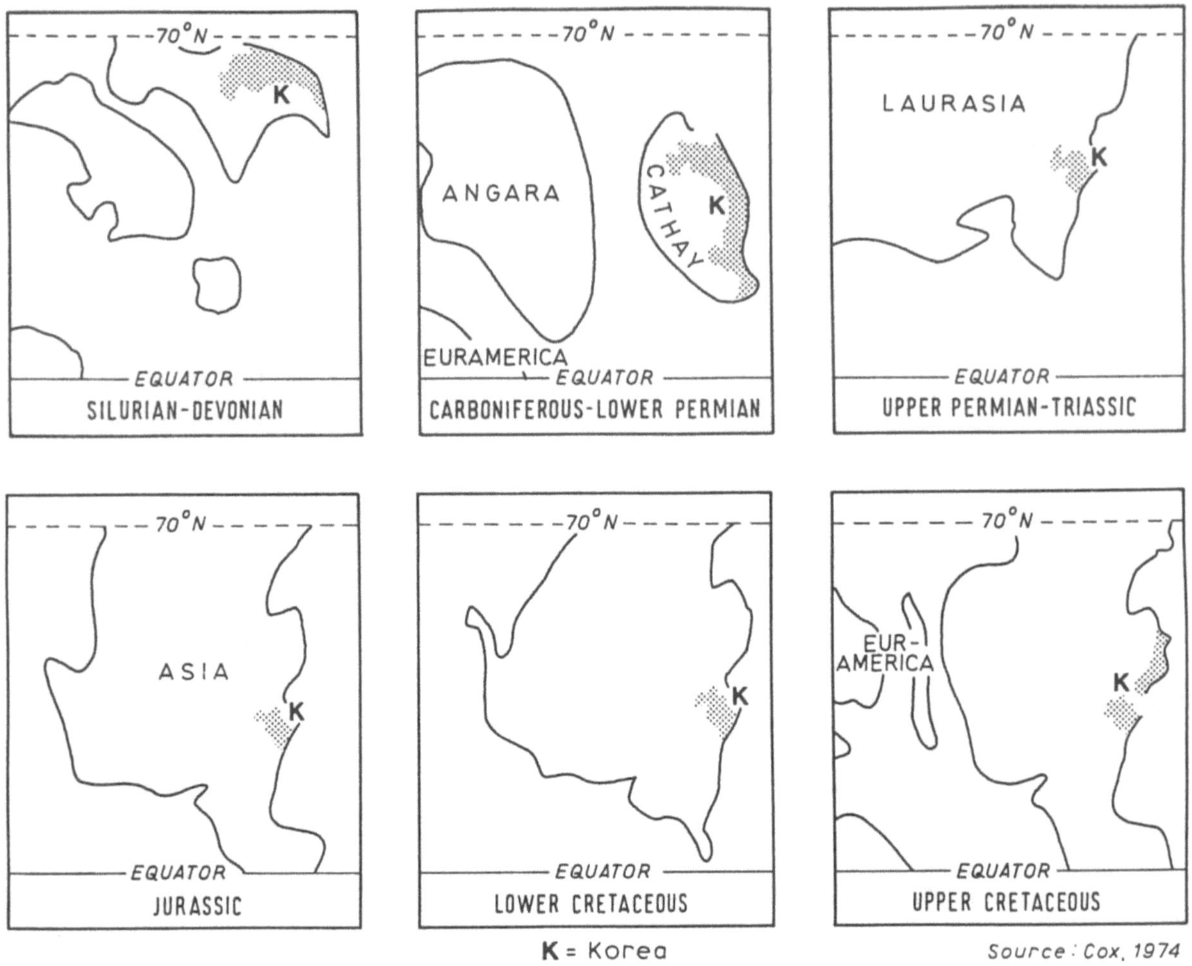

Fig. 3.1. The palaeography of Korea.

(7 spp.). Such groups suggest the existence of a further relatively warm and humid climate within this period.

Jurassic Period (*ca.* 180–135 *my B.P.*)
One hundred and twenty-one species of plant macrofossils have been located in the Seonyeon layer of the Upper Triassic and Lower Jurassic, viz. Filicales and Pteridospermae (32 spp.), Ginkgoites and Ginkgoales (23 spp.), Equisetales (15 spp), Bennettitales (13 spp.), Cycadales (10 spp.), Coniferales (15 spp.) and Incertae Sedis (13 spp.). From the Upper Jurassic ten species are present within the Myogog layer, viz. *Cladophebis* (3 spp.), *Adiantites* (1 sp.), *Onchyopsis* (2 spp.), Equitites (1 sp.), Ginkgoales (1 sp.), *Nilssonia* (1 sp.) and Podozamites (1 sp.). The presence of ferns and gymnospermae again indicate that the climate of the period was similar to that of the Triassic.

Cretaceous Period (*ca.* 135–70 *my B.P.*)
Seventy species have been described for the Sub-Nagdong layer of the Lower Cretaceous, viz. Filicales and Pteridospermae (17 spp.), Equisetales (3 spp.), Bennetitales (9 spp.), Nilssoniales (4 spp.), Ginkgoales (5 spp.) and Coniferales (7

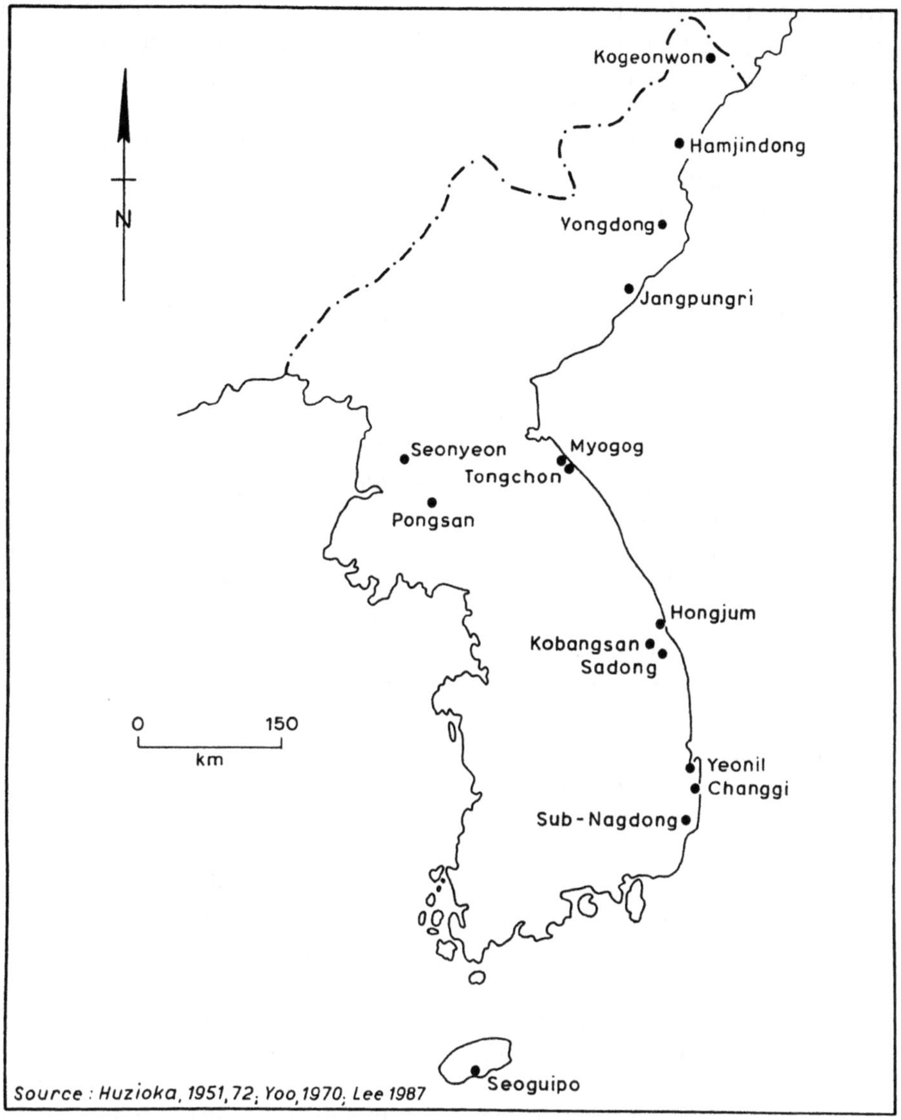

Fig. 3.2. Pre-Pleistocene fossil sites in Korea.

spp.). No drastic climatic change from that of previous periods is envisaged.

The Early Cenozoic Era (mainly Tertiary) (ca. 70–3 my B.P.)

The Cenozoic era is normally taken to include seven major geological periods, i.e. the Palaeocene, the Eocene, the Oligocene, the Miocene the Pliocene, the Pleistocene and the Holocene. The period from the Palaeocene to the Pliocene is also referred to as the Tertiary Era by some authorities, with the Pleistocene and the Holocene referred to as the Quaternary, and this separation is also emphasised herein.

Palaeocene-Eocene Periods (ca. 70–40 my B.P.)

In the Korean peninsula, no floristic data are available for these two periods. However, some indications as to what types of vegetation were present are available from East Asian material collected over a somewhat broader range of territory. In these years, it is clear that, due to the existence of a generally warm climate, an evergreen broadleaved vegetation dominated over most of the world, extending to latitude 60° N during the Palaeocene and 61° N during the Eocene (Wolfe, 1980; Shackleton, 1984). Polewards of this, coniferous forests occupied the northernmost latitudes; and an intermediate formation of deciduous broadleaved forests, along with conifers, were common on high mountains in mid-latitudes (Kryshtofovich, 1929, 1935; Chaney, 1940; Dorf, 1942; Wang, 1961; Wolfe, 1980; Xu, 1984). Bearing this in mind, a typical intermediate floral association of East Asia was *Fagus*, *Ulmus*, *Betula*, *Corylus*, *Populus*, *Juglans*, *Myrica*, *Picea*, *Pinus*, *Abies*, and *Larix* along with related species, but this gradually came to be restricted into more localised climatic zones. Its descendants now survive in the temperate forests of East Asia, Europe and North America (Kryshtofovich, 1929; Chaney, 1936, 1940, 1947, 1964; Axelrod, 1952; Wolfe, 1978). According to Tanai (1972), many other present-day Korean genera also already were present in East Asia during the Upper Eocene, among which were *Platycarya*, *Castanopsis*, *Quercus*, *Ficus*, *Machilus*, *Pittosporum*, *Ilex*, *Zizyphus*, *Ternstroemia*, *Xylosma* and *Ligustrum*.

Oligocene Period (*ca.* 40–25 *my B.P.*)
The macrofossils of 17 broadleaved plants, viz. *Populus* (3 spp.), *Myrica* (1 sp.), ?*Juglans* (1 sp.), *Ficus* (2 spp.), *Platanus* (3 spp.), *Celastrus* (1 sp.), *Acer* (1 sp.), *Sapindus* (1 spp.), *Zizyphus* (1 spp.), *Hedera* (1 sp.) and *Viburnum* (2 spp.) have been identified in the Pongsan layer of the Oligocene. The appearance of such relatively thermophilous genera at c. 38°30′ N, along with a comparison of the present-day northern limits of these genera, *viz*. *Myrica* at 34°30′ N, *Ficus* at 35° N and *Hedera* at 37°30′ N, is indicative of the fact that the Oligocene climate was somewhat warmer than today, to the extent that evergreen broadleaved genera were able to reach the latitude they did.

Miocene Period (*ca.* 25–12 *my B.P.*)
During the Miocene period, world climates cooled considerably, outside the tropics, prior to the full onset of the Plio-Pleistocene cold phases, and reflecting the onset of major world orogenesis. This is reflected in the lists of species identified from beds of this age in several localities in Korea. The effect of this on the survival of floral groups in the peninsula is presented graphically in Table 3.2.

In the Lower Miocene, the Changgi layer displays three ferns, five conifers and 36 dicotyledons, viz. *Pteridium* (1 sp.), *Polypodites* (1 sp.), *Salvinania* (1 sp.), *Picea* (1 sp.), *Pseudotsuga* (1 sp.), *Glyptostrobus* (1 sp.), *Metasequoia* (1 sp.), *Sciadopitys* (1 sp.), *Salix* (1 sp.), *Platycarya* (1 sp.), *Pterocarya* (1 sp.), *Betula* (1 sp.), *Carpinus* (4 spp.), *Ostrya* (1 sp.), *Fagus* (1 sp.), *Quercus* (1 sp.), *Ulmus* (1 sp.), *Zelkova* (1 sp.), *Hydrangea* (1 sp.), *Magnolia* (1 sp.), *Sorbus* (2 spp,), *Cladrastis* (1 sp.), *Rhus* (1 sp.), *Acer* (1 sp.), *Aesculus* (1 sp.), *Tilia* (1 sp.), *Alangium* (1 sp.), *Hemitrapa* (1 sp.), *Cornus* (1 sp.), *Rhododendron* (1 sp.), *Diospyros* (1 sp.), *Fraxinus* (1 sp.), and *Viburnum* (1 sp.). The largest family appears to have been the Betulaceae (8 spp.); the Aceraceae has five species as well. Only 25 genera (80%) of this Lower Miocene flora of Korea still have representatives in the peninsula but 29 genera (94%) may be seen in the Japanese Islands. Six genera viz. *Pseudotsuga*, *Glyptostrobus*, *Metasequoia*, *Sciadopitys*, *Pterocarya* and *Cladrastis* do not grow naturally in present-day Korea. The Changgi flora contains 33 species (77%) common with the known Miocene flora of Japan, but only seven species (16%) are common with that of the Shantung Province of China (Huzioka, 1951, 1972), immediately to the west across the Yellow Sea. The dominance of deciduous broadleaved plants suggests the existence of a warm temperate climate at this time, possibly slightly warmer than that of the present.

Another Lower Miocene flora (the Yongdong flora) consists of 12 families and 15 genera, viz. *Picea* (1 sp.), *Glyptostrobus* (1 sp.), *Metasequoia* (1 sp.), *Pterocarya* (1 sp.), *Betula* (1 sp.), *Carpinus* (1 sp.), *Ostrya* (1 sp.), *Fagus* (1 sp.), *Zelkova* (1 sp.), *Cercidiphyllum* (1 sp.), *Rosa* (1 sp.), *Acer* (4 spp.), *Aesculus* (1 sp.), *Tilia* (4 spp.) and *Alangium* (1 sp.). All these were recorded in the Changgi flora as well, with the exception of *Cercidiphyllum* and *Rosa*. Overall, the evidence suggests that northern Korea (Yongdong) and southern Korea (Changgi) together were under similar, warm temperate climatic conditions at this time. Perhaps it is also worth noting that

Table 3.2. The Miocene flora of Korea.

Period/Sites	Lower Miocene		Middle Miocene			Upper Miocene	
	Changgi	Yongdong	Tongchon	Jangpungri	Kogeonweon	Hamjindong	Yeonil
Plant Names							
Acer	×	×	×	×	×	×	×
Fagus	×	×	×	×	×	×	×
Glyptostrobus	×	×	×	×	×	×	
Zelkova	×	×	×		×	×	×
Metasequoia	×	×	×		×	×	
Alangium	×	×	×		×		×
Betula	×	×	×		×		
Pterocarya	×	×	×			×	×
Carpinus	×	×		×	×	×	×
Aesculus	×	×				×	
Picea	×	×				×	
Tilia	×	×				×	
Ostrya	×	×					
Salix	×			×	×		
Hemitrapa	×		×				×
Quercus			×	×	×		
Sciadoptitys	×				×		×
Pseudotsuga	×				×		
Salvinania	×				×		
Fraxinus	×					×	×
Rhodendron	×					×	×
Diospyros	×					×	
Ulmus	×					×	
Cladrastis	×						
Cornus	×						
Hydrangea	×						
Magnolia	×						
Platycarya	×						
Polypodites	×						
Pteridium	×						
Rhus	×						
Sorbus	×						
Viburnum	×						
Cercidiphyllum		×	×		×	×	
Rosa		×					
Alnus			×		×	×	
Populus			×			×	
Pinus			×				×
Comptonia			×				
Juglans				×		×	
Prunus				×		×	
Cinnamomum				×			
Comptoniphylla				×			
Rhamnus				×			
Platanus					×	×	×
Liriodendron					×	×	
Castanea					×		×
Cyclobalanopsis					×		×
Calocedrus					×		
Engelhardtia					×		
Euonymus					×		
Gleditsia					×		
Ilex					×		
Kalopanax					×		
Keteleeria					×		
Nelumbo					×		

Table 3.2. Continued.

Period/Sites	Lower Miocene		Middle Miocene			Upper Miocene	
	Changgi	Yongdong	Tongchon	Jangpungri	Kogeonweon	Hamjindong	Yeonil
Porana					×		
Thujopsis					×		
Vitis					×		
Liquidambar						×	×
Parrotia						×	×
Cercis						×	
Dryopteris						×	
Equisetum						×	
Hamamelis						×	
Sassafraga						×	
Styrax						×	
Carpites							×
Castanopsis							×
Cryptocarya							×
Entada							×
Lidera							×
Paliurus							×
Passania							×
Phoebe							×
Sapindus							×

× indicates the presence of relevant plant. (Compiled from Huzioka, 1951, 1972; Lee, 1987.)

some other present-day Korean genera were present in adjacent East Asia, although they are not recorded in Korea at this time. These include *Ginkgo*, *Smilax*, *Myrica*, *Corylus*, *Koelreuteria*, *Tilia* and *Viburnum* (Tanai, 1972).

Twenty species were present in the Tongchon flora of the Middle Miocene, viz. *Pinus* (1 sp.), *Glyptostrobus* (1 sp.), *Metasequoia* (1 sp.), *Populus* (1 sp.), *Salix* (2 sp.), *Comptonia* (1 sp.), *Pterocarya* (1 sp.), *Alnus* (1 sp.), *Betula* (1 sp.), *Fagus* (1 sp.), *Zelkova* (1 sp.), *Cercidiphyllum* (1 sp.), *Acer* (5 spp.), *Hemitrapa* (1 sp.) and *Alangium* (1 sp.). The admixture of conifers and deciduous broadleaved plants again suggests a warm temperate climate.

Also from the Middle Miocene, ten species have been described from the Jangpungri flora, viz. *Glyptostrobus* (1 sp.), *Salix* (1 sp.), *Comptoniphylla* (1 sp.) *Juglans* (1 sp.) *Carpinus* (1 sp.), *Cinnamomum* (1 sp.), *Prunus* (1 sp.), *Acer* (1 sp.), and *Rhamnus* (1 sp.). The most significant feature here, perhaps, is the appearance of the evergreen broadleaved *Cinnamomum*, which is indicative of a slightly warmer climate in this locality as compared to that of Tongchon.

Another Middle Miocene flora (the Kogeonweon flora) contains 21 families, 30 genera and 42 species, viz. *Salvinia* (1 sp.), *Keteleeria* (1 sp.), *Pseudotsuga* (1 sp.), *Glyptostrobus* (2 spp.), *Metasequoia* (1 sp.), *Sciadopitys* (1 sp.), *Calocedrus* (1 sp.), *Thujopsis* (1 sp.), *Salix* (1 sp.), *Engelhardtia* (1 sp.), *Alnus* (1 sp.), *Betula* (2 spp.), *Carpinus* (1 sp.), *Castanea* (1 sp.), *Fagus* (2 spp.), *Quercus* (3 spp.), *Cyclobalanopsis* (2 spp.), *Zelkova* (1 sp.), *Nelumbo* (1 sp.), *Cercidiphyllum* (1 sp.), *Liriodendron* (1 sp.), *Platanus* (1 sp.), *Gleditia* (1 sp.), *Ilex* (1 sp.), *Euonymus* (1 sp.), *Acer* (4 spp.), *Vitis* (1 sp.), *Kalopanax* (1 sp.), *Porana* (1 sp.) and *Alangium* (1 sp.). The appearance of the evergreen broadleaved plants *Cyclobalanopsis* and *Ilex* in northern Korea (Kogeonweon, ca. 42° N) and neighbouring Jangpungri suggests the presence of a major climatic amelioration at this time, in which occurred possibly the maximum northward expansion of evergreen broadleaved vegetation in Korean vegetation history, for these genera then existed some 6 degrees of latitude further to the north than they do today.

In the Upper Miocene, the Hamjindong layer in the north of the peninsula comprises 23 families, 31 genera and 51 species, viz. *Equisetum* (1 sp.), *Dryopteris* (1 sp.), *Picea* (2 spp.), *Glyptostrobus* (1 sp.), *Metasequoia* (1 sp.), *Populus* (1 sp.), *Juglans* (1 sp.), *Pterocarya* (2 spp.), *Alnus* (1 sp.) *Carpinus* (2 spp.), *Fagus* (4 spp.) *Quercus* (2 spp.), *Ulmus* (2 spp.), *Zelkova* (1 sp.), *Cercidiphyllum* (1 sp.), *Liriodendron* (1 sp.), *Lindera* (1 sp.), *Sassafraga* (1 sp.), *Platanus* (1 sp.), *Hamamelis* (1 sp.), *Liquidambar* (1 sp.), *Parrotia* (1 sp.), *Prunus* (1 sp.), *Cercis* (1 sp.), *Acer* (10 spp.), *Aesculus* (1 sp.), *Tilia* (2 spp.), *Rhododendron* (2 spp.), *Diospyros* (1 sp.), *Styrax* (1 sp.) and *Fraxinus* (1 sp.).

In the south, another Upper Miocene flora (the Yeonil flora) contains 17 families, 27 genera and 35 species, viz. *Pinus* (1 sp.), *Keteleeria* (1 sp.), *Sciadopitys* (1 sp.), *Pterocarya* (1 sp.), *Carpinus* (3 spp.), *Castanea* (1 sp.), *Fagus* (1 sp.), *Castanopsis* (1 sp.), *Cyclobalanopsis* (3 spp.), *Lindera* (1 sp.), *Phoebe* (1 sp.), *Passania* (2 spp.), *Zelkova* (1 sp.), *Cinnamomum* (1 sp.), *Cryptocarya* (1 sp.), *Platanus* (1 sp.), *Liquidambar* (1 sp.), *Parrotia* (1 sp.), *Entada* (1 sp.), *Acer* (1 sp.), *Sapindus* (1 sp.), *Paliurus* (1 sp.), *Alangium* (1 sp.), *Hemitrapa* (1 sp.), *Rhododendron* (1 sp.), *Fraxinus* (1 sp.) and *Carpites* (1 sp.). The presence of such a relatively rich flora, in which are the evergreen broadleaved plants *Castanopsis*, *Cyclobalanopsis* and *Cinnamomum*, suggest the continuation in this district of a warm climate; however, the absence of these indicator species at Hamjindong suggests that some significant cooling had already begun in the north.

Overall, ten genera occur continuously from the Lower Miocene to the Upper Miocene in the Korean peninsula as a whole, namely *Glyptostrobus*, *Metasequoia*, *Pterocarya*, *Carpinus*, *Fagus*, *Zelkova*, *Acer*, *Alangium*, *Hemitrapa* and *Cercidiphyllum*. Out of these, *Carpinus*, *Fagus*, *Zelkova*, *Acer* and *Alangium* are still growing naturally in present-day Korea: the other genera have retreated subsequently to China and Japan. The intermittent occurrence of numerous other representatives of the present-day flora from the Oligocene and within the Miocene periods also suggests, bearing in mind the general paucity of records, that they were a stable feature of Korean vegetation, ensuring its continuity over a very long period of time. Among these latter are *Picea*, *Pinus*, *Populus*, *Salix*, *Myrica*, *Juglans*, *Platycarya*, *Carpinus*, *Ostrya*, *Alnus*, *Betula*, *Ficus*, *Celastrus*, *Castanopsis*, *Castanea*, *Cyclobalanopsis*, *Fagus*, *Quercus*, *Ulmus*, *Zelkova*, *Nelumbo*, *Hydrangea*, *Magnolia*, *Cinnamomum*, *Sorbus*, *Rosa*, *Prunus*, *Cladrastis*, *Rhus*, *Acer*, *Ilex*, *Euonymus*, *Sapindus*, *Paliurus*, *Vitis*, *Tilia*, *Kalopanax*, *Alangium*, *Cornus*, *Rhododendron*, *Fraxinus*, *Styrax*, *Viburnum*, *Equisetum* and *Dryopteris*. In summary, one may say that the temperate climate of the Lower Miocene of Korea subsequently became even milder in the Middle Miocene, but later during the Upper Miocene, regional differences in vegetation type and climate began to occur, as temperatures in the north were lowered relative to those in the south.

Pliocene Period (*ca.* 12–13 *my. B.P.*)

Though no floristic data are available for Korea from this period, planktonic foraminiferal assemblages from Seoguipo, mainly of *Globigerina pachyderma*, show a sequence of warm and cold phases, the details of which are as yet not clearly differentiated (Yoo, 1970).

Vegetation of the Pleistocene Period (ca. 3 my–10 000 y B.P.)

For Europe and East Asia, at least four glacial epochs occurred within this period (Table 3.3; Nilsson, 1982; Chiu, 1984; Sibrava *et al.*, 1986).

In Korea, however, an almost total lack of data from the Lower Pleistocene means that only the Middle and Upper Pleistocene are considered herein. The faunal and floral fossil assemblages compiled by Choi (1983, 1986), Park (1983) and Sohn (1984), are essential information, and they have been utilised widely in this section.

Middle Pleistocene (*ca.* 700 000–110 000 *y B.P.*)

Middle Pleistocene faunal and floral data have been collected from only three Early Palaeolithic sites. So far, one (Sangwon Kumunmoru cave) from the north of the peninsula, and the other two (Chewon Chummal Yonggul cave and Cheongwon Turubong cave) from the central area, contain faunal macrofossils, and the first two also contain a range of floral macrofossils (Fig. 3.3; Table 3.4). No absolute dating is available for any of these sites.

In Sangwon Kumunmoru cave, 7 orders, 16 families, 23 genera, 1 sub-genus and 29 species of faunal macrofossils are present. The small

Table 3.3. Correlation chart for the European and East Asian Pleistocene.

Estimated Age (B.P.)	General Division	Glacial/ Interglacial	North-West Europe	Britain	Alps	West Siberia	South China
0							
10 000	Holocene		Holocene	Holocene	Holocene	Holocene	Holocene
75–110 000		G.	Weichsel	Devensian G.	Würm	Zyrianka	Dali
? 125 000	Upper	I.G.	Eem	Ipswichian I.G.	R/W	Kazantsevo	L/D
		G.	Saale	Wolstonian G.	Riss	Samarov	Lushan
		I.G.	Holstein	Hoxonian I.G.	M/R	Tobol	Dagu
400 000	Pleistocene Middle	G.	Elster	Anglian G.	Mindel		
700 000		I.G.	Cromer	Cromerian I.G.			
			Complex				
900 000			Menapian				
			Waal				
1.7 million	Lower	G.	Eburonian	Baventian G.			
		I.G.	Tiglian	Antian I.G.			
2–2.5 million		G.	Praetiglian	? Waltonian G.			
	Pliocene						

G: Glacial Period. I.G.: Interglacial Period.
(Modified from Nilsson 1982; Chiu, 1984; Sibrava *et al.*, 1986.)

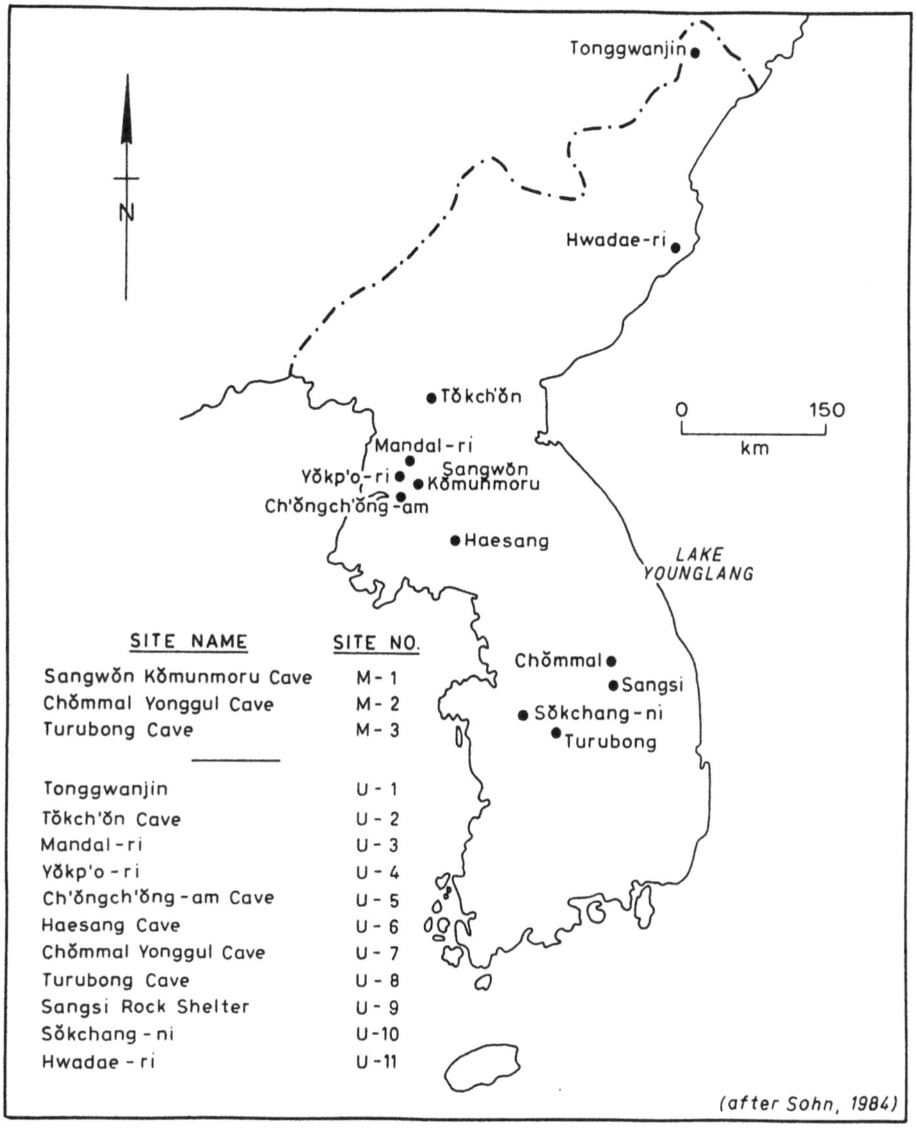

Fig. 3.3. Middle- and Upper-Pleistocene human-occupied sites in Korea.

mammals from this site (ca. 41.4% of the total species number or ca. 38.7% of the total number of individuals) are also found, at the same period, in southeastern Siberia, northeastern China and Mongolia. A large number of large mammals further are present (ca. 58.6% of the total species number or ca. 61.3% of the total number of individuals). Charcoal of *Pinus* has been recorded as well.

In Chewon Chummal Yonggul cave, 2 classes, 7 orders, 12 families, 12 genera, 1 sub-genus and 20 species of vertebrates are present and there are also seeds of Juglandaceae, *Celtis* and leaves of *Benzoin obtusilobum*.

In Cheongwon Turubong cave, 3 classes, 3 orders, 3 families, 3 genera, 1 sub-genus and 4 species of vertebrates have been recorded, but there are no plant macrofossils.

In respect of microfossils, arboreal pollen of Juglandaceae, *Pinus* and *Thuja*, and non-arboreal pollen of *Equisetum* is present in the Sangwon Kumunmoru cave.

In the Chewon Chummal Yonggul cave, arboreal pollen of Aquifoliaceae, *Betula*, *Castanea*,

Table 3.4. Faunal and floral assemblages of the Middle Pleistocene in Korea.

Fauna	M-1	M-2	M-3*
Bos primigenius	×		
Bubalus sp.	×		
Sus scrofa	×		
Pseudaxis grayi	×		
P. grayi var.			×
Cervus elaphus	×		
C. sp.		×	
Megaceros flabellatus	×		
M. sangwonensis	×		
Capreolus capreolus		×	
Moschus moschiferus		×	
Mustela sibirica manchurica		×	
Meles meles	×	×	
Vulpes vulpes		×	
Cuon alpinus	×		
Panthera tigris	×		
P. pardus		×	×
Ursus arctos	×		
U. spelaeus	×		
Hyaena brevirostris	×		
Hyaena sp.		×	
Rhinolophus		×	
Murina leucogaster		×	
Plecotus sp.		×	
Miniopterus schreibersii		×	
Equus sangwonensis	×		
E. sp.		×	
Dicerorhinus	×	?	
Elephantinus	×		
Microtus brandtioides		×	
M. epiratticeps	×	×	
M. brandti	×	×	
M. sangwonensis	×		
M. sp.	×	×	
M. fortis	×		
Mimoys sp.	×		
Pitymys simplicidens	×		
Rattus norvegicus	×		
Apodemus sylvaticus	×		
Myospalax	×		
Castor	×		
Talpa		×	
Macaca sp.	×		
Rana asiatica		×	
Bufo asiaticus		×	×
(Charcoal)			
Pinus	×		
Leaves			
Benzoin obtusilobum		×	
Juglandaceae		×	
Celtis		×	
(Pollen: AP)**			
Aquifoliaceae		×	
Betula		×	
Castanea		×	
Fagus		×	
Quercus		×	
Juglandaceae	×		
Salix		×	
Tilia		×	
Taxus		×	
Abies		×	
Pinus	×	×	
Thuja	×		
(Pollen: NAP)**			
Artemisia		×	
Equisetum	×		

× indicates the presence of relevant plants (Modified from Sohn, 1984).
* M-1, Sangwon Kumunmoru cave; M-2, Chewon Chummal Yonggul cave; M-3, Cheongwon Turubong cave. For reference see Fig. 3.3.
** AP: Arboreal Pollen
NAP: Non-Arboreal Pollen

Fagus, *Quercus*, *Salix*, *Tilia*, *Taxus*, *Abies*, *Pinus* and non-arboreal pollen of *Artemisia* has been found.

While one may not draw any firm conclusions chronologically or spatially about vegetation in the Middle Pleistocene from these data, it is clear that many deciduous broadleaved plants and evergreen coniferous plants were present in the peninsula for at least part of this period. Also, the appearance of *Pinus* charcoal suggests that some anthropogenic disturbance of natural vegetation already was taking place during this period.

Upper Pleistocene (*ca.* 110 000–10 000 *y B.P.*)

Upper Pleistocene faunal and floral data have been collected by Sohn (1984) from 11 sites (Fig. 3.3). The complete faunal and floral assemblages of Upper Pleistocene age are given in Table 3.5.

Macrofossil data. Faunal macrofossils of this age are present in nine sites, and two of these sites contain floral macrofossils as well (Table 3.5).

In Tonggwanjin, 6 orders, 10 families, 14 genera, 17 sub-genera and 2 sub-species of faunal macrofossils are present. One species, *Mam-*

Table 3.5. Faunal and floral assemblages of the Upper Pleistocene in Korea.

Fauna	U-1	U-2	U-3	U-4	U-5	U-6	U-7	U-8	U-9*
Bison priscus			×				×		
Bos primigenius		×	×					×	
Bubalus sp.			×						
Nemorhadeus goral		×	×		×	×			
Sus scrofa			×		×	×		×	
Pseudaxis grayi								×	×
P. grayi var.								×	
Cervus elaphus		×				×		×	
C. nippon			×		×	×			
C. sp.			×				×	×	×
Megaceros sp.		×	×						
Capreolus capreolus		×	×		×	×	×	×	×
Hydropotes inermis						×			
Moschus moschiferus			×				×		×
Mustela sp.		×	×					×	
M. sibirica manchurica							×		
Lutra lutra			×				×		
Martes melampus							×		×
M. cf. zibbellina			×				×		
M. sp.			×						
Meles meles		×			×	×		×	
Vulpes vulpes			×				×		×
Canis lupus								×	×
Cuon alpinus			×						
Nyctereutes			×				×	×	×
Panthera tigris			×			×	×	×	
P. cf. *leo*							×	×	
P. paruds							×	×	
P. leo spelaea			×						
Ursus thibetanus						×	×	×	
U. arctos			×		×	×	×	×	
U. spelaeus			×			×	×	×	
Hyaena ultima	×	×						×	
H. sp.							×	×	×
Rhinolophus							×	×	
Murina leucogaster							×	×	
Plecotus sp.							×	×	
Myotis nattereri								×	
Lasionycteris noctivagans								×	
Miniopterus schreibersii							×	×	
Equus caballus		×					×		×
E. sp.	×							×	
Dicerorhinus		×						×	
Coelodonta antiquitatis	×						×		
Mammuthus sp.	×	×							
Sciurus sp.	×						×	×	
Siphenus cf. *wongi*								×	
Clethrionomys sp.							×	×	
Microtus brandtioides							×	×	
M. epiratticeps							×		
M. brandti							×		
M. sp.							×		
M. fortis		×					×	×	
M. mandarinus								×	
Allocricetus cf. *bursae*							×	×	
Cricetinus sp.						×			
Apodemus agrarius								×	
Rattus norvegicus								×	
Apodemus sylvaticus							×		
Castor sp.		×							

Table 3.5. Continued.

Fauna	U-1	U-2	U-3	U-4	U-5	U-6	U-7	U-8	U-9*
Talpa sp.							×	×	
Sorex sp.							×	×	
Crocidura sp.								×	
Erinaceus sp.							×	×	
Evotomys rufocanus							×		
Lepus sp.					×		×	×	×
Ocotona sp.							×	×	
Macaca cf. *robusta*							×	×	
M. sp.							×		
Homo sapiens		×		×					×
H. sapiens sapiens		×	×				×	×	×
Rana nigromaculata								×	
R. asiatica							×	×	
B. raddei							×	×	
B. bufo							×	×	
Geoclemys reevesii							×	×	

Flora	U-5	U-6	U-7	U-8	U-10	U-11
(Charcoal)						
Sambucus			×			
Prunoideae			×			
Papilionoidae			×			
Prunus			×			
Eleaganaceae			×			
Araliaceae			×			
Fagaceae			×	×		
Taxaceae			×			
Pinus			×	×		
Cupressaceae			×			
Aceraceae				×		
(Leaves)						
Smilax china var. *microphylla*			×			
Juglandaceae			×			
Celtis			×			
(Pollen: AP)**						
Aceraceae				×	×	
Aquilfoliaceae			×			
Alnus			×	×	×	×
Betula			×	×	×	×
Carpinus			×	×		
Castanea			×			
Fagus			×			
Quercus			×	×		
Juglans sp.	×					
Juglandaceae			×			
Lauraceae			×			
Magnoliaceae					×	
Nymphaceae					×	
Populus			×			
Salix			×			×
Ligustrum				×	×	
Tilia			×	×		
Ulmus			×			
Ginkgoaceae			×			
Taxus			×	×		
Abies		×	×	×	×	×

Table 3.5. Continued.

Flora	U-5	U-6	U-7	U-8	U-10	U-11
Larix			×	×		×
Picea			×			×
Pinus		×	×	×	×	×
Pinus koraiensis						×
Pinus pumila						×
Pinus sibirica						×
Pseudotsuga			×			
Tsuga			×			
Thuja	×					
Juniperus dahurica						×
(Pollen: NAP)**						
Lycopodiaceae					×	
Chenopodium			×			
Artemisia			×	×		
Drosera			×			
Saxifraga						×
Oxytropis						×
Bryales						×
Sphagnum						×
Iridaceae						×
Urtica						×
Gramineae			×	×		×
Cyperaceae				×	×	×
Liliaceae				×	×	
Typha				×		
Scrophulariaceae				×		
Compositae				×		×
Pteridaceae	×	×				
Osmunda		×				×

× indicates the presence of relevant plant. (Modified from Sohn, 1984)
*U-1: Tonggwanjin
U-2: Tokchon cave
U-3: Mandal-ri
U-4: Yokpo-ri
U-5: Sangwon Cheongchong-am cave
U-6: Pyongsan Haesang cave
U-7: Chewon Chummal Yonggul cave
U-8: Cheongwon Turubong cave
U-9: Sangsi rock shelter
U-10: Sokchang-ni cave
U-11: Hwadae peat layer
**AP: Arboreal Pollen NAP: Non-Arboreal Pollen
For site locations, see Fig. 3.3.

muthus (*Elephas*) *primigenius*, is a cryophilous animal.

The macrofossil fauna of the Tokchon cave can be divided into 2 orders, 13 families, 25 genera, 1 sub-genus, 31 species and 2 sub-species of vertebrates, including human fossils, viz. *Homo sapiens* and *H. sapiens sapiens*. Two human fossils are also found at adacent Mandal-ri and Yokpo-ri respectively.

In Sangwon Cheongchong-am cave, 3 orders, 5 families, 7 genera and 7 species of vertebrates have been discovered. All of these genera are still present in Korea.

In Pyongsan Haesang cave, 3 orders, 7 families, 9 genera, 1 sub-genus and 12 species of mammals are present.

In Chewon Chummal Yonggul cave, there are 2 classes, 9 orders, 22 families, 33 genera, 1 sub-genus, 45 species and 1 sub-species of vertebrates. These species include *Homo sapiens sapiens*.

From there, 2 classes, 3 sub-classes, 18 orders, 22 families, 2 sub-families, 24 genera and 1 species (leaves of *Smilax china* var. *macrophylla*) of macrofossil flora have been described. The cave also includes charcoal of *Sambucus*, Prunoideae, Papilionoidae, *Prunus*, Elaeagnaceae, Araliaceae, Fagaceae, Taxaceae, *Pinus* and Cupressaceae, and seeds of Juglandaceae and *Celtis*.

Cheongwon Turubong 2 cave has 7 classes, 15 orders, 28 families, 37 genera, 1 sub-genus and 46 species of vertebrates. The existence of *Bos primigenius*, *Dicerorhinus kirchbergensis*, *Hyaena ultima*, *Hyaena* spp. and *Macacus robustrus* indicates a comparatively warm climate. In Cheongwon Turubong 9 cave, the macrofossil fauna, consisting of 4 classes, 9 orders, 22 families, 27 genera, 1 sub-genus and 31 species of vertebrates, with mammals like *Macacus* sp., *Panthera* cf. *leo*, *Sus scrofa*, being present, along with its Chiroptera and Mollusca, also indicates a warm climate. This cave also contains 22 species of macrofossil flora, including charcoal of Fagacene, *Prunus* and Aceraceae.

In the Sangsi rock shelter, 12 species of mammal fosssils, including *Homo sapiens* and *H. sapiens sapiens* are present.

Microfossil data. All microfossils of this age consist of pollen grains, and have been collected from seven different sites, as indicated below. These include several cave sites, a peat layer and one pollen sequence from bore-hole material (Lake Younglang). They are of variable quality and only the bore-hole has a dated sequence (p. 68).

In the Sangwon Cheongchong-am cave, pollen of 1 family and 2 genera, i.e. *Juglans* sp., *Thuja* and Pteridaceae, are present.

In the Pyongsan Haesang cave, pollen of 1 family and 3 genera, e.g. *Abies*, *Pinus*, *Osmunda*, along with Pteridaceae, are present.

The flora of the Chewon Chummal Yonggul cave contains pollen of *Taxus*, *Abies*, *Larix*, *Picea*, *Pinus*, *Pseudotsuga*, *Tsuga*, *Alnus* and *Betula*.

The flora of Cheongwon Turubong cave also contains pollen of *Taxus*, *Abies*, *Larix*, *Pinus*, *Betula*, Ericaceae, Aceraceae, *Alnus*, *Carpinus*, *Quercus*, *Ligustrum* and *Tilia*.

In the Sokchang-ni cave, 4 classes, 3 sub-classes, 9 orders, 10 families and 6 genera of pollen types have been collected. In this, deciduous broadleaved trees of Aceraceae, *Alnus*, *Betula* and Magnoliaceae, along with the hydrophilous Lycopodiaceae, Cyperaceae, Liliaceae and *Osmunda*, are common.

From the Hwadae peat layer, 5 classes, 5 sub-classes, 11 orders, 17 families, 16 genera and 7 species of flora, are present. The occurrence therein of the cryophilous *Pinus koraiensis*, *P. pumila*, *P. sibirica*, *Juniperus dahurica*, *Abies*, *Picea*, *Pinus* (Haploxylon), *Larix*, *Salix*, *Betula*, *Alnus*, *Saxifraga* and *Oxytropis* indicates the existence at this time of a rich floristic association, but also a cold climate. Non-arboreal pollen of Bryales, *Sphagnum*, *Osmunda*, Gramineae, Cyperaceae, Iridiaceae, *Urtica* and Compositae are also present.

For the Upper Pleistocene, indications of a diverse flora accordingly are found from the evidence of charcoals (*Sambucus*, *Prunus*, *Pinus* and others), fossil leaves of *Smilax*, Juglandaceae and *Celtis* and pollen grains of *Alnus*, *Betula*, *Carpinus*, *Castanea*, *Fagus*, *Quercus*, *Juglans*, *Populus*, *Salix*, *Ligustrum*, *Tilia*, *Ulmus*, *Taxus*, *Abies*, *Larix*, *Picea*, *Pinus*, *Pseudotsuga*, *Tsuga*, *Thuja*, *Chenopodium*, *Artemisia*, *Drosera*, *Typha*, *Osmunda* and others. The vegetation of the Upper Pleistocene in other words, at particular times, was a mixture of deciduous broadleaved plants and cryophilous evergreen coniferous plants. The clearest evidence comes from the Hwadae peat layer of the north, in which *Pinus koraiensis*, *P. pumila*, *P. sibirica*, *Abies*, *Thuja*, *Taxus*, *Picea*, *Pseudotsuga*, *Tsuga* and *Juniperus* occur, along with some deciduous broadleaved species, suggesting the presence of a cold episode in this period. How these features relate generally to glacial episodes in East Asia as a whole is not yet clear. The bore-hole data from Lake Younglang has particular relevance for the end of the Upper Pleistocene, and is discussed later (p. 68).

Pleistocene glaciation in East Asia

There is considerable controversy regarding the exact numbers and extent of glacial cycles in East Asia (Hultén, 1937; Dahl, 1946; Charlesworth 1957; Florin, 1963; Bowen, 1978; Imbrie & Imbrie, 1979; Nilsson, 1982). In 1922, Brooks suggested that two glacial epochs had occurred in eastern Siberia; in 1933, Obruchev proposed three such epochs for the Verkhoyansk-Chersky mountains; and later (1939) two in the Chukotsk peninsula. In 1939, Gerasimov & Markov described the existence of two glacial epochs in the Kamchatka peninsula. Then in 1945, Flint & Dorsey proposed three glacial episodes for the Kolyma mountains, two for the highlands near Lake Baikal, and at least two in the Sayan mountains. More recently, four glaciations have been determined for eastern Siberia by Golubeva (1964), and five for the north Pacific region of East Asia generally by Karlstrom (1966). However, no precise chronology has been established so far. In the case of China, at least four glaciations currently are recognised (Young, 1950; Chang, 1968a,b; Wang, 1984). Kung (1976), however, has suggested five glacial episodes for this country, which correlate with the Donau, Günz, Mindel, Riss and Würm glaciations of the Alps. In Japan, two to five glaciations have been suggested by Minato (1972) and Okamoto (1972).

For many places on the earth's surface, the extent of maximum ice development during the Pleistocene was reached during the Wolstonian (Riss in the Alps, Samarov in West Siberia). The evidence in East Asia, however, suggests that only a few places were occupied by a glaciation of maximum extent at this time (Sharp, 1956; Griffin, 1960; Lindroth, 1970; Flint, 1971; Frenzel, 1973a; Nilsson, 1982; Fig. 3.4). Instead, many regions in East Asia then either were unglaciated or poorly glaciated. The possible reasons for this are as follows: first, there might have been inadequate precipitation during the appropriate period (Cressey, 1926; Flint & Dorsey, 1945; Klein, 1971; Chard, 1974; Goudie, 1979; Aleksandrova, 1980; Nilsson, 1982); second, a variation in the season of precipitation might have occurred, so that most fell during the summer rather than in cooler seasons (Berkley & Morris, 1927); and finally, linked with both of these, a semi-permanent anticyclonic mass seems at that time to have extended over the region (Giterman, 1984; Velichko, 1984a).

What seems to have happened then in East Asia, certainly in the Walstonian and possibly at other times as well, was that while most regions were extremely cold during the glacial phases, with intense cryogenic activities, no great ice-sheet development took place; however it must be said that where humidity was generally higher than normal, as along the Pacific fringe mountains, local glaciations did occur. This feature is discussed further below and in the succeeding chapter.

Pleistocene glaciation in Korea

A report on the existence of Pleistocene glaciation in northern Korea by Berkeley & Morris (1927) was one of the earliest works to have examined the possibility of glacial activity within the peninsula. Subsequently, Kano (1937) and Sasa & Tanaka (1938) concluded that Mts. Paikdoo (2744 m a.s.l.), Kwanmo (2451 m a.s.l.), Solryong (2442 m a.s.l.), Nampodae (2435 m a.s.l.) and Pukpodae (2289 m a.s.l.) in northern Korea, all had possessed a few glaciers during the Pleistocene (Fig. 3.5). Their evidence was based on the presence of 18 glacial cirques, lying at about 2000 m a.s.l., and moraines, found as low as 1700 m a.s.l. Sasa & Tanaka (1938) were of the opinion that two stages of cirque formation could be detected for northern Korea within the Pleistocene, one at 1900 to 2000 m a.s.l. and the other at 2000 to 2100 m a.s.l. The first of these displayed the greatest extent of glacial activity and may be contemporaneous with the penultimate major world glaciation (Wolstonian, Riss, Samarov), whereas the latter has a much more restricted pattern of development, probably being associated with the last glaciation (Devensian, Würm, Zyrianka). During these glaciations, the climatic snowline in Korea was at ca. 2000 m in northernmost Korea, ca. 2250 m in northwestern Korea, ca. 2500 m in the midland mountains and ca. 2750 m in southern Korea (Frenzel, 1973a; Nilsson, 1982). For all practical purposes, this means that glaciation *per se* within the peninsula always was restricted to the northernmost parts of Korea.

For the last glaciation at least, there is also a good deal of additional evidence to suggest that, while glaciers were not present in mid- and southern Korea, climates were nevertheless extremely cold during these phases throughout the peninsula, except possibly on the southern lowland coastal fringes. Thus, patterned ground is present on Mt. Halla, Cheju Island (Kim, 1970); solifluction phenomena are found on Mt.

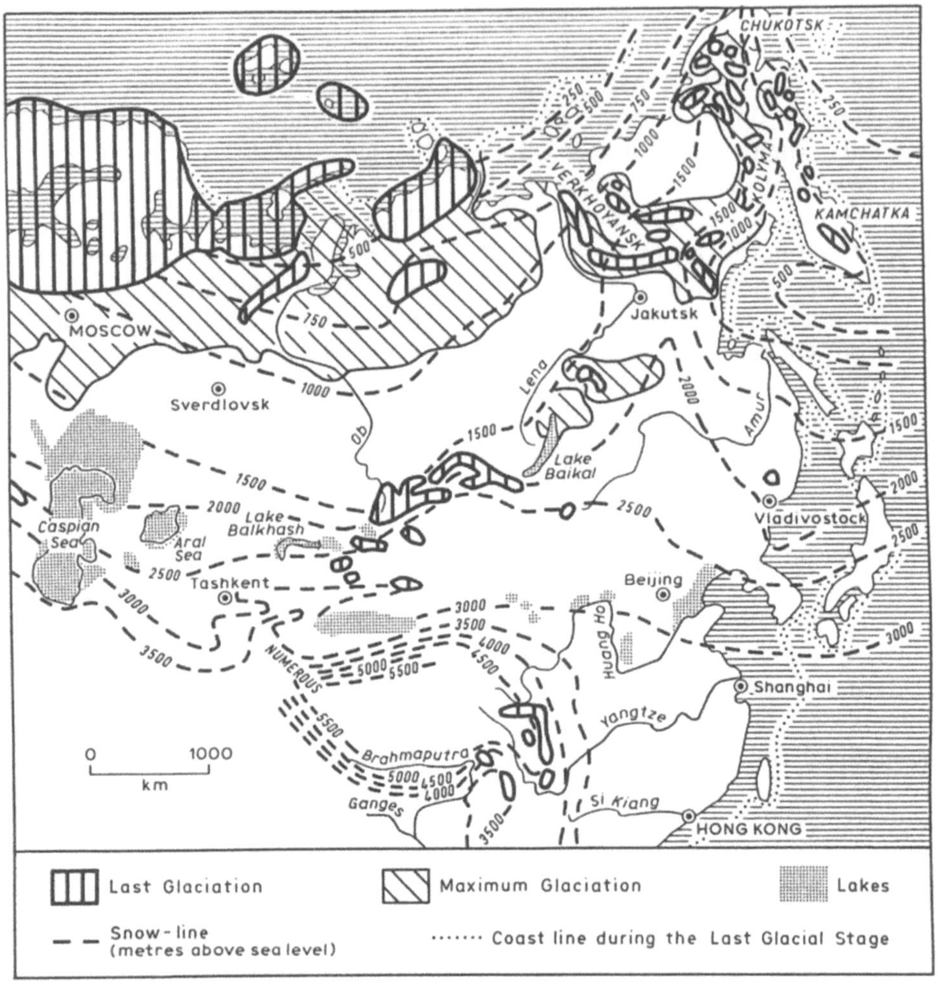

Fig. 3.4. Pleistocene glaciations, current snow-line data and lake sites in Eurasia.

Keumjung, Pusan (Kwon, 1980); and other periglacial landforms have been described for Mt. Chiri (Chang, 1983a). Elsewhere, on low ground, cryogenetically disturbed beds occur near Chongju (Kwon, communication with Watts & Kong, 1987; Kwon, 1987).

Palaeoclimates of the Upper Pleistocene

For the penultimate interglacial, at ca. 122 000 years B.P., the February sea surface temperatures near Korea have been estimated to be 4° to 10°C (N. to S.) in the West Sea (Yellow Sea), 10° to 14°C (N. to S.) in the South Sea and 2° to 10°C (N. to S.) in the East Sea (Sea of Japan). For August, the respective values were 24° to 26°C, 26° to 27°C and 21° to 26°C (CLIMAP, 1984). In reality, these sea surface temperatures are quite similar to the present sea surface temperatures near Korea, as specified by Jo (1980a) and Lee (1982). Then, during the penultimate glaciation (Walstonian), the mean annual temperature of the peninsula is thought to have been lowered by ca. 8° to 12°C, as compared to present-day temperatures. This figure has been derived from temperature data for China and Japan, and from the general thermal indicators of the last glaciation (Minato, 1972; CLIMAP, 1976, 1981; Zhang, 1984).

At around 40 000 years B.P., the Devensian (Würm in the Alps, Zyrianka in Siberia) glaciation was expanding toward the south, and lasted up to 10/12 000 years B.P. (Kind, 1967). According-

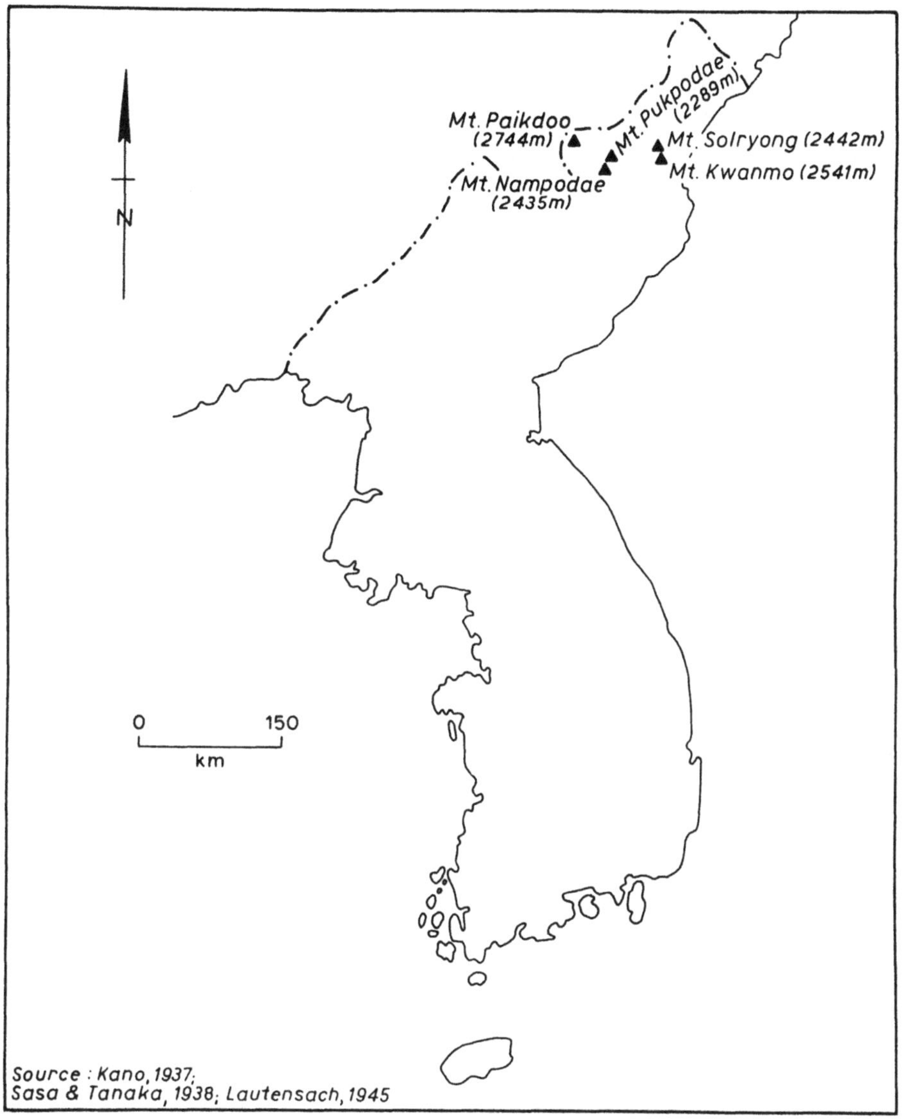

Fig. 3.5. Pleistocene-glaciated mountains in Korea.

ing to CLIMAP (1981) and Ruddiman & Dempsey (1985), the maximum cold peak on a global scale for the last glaciation was ca. 18 000 years B.P. In the case of East Asia however, the period of maximum cold varies widely from region to region. Thus, it occurred at ca. 22 000 years B.P. in West Beringia (Lozhkin, 1984); at ca. 22 000 to 18 000 years B.P. in eastern Siberia (Frenzel, 1973a; Powers, 1973; Howe, 1986); and at ca. 20 000 to 18 000 years B.P., in the area of the northeastern China Sea (Qin *et al.*, 1981). There is, as yet, no indication as to the date of the maximum cold period in Korea during the Devensian, though it must have occurred between 22 000 to 18 000 years B.P.

In East Asia, the temperatures themselves at the Devensian peak, both of the sea surface and the atmosphere, also varied widely from region to region. Thus in Siberia, the annual temperatures are thought to have fallen by about 8°C (Velichko, 1984b); and they dropped by ca. 7°C in Japan, with a fall of summer temperatures of ca. 8° to 9°C in Hokkaido (Tsukada, 1983a; Heusser & Morley, 1985). There was a reduction of ca. 7°C near Beijing, China (Kong *et al.*, 1976); and of ca. 4° to 6°C in Taiwan (Chang, 1970).

However, it should not be forgotten that a good deal of land was exposed during this period in the West Sea (Yellow Sea), and this would accordingly have further reduced any ameliorating maritime effects on temperatures near Korea. Thus, where ocean water was maintained, August sea-surface temperatures near Korea fell by ca. 8°C, while February sea-surface temperatures fell by only ca. 2° to 5°C; however, the annual air temperatures in the peninsula are thought to have fallen by ca. 6°C overall (CLIMAP, 1976, 1981; Rind & Peteet 1985). The vegetation indicators for this period have been indicated previously (pp. 56–60).

Archaeology

Archaeological evidence (settlement sites, artefacts, tomb sites and others) suggests that Palaeolithic man certainly existed in the Korean peninsula between ca. 50 000 and 30 000 years B.P., and may even have been present prior to 110 000 years B.P. (pp. 56–58). Such groups appear to have lived mostly along river banks (Osgood, 1951; Butzer, 1977; Yi, 1984). Their diet included fruits, berries, edible roots and other plants. There are few indications that they were able to hunt and fish, at least to begin with. The existence of hearths for cooking and heating, which date back to 20 830 ± 1880 years B.P. (Sohn, 1978), suggest both that hunting and fishing had in fact begun by this time, and also that early-stage deforestation for firewood was under way. An archaeological hiatus for data exists for Korea between 13 700 ± 180 and 6430 ± 180 years B.P., during which time the Palaeolithic age ended (Kim, 1981; Chang, 1983b).

Quarternary vegetation of East Asia

In the extreme northeast of Asia (viz. the Chukchi and Kamchatka peninsulas), all the evidence suggests that vegetation below the level of glaciers during the last two glacial periods was dominated by alpine tundra, with *Pinus pumila* and *Betula* dominant. It is likely that the *P. pumila* communities adopted a krummholz form, as they do today. The southern parts of the maritime territory of eastern Siberia were also occupied by tundra and shrub tundra. Farther to the south, the cover of mixed coniferous and broadleaved forests, present during the interglacial, became more restricted spatially, and *Larix* and shrubby forms of *Betula* and *Alnus* became important instead (Frenzel 1973b; Powers, 1973; Wright & Barnosky, 1984), with further alpine tundra communities in the high mountain land.

According to Grichuk (1984), vegetation in the Far East of the USSR, close to the Korean border, also displayed regional variations during the Devensian: open *Betula* and *Picea* forests were present in the Lower Amur; a montane glacier, montane subarctic tundra, subalpine and alpine meadows in the Middle Amur; a light coniferous montane forest in the Upper Amur; a montane tundra, subalpine and alpine meadows in the Upper Vladivostok; a dark coniferous and *Betula* montane forest in the Middle Vladivostok; and coniferous and broadleaved montane forests in the Lower Vladivostok.

At the same time, in East China, the southern limit of the boreal forest moved southward to extend from the southern part of the Liadong peninsula to the mountain region of northern Hebei. In turn, to the south of this line, a coniferous and deciduous temperate forest, or steppe forest, extended southward to the Yangtze delta (Zhang, 1984). The present loess areas inland along the Hwang He River (Yellow River) were occupied by grassland (Nilsson, 1982).

In Japan, most of Hokkaido was influenced by periglacial controls (Suzuki, 1962; Kaizuka, 1965) in the Devensian. Tundra vegetation occupied the area to the north of 44°30′ N, and was present also in the south above about 900 m a.s.l. The rest of the area was covered generally with more specifically arctic-alpine plants, dominated by *Pinus pumila* and *Betula ermanii*. However, in climatically more favourable locations, *Picea* spp., *Betula ermanii*, *Pinus* spp., *Larix* spp., boreal and bog taxa were a common feature of the pollen data (Yamanaka, 1978; Igarashi & Kumano, 1981; Tsukada, 1983b; Heusser & Morley 1985). Northern and central Honshu were characterised by a subalpine coniferous forest, mainly comprised of *Picea*, *Abies* and *Tsuga* (Quaternary Flora Research Group, 1974; Yasuda, 1978; Sakai, 1981; Yasuda & Narita, 1981; Tsukada, 1985).

Quaternary vegetation of Korea

The idea of the existence of a mixed forest, put forward by Chard (1974) for the maximum cold-phase Pleistocene of Korea, seems to be a simplistic one. On the contrary, Wissmann's (1956) proposal of five zones of vegetation appears to be more realistic, bearing in mind the known present-day biogeographic divisions of Korean vegetation (Fig. 2.2). Wissmann's zones are as follows; first, glaciers existed on the northernmost mountain tops; second, tundra or alpine vegetation occurred below the glacier areas; third, steppes were located in the northwestern region; fourth, a coniferous forest had developed at lower altitudes in both the northern and midland mountains, with alpine vegetation above; and finally, a mixed temperate forest was present from central to southern Korea. Another alternative is that of Gellert (1976), who proposed that taiga existed in the north, steppe in the northwestern region and mixed evergreen and summer-green forests in central and southern Korea.

Our own work suggests that, despite the lack of adequate pollen data from the peninsula, it is possible to present a schema for vegetation in Korea during the last glaciation, which is more detailed than any of the above. This is based on our knowledge of present-day Korean biogeographic divisions, along with the present-day altitudinal range of species found within them (already explained in Chapter 2), with supporting evidence from Wissmann (1956) and Gellert (1976), palynological results from neighbouring countries (e.g. USSR, China and Japan) and also from Lake Younglang in Korea (Fig. 3.7). From all of this, it is possible to divide Korean vegetation during the last glacial period into six tentative zones (Fig. 3.6, Table 3.6).

First, it is clear that local alpine glaciers, and intensive periglacial zones existed in the northernmost high mountains, above 2000 m a.s.l., including Mts. Paikdoo, Kwanmo, Solryong, Nampodae and Pukpodae (Fig. 3.5). Below, second, was an alpine tundra or montane-subarctic vegetation, which occupied the northernmost mountains above ca. 500 or 1000 m a.s.l. Third, an arctic-alpine vegetation also covered many central and southern high mountains. This probably occurred even on the highlands of Mt. Halla, Cheju Island (33°30′ N), where numerous arctic-alpine plants even now are present. Common arctic-alpine plants which would have been present in this community include *Sabina sargentii*, *Empetrum nigrum* var. *japonicum*, *Diapensia lapponica* subsp. *obovata*, *Rhododendron* spp., *Vaccinium* spp., *Silene* sp., *Sieboldia* sp., *Geranium* sp., *Adenophora* sp. and *Senecio* sp. From northwestern Korea to northeastern China, fourth, steppe vegetation was present, in a broad area, which probably included the exposed continental shelf of the West Sea (Yellow Sea). The presence of this steppe grassland might have been due to the development of extremely low humidity, low precipitation and a marked increase of continentality within this district at this time. Southwards of this, fifth, lay coniferous vegetation, which occupied the central and southern high mountains, below the arctic-alpine vegetation. Finally, a mixed coniferous and deciduous broadleaved vegetation was widespread in the central lowland areas and in the south of Korea.

Of vegetation changes subsequent to the Devensian glacial maximum, little is known, though the bore-hole sample from Lake Younglang (mid-eastern coast, 128°35′ E, 38°12′ N, 0 m a.s.l.) is of importance in this respect, since it contains pollen grains and an absolute ^{14}C dating record which goes back to 17 000 years B.P. (Kim, 1980; Yasuda *et al.* 1980; Fig. 3.7). This, however, is the only such sample available from Korea at present.

An interpretation of its data suggests that from 17 000 to 15 000 years B.P. (pollen zone I), cryophilous coniferous plants, such as *Picea*, *Larix*, *Abies* and *Pinus* (Haploxylon), still were dominant in this region with a large amount of open-land *Artemisia* and hydrophilous Cyperaceae. Small amounts of *Tilia*, *Quercus*, Gramineae, Umbelliferae and Chenopodiaceae also were present.

Then, from 15 000 to 10 000 years B.P. (pollen zone II), as the climate ameliorated, the coniferous tree element declined in abundance, and was replaced largely by deciduous broadleaved trees like *Tilia* and especially *Quercus*. There was a decrease in the number of *Artemisia* and hydrophilous Cyperaceae, and an increase in Gramineae. Other non-arboreal pollens of Oleaceae, Chenopodiaceae, Umbelliferae, Compositae and Typhaceae were also present at this time.

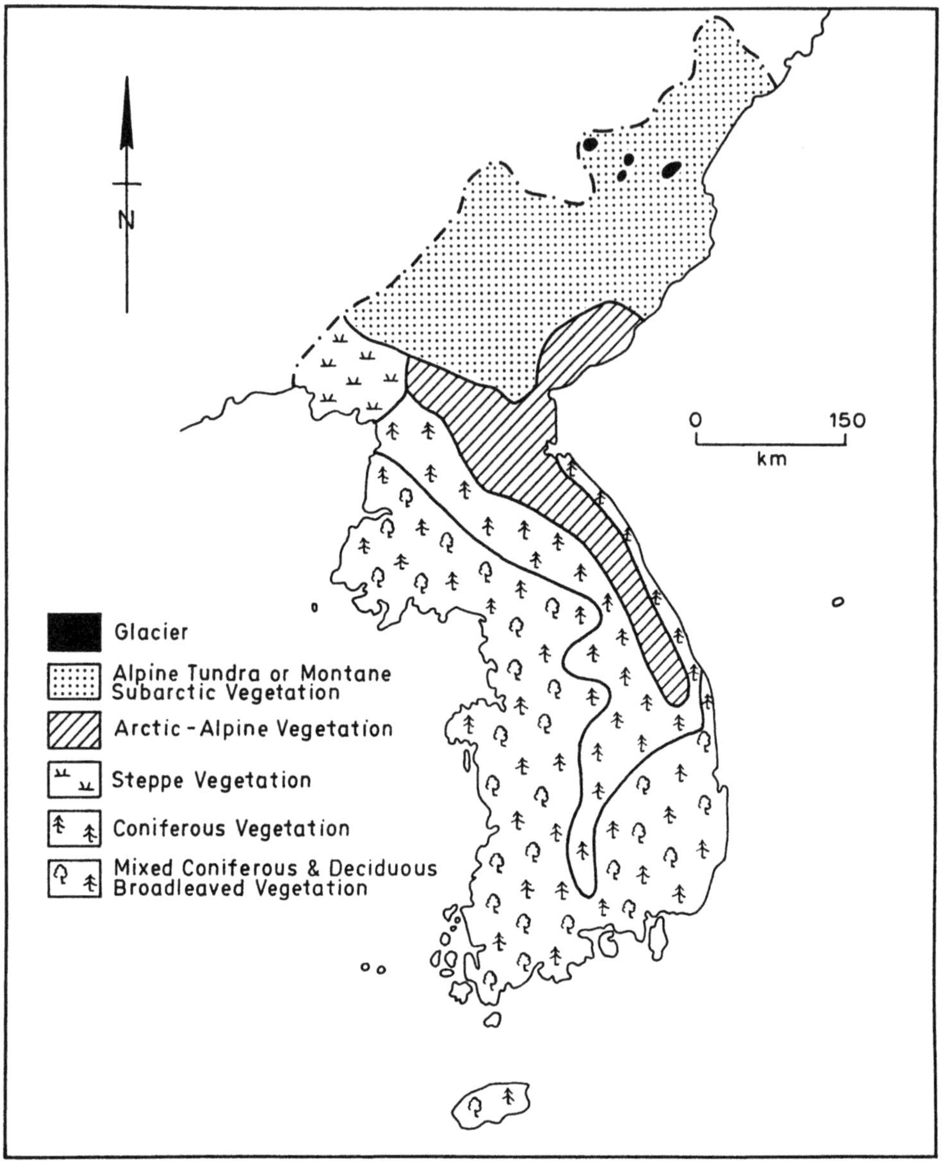

Fig. 3.6. Tentative vegetation zones during the last glacial maximum (Devensian).

Vegetation of the Holocene Period (ca. 10 000 y B.P. – present)

The boundary between the Pleistocene and the Holocene has been placed at ca. 10 000 years B.P. for most parts of the world (Godwin, 1966; Wendland & Bryson, 1974; Bowen, 1978; Nilsson, 1982), and the same division is employed in this work. For ease of analysis, we have also divided the Holocene of Korea into two, viz. the Lower Holocene and the Upper Holocene (Recent), with the division placed at ca. 2 000 years B.P., when some early documentary climatic records become available.

Lower Holocene (ca. 10 000–2 000 *y B.P.)*

Microfossil (pollen) evidence from Lake Younglang (Fig. 3.7; Table 3.6) indicates that along the lowland central east coast region, deciduous broadleaved trees (e.g. *Juglans*, *Carpinus*, *Alnus*, *Ulmus* and *Celtis*) gradually increased in numbers from 10 000 to 6700 years B.P. (pollen zone III), contemporaneously with the disappearance of

Table 3.6. Tentative vegation history since 17 000 y B.P. in Korea

Dating	Vegetation			Archaeology		Human		Climate			Geology
(B.P.)	Pollen zone	AP:	NAP	Cultural age	Agriculture	Habitation	Deforestation	Temperature	Humidity	Period	Sea level
Present 1000	VI	Pinus Ouercus Alnus Corylus	Artemisia Gramineae Chenopodiaceae Oryza	Iron				Cool	? Humid		+0.426 mm/yr.
3000	V	Quercus Pinus Carpinus Corylus Alnus Ulmus Juglans	Artemisia Oleaceae Gramineae Spore	Iron Brozne Neolithic				Mild	? Dry		–1 m (Korea)
5000	IV	Pinus Ouercus Carpinus	Artemisia Oleaceae Gramineae	Neolithic				Warm	Dry		+1.4 mm/yr.
7000	III	Quercus Pinus Carpinus	Gramineae Spore Artemisia Cyperaceae	?				Mild Dry	Post Glacial		–10 m (Korea)
10000 13000	II	Ouercus	Cyperaceae Spore Gramineae Artemisia	Palaeolithic				? Cold	Dry	Late Glacial	–25 m (Korea) –40 m (E.China) –70 m (E.China)
17000	I	Picea Larix Abies Pinus (Haploxylon)	Cyperaceae Artemisia Spore					Cool	? Dry		–80 m (Japan)

(Compiled from Chard, 1960a; Sample, 1974; Oh, 1976; Park, 1977; Kim 1980; Jo, 1980a,b; Scoul National University, 1980; Yasuda *et al.*, 1980; Nelson, 1982; Kim 1985; Kang, 1987)

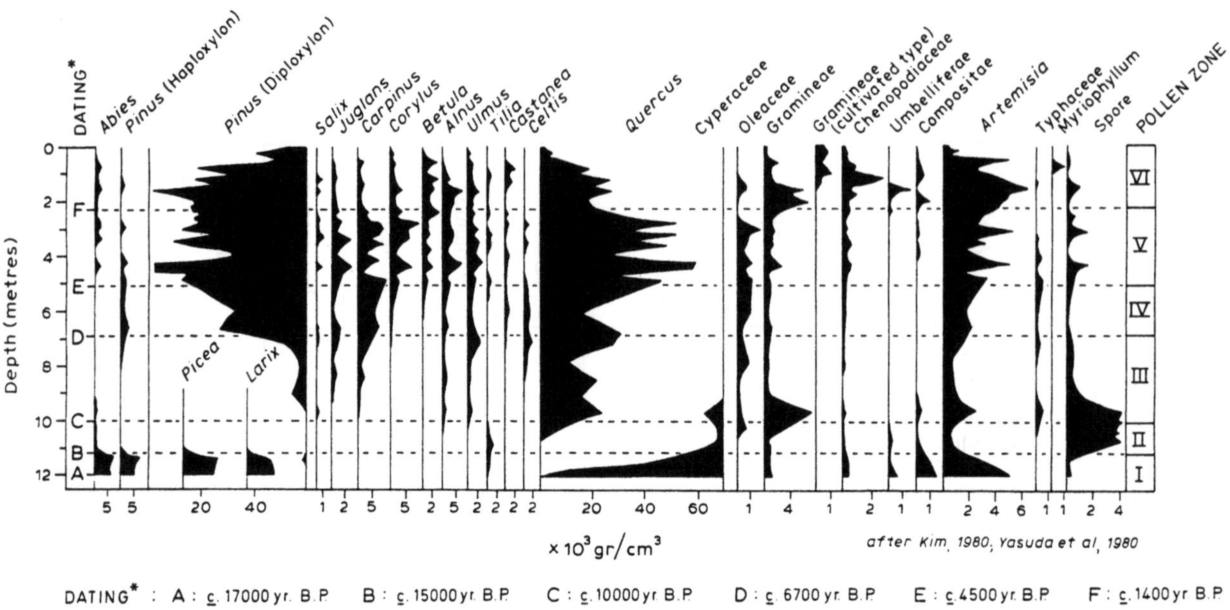

Fig. 3.7. Pollen diagram from Lake Younglang, Korea.

cryophilous trees, such as *Picea*, *Larix* and *Betula*, notably at ca. 8000 years B.P. However, the dominant trees at this stage were *Quercus* and *Pinus* (Diploxylon). Also, the non-arboreal pollen of Cyperaceae, Gramineae and spores was sharply decreased.

From 6700 to 4500 years B.P. (pollen zone IV), the vegetation remained dominated by *Pinus* (Diploxylon) and *Quercus*, along with *Betula*, *Alnus*, *Tilia* and *Castanea*. A gradual increase of *Artemisia*, Oleaceae, Gramineae, Chenopodiaceae, Typhaceae and spores over this period is also noticeable. At the same time, pollen evidence from elsewhere (Yasuda *et al.*, 1980), which unfortunately has no precise dating, suggests that thermophilous evergreen broadleaved trees, such as *Cyclobalanopsis* and *Castanopsis*, appeared in Bangojin, southeastern Korea (129°25′ E, 35°29′ N, 1.5 m a.s.l.) and in Kaheung, southwestern Korea (126°38′ E, 34°59′ N, 3 m a.s.l.). It should be noted that these two trees do not exist anymore in these areas today, though whether this is due to deforestation or to a minor climatic fluctuation is not yet clear.

After 4500 years B.P. until 1400 years B.P. (pollen zone V), *Pinus* (Diploxylon) and *Quercus* continued to dominate, along with deciduous broadleaved trees, such as *Salix*, *Juglans*, *Carpinus*, *Corylus*, *Betula*, *Alnus*, *Ulmus*, *Tilia*, *Castanea* and *Celtis*. However, in contrast to the preceding period, cryophilous coniferous trees, such as *Abies* reappeared. In addition, a sharp decline of *Pinus* (Diploxylon) and *Quercus* pollen is reported from the Pyungtaek peat deposit (mid-western coast of Korea) at ca. 3500 years B.P. (Oh, 1976). At the same time, according to Yasuda *et al.* (1980), thermophilous evergreen broadleaved trees (e.g. *Cyclobalanopsis*, *Castanopsis*, *Myrica* and *Camellia*) began to occur from Bangojin (southeastern Korea), and the first three of these were present in Kaheung (southwestern Korea) as well. The reappearance of cold-tolerant species at Lake Younglang, coupled with the presence of thermophilous evergreen broadleaved trees in the southeast, seems to imply that the latitudinal divisions of climate in the peninsula were becoming more exaggerated than formerly. Additionally, an expansion in the frequency of non-arboreal pollen and spores suggests an increased rate of deforestation for this region within this period. Some three degrees latitude to the south of Lake Younglang, early agriculture clearly was well under way at this time, for Kim (1980) has reported the presence of cultivated rice pollen (*Oryza*) from Naju (southwestern Korea) at ca. 3500 years B.P. and from Kimhae (southeastern Korea) at ca. 3000 years B.P. (see also p. 70).

Sea-levels
Continuing the trend of the previous 10 000 years (Table 3.6), a major rise of sea-level took place during the period between 10 000 and 4000 years B.P. Sea level is thought to have been 25 m below present at ca. 10 000 years B.P., 10 m below present at ca. 7000 years B.P., and 1 m below present at ca. 4000 years B.P. (Jo, 1980b,c), and according to Park (1977), land on the western coasts has been further submerged at the rates of 1.4 mm per year from ca. 6500 to 3820 years B.P., and 0.426 mm per year since then. A further expansion in the area of land covered by the West Sea (Yellow Sea) therefore was conspicuously maintained over these years.

Palaeoclimates

The postglacial hypsithermal period in Korea appears to have lasted from ca. 8000 to 5000 years B.P., or possibly to ca. 4000 years B.P. (Chard, 1974; Pearson, 1977; Kim, 1983). According to Sohn (1984), the mean annual temperature at ca. 5000 years B.P. was 2°C warmer than that of today. The dominance of *Pinus* (Diploxylon) and *Quercus*, as well as a gradual increase of deciduous broadleaved trees in central Korea, and the appearance of thermophilous evergreen broadleaved trees in southern Korea, also suggest the postglacial hypsithermal from ca. 8000 to 4500 years B.P. Following this, increased cooling in the north, and a greater climatic differentiation between north and south in the peninsula (see above), marked the end of this phase.

Archaeology
No cultural and archaeological records are reported from ca. 10 000 to 6700 years B.P. for the Korean peninsula, and this clearly makes the determination of the cultural sequence extremely difficult in detail. According to Lee (1975), the Neolithic Age extended from ca. 7390 to 3040 years B.P. Later, Choi (1982) suggested that it ranged from ca. 7000 to 3300 years B.P., and Yi (1984) reported it as existing from ca. 5950 to 2850 years B.P. This present research is based on the cultural chronology of Korea as put forward by Yi (1984).

It is fairly clear that Neolithic man appeared in Korea at some time between ca. 7500 and 6000 years B.P., and continued until between ca. 3500 and 2000 years B.P. During this time, his active presence was noticed throughout the peninsula. The first dwelling sites of Neolithic man are mainly associated with the banks of rivers, streams and seas. With time, these people then seem to have occupied inland areas as well. This inland expansion may have given rise to early deforestation, mainly for agricultural and domestic (fuel and building) purposes. Evidence for agriculture comes from the discovery of tools, such as sickles, mattocks, hoes, querns, antler objects, stone spade-like tools, storage pits, huge storage vessels, and grains (e.g. millet and barnyard grass), which were grown at ca. 5000 years B.P. (Chard, 1960a; Sample, 1974; Seoul National University, 1980; Yi, 1984; Scarre, 1988). A similar pattern has been found in Manchuria and eastern Russia (Chard, 1960b).

From ca. 4500 to 1400 years B.P., intensive agricultural activity took place in the Late Neolithic, Bronze and Early Iron Ages (see Table 3.6). This was marked most noticeably by the cultivation of rice. According to Grigg (1984), the introduction of rice to Korea took place at ca. 5000 years B.P.; in turn, Nelson (1982) proposed that rice cultivation had been well established by ca. 3500 years B.P. In southwestern Korea, the cultivated rice pollen *Oryza* dates back to ca. 3500 years B.P., and to ca. 3000 years B.P. in southeastern Korea (Yasuda *et al.*, 1980; Scarre, 1988). In central Korea, rice cultivation began between ca. 2850 and 2350 years B.P. (Yi, 1984). According to Akazawa (1983), irrigated rice cultivation was transferred *to* Japan from Asia, possibly via southern Korea, over this period. Clearly, by this time, a sophisticated agricultural technology centred on rice had emerged in the peninsula.

From the Bronze Age, the Korean population began to dwell on higher ground more widely, as its density increased. This was a classic age for metal-making skills, especially in the preparation of tools and ornaments. In the process of expansion, forest must have been cleared quickly to provide agricultural and residential space, as well as for a source of charcoal in the smelting of ores for metallurgical purposes (Turner, 1970; Jahn, 1977). Further deforestation might have been caused by the development of the classic Korean heating system of 'Ondol' (Korean-style underfloor flues for heating), which commenced in this period and which in itself, in its operation, requires a large amount of fuel, the building of wooden houses, and the use of wood in pottery kilns. It is known that deforestation for rice cultivation in central Korea dates back to ca. 2500 and 2300 years B.P. (Kim, 1985; Kang, 1987): elsewhere the dates are not yet clarified.

Upper Holocene (ca. 2000 yr B.P.–present)

A wide range of information is available for this period for the reconstruction of climate and vegetation, and this includes pollen evidence, archaeological data and documentary materials.

Microfossil data
The period between 2000 to 1400 years B.P. has been dealt with in the preceding section (p. 68). From 1400 years B.P. (Pollen Zone VI), the available pollen evidence (see Fig. 3.7) suggests that there was a sharp decrease in the representation of *Pinus* (Diploxylon) and *Quercus* within the pollen spectra. Also, pollen from *Alnus*, *Corylus*, *Ulmus*, *Carpinus*, *Castanea*, *Juglans*, *Salix* and *Tilia* further was reduced. The presence of the cryophilous genera *Abies* and *Pinus* (Haploxylon) is noticeable. A further decline in *Artemisia* and Gramineae takes place during this period. From ca. 1000 years B.P., there was moreover a gradual increase of *Oryza* pollen in the central east coast area of Korea, with a sudden and pronounced increase of *Oryza* at ca. 1500 years B.P. in Kaheung (southwestern Korea) and a similar sharp increase at ca. 1000 years B.P. in Bangojin (southeastern Korea) as well (Yasuda *et al.*, 1980). Indications from the pollen data accordingly suggest that, certainly in the central east coast region, from which the pollen data were derived, the climate was getting colder over this period; additionally, agricultural activity was becoming more intensive, and this was also the case in southern regions.

Palaeoclimates and vegetation: pollen data and historical records of the Upper Holocene
From 2000 years B.P., the palaeoclimate of the Korean peninsula can be reconstructed by means of reference to the palynological record and also from a series of written historical documents which are unique to Korea. Both complement each other, and the historical records are used in particular as a means of clarifying both the anthropogenic influences suggested by the pollen diagram, and the nature of the recent climatic changes. It is clear that the latter are complex both spatially and temporally. A series of very early references to climatic conditions in Korea are found in the Samguksagi (History of the Three Kingdoms), which covers a period from 51 B.C.E. to C.E. 921. Other similar records include the Munhonbigo, Seowoonkwan Ji (Diary of the Seowoonkwan), the Chosun Wangjo Sillok (Annals of the Yi Dynasty) and the Sungjungwon Ilgi (Diary of the Royal Secretariat of the Yi Dynasty), which cover the period from 1413 to 1894. All these were transcribed into modern Korean script by Kim (1983, 1984a,b), and they have been used in this work to provide indications of relative rainfall and humidity, snow, fog, frost, hail, drought, flood and famine. From them, the climatic fluctuations since 50 B.C.E. may be elucidated, and the patterns which arise are presented in Table 3.7.

In this table, climatic divisions have been indicated, with a 100-year time interval, on the basis of relative temperature and humidity, so that they may be categorised as cold or warm, dry or humid. Overall, major cold phases may be detected between 50 B.C.E. to C.E. 100 and C.E. 301 to 600; dry phases from C.E. 1 to 100, 201 to 400 and 601 to 1100; and humid phases from C.E. 101 to 200 and 1101 to 1900. The period between C.E. 601 to 800 also shows a strongly sequential succession of cold and warm years; and sequences of dry and humid years existed from C.E. 401 to 600. Famine seems to have occurred especially during the cold stages, and mainly between C.E. 101 to 200, 801 to 1400 and 1601 to 1800; however, famine was also recorded during the relatively warm and dry years between C.E. 300 to 600, perhaps because of an excess of heat or aridity during the main crop-growth season. The periodic synchronous occurrence of cold and

Table 3.7. Climatic fluctuations since 50 B.C.E. in Korea.

Period	B.C.E.	C.E.																		
No. Climatic	50	1	101	201	301	401	501	601	701	801	901	1001	1101	1201	1301	1401	1501	1601	1701	1801	
Events	0	100	200	300	400	500	600	700	800	900	1000	1100	1200	1300	1400	1500	1500	1600	1700	1800	1900 Total
Warm	1	4	3	3	4	4	3	1	4	5	1	2	6	3	8	2	4	–	1	–	59
Cold	–	4	9	3	2	3	2	5	3	7	5	11	10	16	17	–	3	7	6	1	114
Dry	2	12	6	17	16	14	10	7	16	12	7	27	18	23	8	8	2	5	4	1	215
Humid	–	7	9	5	4	13	4	8	7	6	4	14	40	24	22	22	15	41	41	37	323
Warm & Cold	–	–	–	–	–	–	–	–	–	–	–	–	1	–	4	–	–	–	–	–	5
Dry & Humid	–	–	1	–	–	–	–	–	–	–	–	13	20	11	23	–	2	11	3	1	85
Famine	1	1	6	5	7	7	8	5	8	9	–	6	3	10	16	9	9	35	21	2	169

The number of climatic events for each column represents the number of years associated with such events (Modified from Kim, 1985).

wet years is also thought to have been responsible for the serious famines of C.E. 1200 to 1400 and 1600 to 1800.

During the cold and/or humid stages, an uneven seasonal distribution of rainfall often occurred as well, and this often had the effect of accentuating drought during the sowing season of rice, and frequently also produced unusually severe floods during the main growth season of the crop, so that both of these features accentuated the potential for famine; and at the same time, people then were forced to disturb forest land at a rate which was well above average in their search for substitute foods.

The large number of cold and humid years in Korea between C.E. 1600 and 1800 in particular is also of interest, in that these appear to be related to the incidence of the Little Ice Age, a period which in Korea was marked by an increase in the occurrence of famine, and in a consequent sharp decline of population, which is noted by Kwon & Shin (1977).

Pollen evidence (Fig. 3.7) suggests a sharp decline of the thermophilic coniferous and deciduous broadleaved trees, and a minor increase of the cryophilous coniferous trees, such as *Abies*, from 1400 years B.P. However, the cool phase of the Little Ice Age in Korea is poorly reflected in the vegetation data. In other respects, vegetation change over this period is largely a matter of anthropogenic activity. Thus, the increased presence of *Oryza* indicates an extension of rice cultivation activities during this period, both in the central eastern coast of Korea and elsewhere (pp. 68, 70), and this eventually caused conspicuous deforestation.

Over the last 650 years, seven further historical works are also useful in the reconstruction of these vegetation patterns and changes, even though many of these deal with cultivated plants. These are the Kyungsangdo Chiriji (Regional Geography of Kyungsang Province) of 1425; the Saejong Sillok Chiriji (Annals of King Saejong) of 1454; the Shinzung Dongguk Yeoji Seungram (Augmented Survey of Korean Regional Geography) of 1531; the Takriji (Selection of Suitable Habitat) of 1751; the Kongsa Shinseo (New Bibliography of Administration) of 1771; and the Chosun Mulsan (Products of Korea) of 1928. The records within these works provide information relating to the distribution of 32 trees, four textile plants, four medical herbs, nine cereal plants, and of these, the records of 23 are especially complete, and accordingly are presented herein (Table 3.8).

But before this table was compiled in final form, some interpretation of available data was necessary, in that the old descriptions of trees, herbs, cultivated plants and other products needed to be reassessed for the purpose of species classification. This painstaking work, involving comparisons between present and past floras, was done after the method used in the West Indies by Watts (1966, 1987). Always, one has further to remember the risk that some artificial manipulation of these data, for tax purposes, might have been made by the original recorders, and that this must be discounted as far as possible.

Bearing all this in mind, it is clear that several recurring patterns of distribution of these mainly cultivated species, and of other distributional changes associated with them, occurred in Korea over the years 1425 to 1928. In recent times (1751–1907), eight common spatial distribution patterns may be recognised, as follows (Fig. 3.8).

First, some species grow over the whole of the peninsula, e.g. lacquer (*Rhus verniciflua*), Korean pine (*Pinus koraiensis*) and ginseng (*Panax schinseng*); second, other species occur predominantly in northern Korea, such as mulberry (*Morus alba*), cotton (*Gossypium nanking*) and hemp (*Cannabis sativa*); third, some species are found mainly in northern and central Korea, e.g. paper mulberry (*Broussonetia kazinoki*) and soy bean (*Glycine max*); fourth, others are located mainly in central Korea, e.g. the box tree (*Buxus koreana*) and the jujube (*Zizyphus jujuba*); fifth, species such as the small bamboo (*Pseudosasa japonica*), the persimmon (*Diospyros kaki*) and the ramie (*Boehmeria frutescens*), occur in both central and southern Korea; sixth, other species occur almost solely in southern Korea, e.g. the pomegranate (*Punica granatum*), tea plant (*Thea sinensis* var. *bohea*), nutmeg (*Torreya nucifera*) and large bamboo (*Sinoarundinaria reticulata*); seventh, certain species are found mainly in southern Korea and Cheju Island, e.g. citrus (*Citrus junos*); and finally, some are restricted to the subtropical Cheju Island itself, e.g. machilus (*Machilus thunbergii*) and mandarin (*Citrus* spp.).

Table 3.8. Historical changes of selected Korean plant ranges in Korea.

Plant	Period			
	1425–1531	1751–1787	1864–1907	1910–1928
Mulberry	NW SW	NW SE S	NW S	–
Paper Mulberry	S NW	S NW M	S NW M	S M NW
Lacquer	ME SE SW NW	ME SE SW	SE SW NE NW	SE NE
Pomegranate	S	S	S	–
Tea Plant	SW SE	SW SE	SW SE	SW
Box Tree	ME N	ME N	NE ME	–
Machilus	Cheju	Cheju	Cheju	–
Citrus	Cheju	Cheju	Cheju	–
Pear	SE ME	M S	NE NW SW M	SW MW SE
Chestnut	SE SW	SE SW MW	S NW M	NW M S
Jujube	M SE SW	M S	M S N	M
Korean Pine	N M S	N M S	N M S	–
Persimmon	S M	S M	S M	S M
Nutmeg	SW SE Cheju	SW SE Cheju	SW SE Cheju	–
Mandarin	Cheju	Cheju	S Cheju	–
Large Bamboo	S MW	S MW	S MW ME	S
Small Bamboo	S ME MW NE	S ME MW NE	S ME MW	ME S
Cotton	N M	N M	N M S	M S N
Hemp	N M S	N M S	N M S	N M S
Ramie	ME S	ME S	ME S	MW SW
Ginseng	N ME S	N ME S	NE ME	NW M SW
Tobacco	–	–	M N S	M S N
Soy Bean	–	–	N M S	M S N

The data indicate major areas of distribution for each species, and have been compiled from various records, viz. the Kyungsangdo Chiriji (1425), the Saejong Sillok Chiriji (1454), the Shinzung Dongguk Yeoji Seungram (1531), the Takriji (1751), the Kongsa Shinseo (1771), and the Chosun Mulsan (1928), and analyzed by the authors.

Legend: E: East W: West
S: South N: North
M: Midland

These patterns emphasise the climatic divisions which are prevalent in the peninsula, and which have been detailed in Chapter 2.

On a temporal basis, five patterns of distributional change can be recognised from Table 3.8 and Fig. 3.9.

First, species which have no noticeable change in their range distribution over the last 500 years include paper mulberry, lacquer, pomegranate, box tree, machilus, nutmeg, mandarin and *Citrus junos*. The stability of distribution implied by these records is perhaps surprising, bearing in mind the climatic changes associated with the Little Ice Age, of from ca. C.E. 1600 to 1800, and seem to imply that they were more a matter of degree, than on any major scale. Second, a general decline in range can be seen in the case of the tea plant, large bamboo, small bamboo, ramie and ginseng. This might be the result of a response to climatic deterioration in the case of thermophilous plants, such as tea and bamboo, but it also may relate to a relative decline in range associated with the expansion of other crops (Kong & Chung, 1984, 1985; Kong, 1985). Third, an expansion of range can be noted in the case of some fruit trees, e.g. pear, chestnut, persimmon and jujube. This expansion is probably due to a steady and increasing demand from markets. Fourth, a southward expansion of species can be seen in the case of the mulberry, cotton, hemp, tobacco and soy bean. The sudden expansion of tobacco (*Nicotiana tabacum*) and soy bean (*Glycine max*) in central and southern Korea since 1864 again suggests the active cultivation of cash crops in demand for commercial purposes, since no clear evidence for climatic change over this small period of time is forthcoming. Finally, the sudden appearance and expansion of certain grain

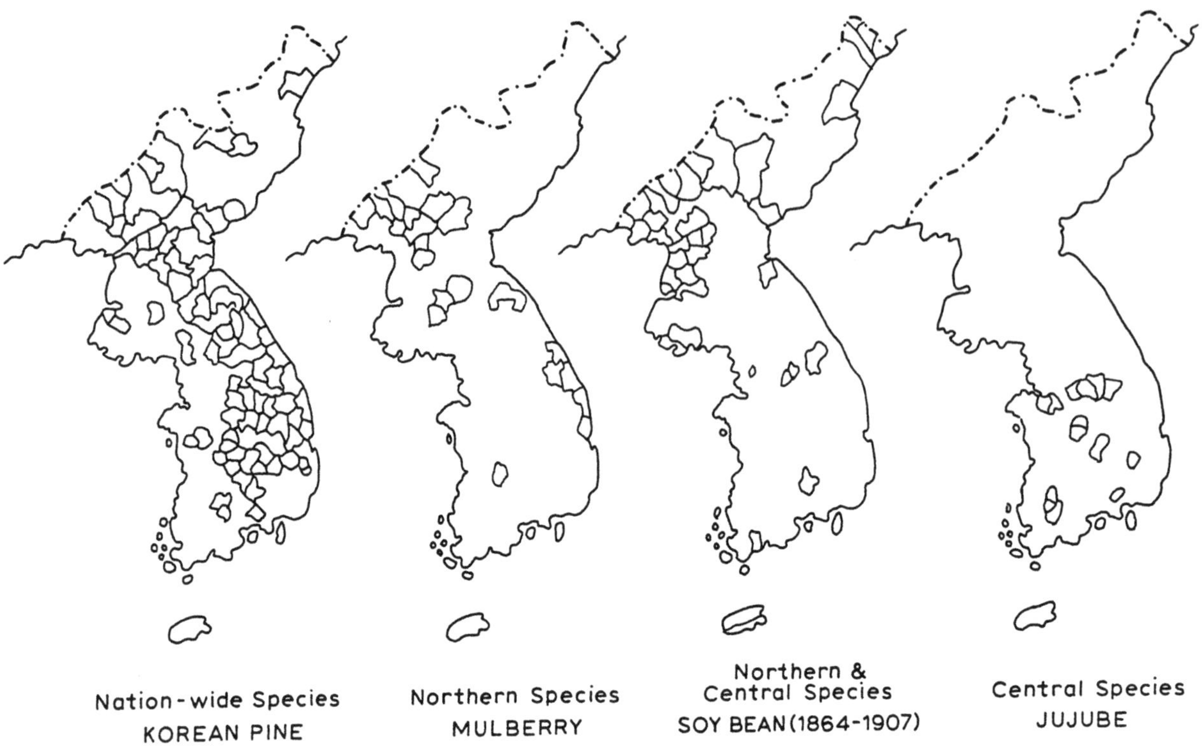

Fig. 3.8.(a) Distributional ranges of selected species from 1751–1787-I.

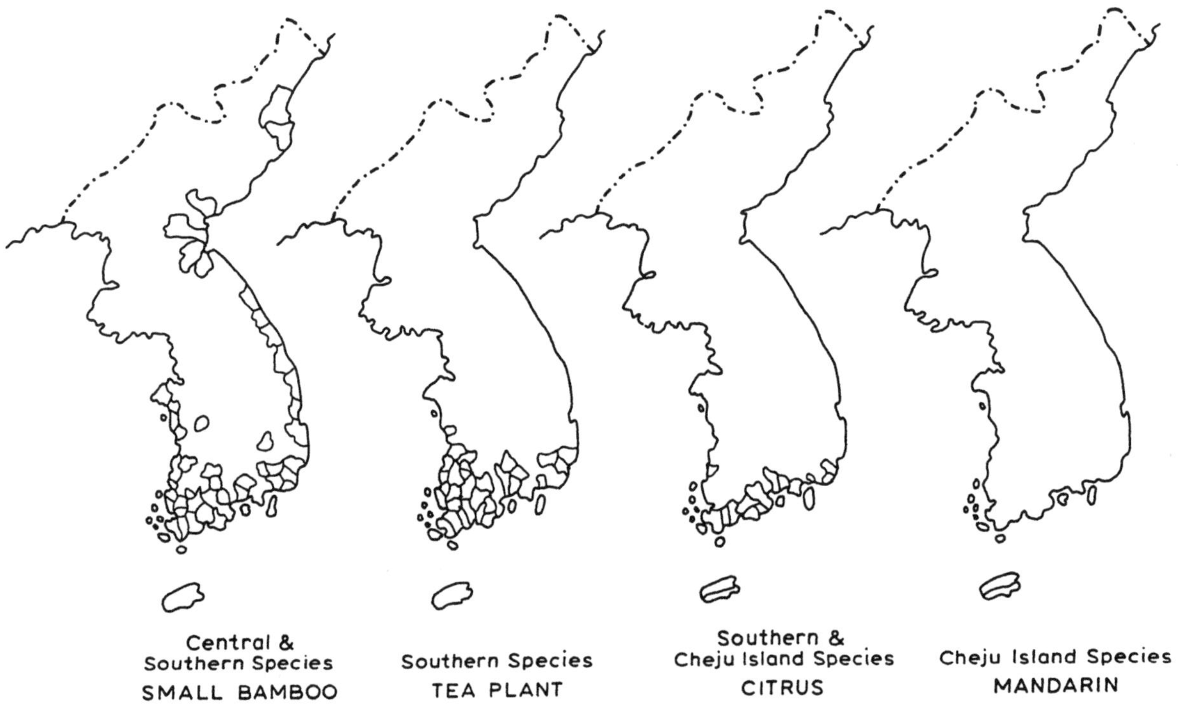

Fig. 3.8.(b) Distributional ranges of selected species from 1751–1787-II.

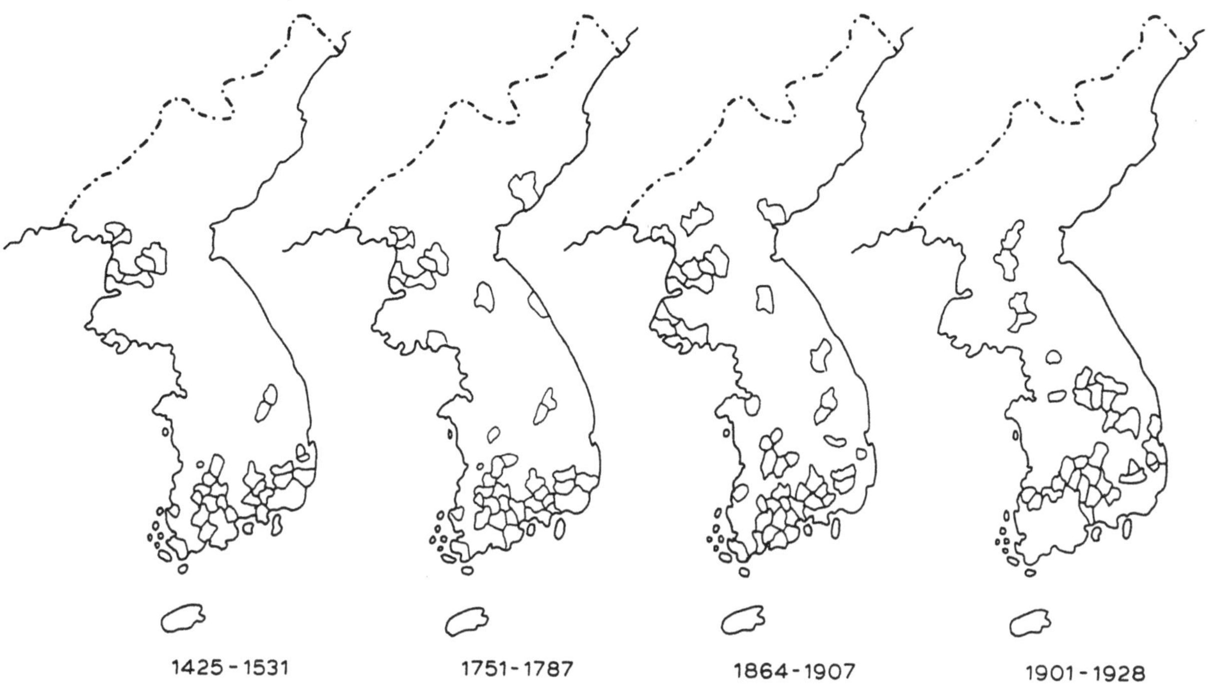

Fig. 3.9.(a) Temporal distributional changes of plants, 1425–1928-I.

crops, such as buckwheat (*Fagophyrum esculentum*), oats (*Avena sativa*), wheat (*Triticum aestivum*), barley (*Hordeum sativum*), maize (*Zea mays*), barnyard grass (*Echinochloa frusanentacea*), millet (*Panicum miliaceum*), and foxtail millet (*Setaria italica*) in northern Korea since 1864 may be the result of the sometimes frequent failure of the rice crop in this region, the cultivation of which appears there to be at the limit of its range, as a consequence of abnormal climatic events, (e.g. drought, flood, late frost, etc.), and the general conversion to more suitable crops such as these.

Overall, there are several factors which are responsible for the rapid deforestation generally found in Korea since 2000 years B.P. The most common method of cultivation in mountain areas until very recently has always been a form of slash-and-burn, and millets, potatoes and oats have been grown on this burnt-over land since the Neolithic. Even though this method of land clearance was banned by law at several times in the past, as in 1469, 1592 and 1736, it was maintained widely even so, and was still noted in 1963 on the upper parts of the Han River in central Korea (McCune, 1956; Korean Commission for Nature and Natural Resources, 1971; Sample, 1974; Kang & Iizumi, 1981). Although the extension of rice cultivation in recent years has generally been confined to the intensification of lowland production, some movement upslope on mountain land did take place, and this involved further deforestation. Fuel was always widely used to maintain the Korean system of heating houses, especially bearing in mind the severely cold Korean winter and the lack of alternatives, coal resources being confined mainly to the Kangwon Province along the East Sea, and being used for the first time quite recently for domestic purposes.

Today, the total forest area in Korea is about 67 per cent of the nation's territory, and the natural components of this have been presented in Chapter 2. However, some mountains have been denuded of timber, mainly due to reckless deforestation during the last half-century, and also as a result of damage induced by the Korean War (1950–1953). During the early 1960's, a national

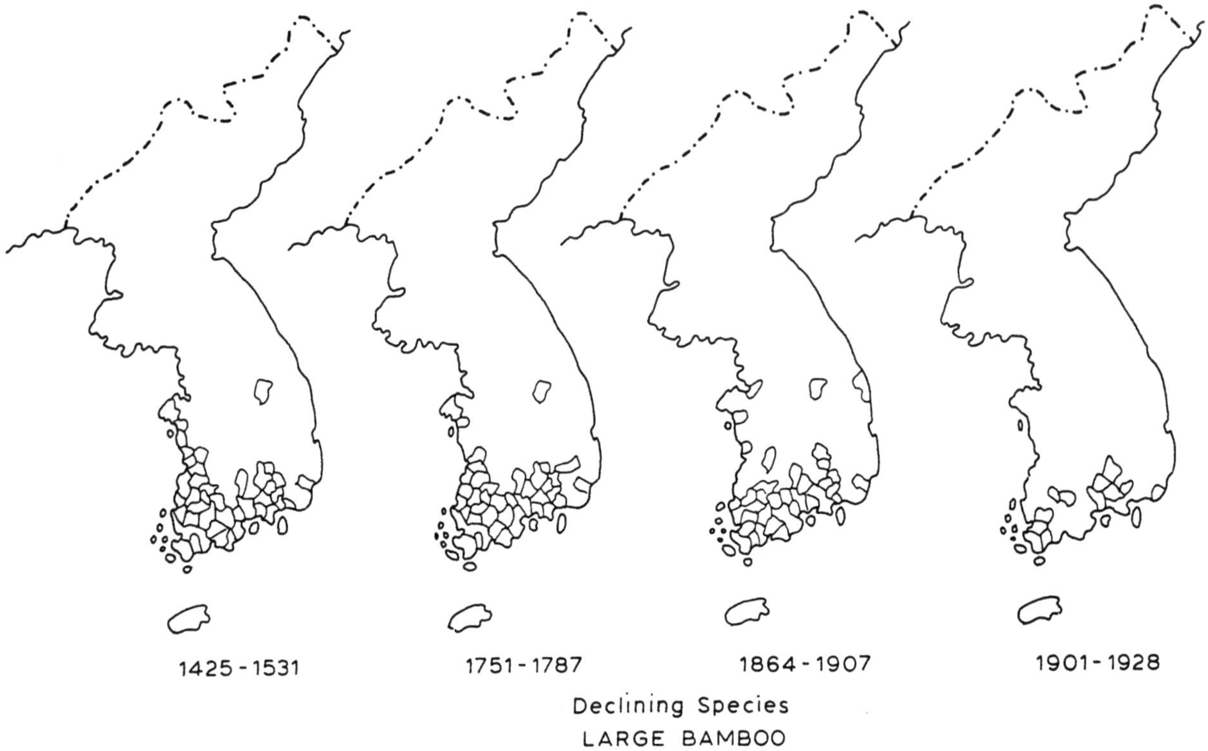

Fig. 3.9.(b) Temporal distributional changes of plants, 1425–1928-II.

reforestation scheme was launched, and laws concerning the preservation of forests then were instituted. Under the afforestation policy, huge numbers of trees were planted. At present, the 5 742 000 hectares of forest in South Korea consists of 3 317 000 hectares of coniferous forest, 1 217 000 hectares of deciduous forest, 1 201 000 hectares of mixed forest and only 7 of bamboo stands (FAO, 1976; Korean Overseas Information Service, 1978).

The growing number of intentional and/or accidental introductions of alien plants also have changed the vegetation of Korea, particularly since 1900, since some have become naturalised. Numerous alien plants have been introduced from China, Europe, America, Asia, Africa and Australia for the purpose of horticulture, industry, food, plantations or other uses (Lee & Kim, 1961; Lee, 1970). Recently, Yim & Jeon (1980) have noted that, from these and arriving weeds, 110 have become naturalised. These are predominantly herbaceous plants, including representatives from the Compositae (13 spp.), Polygonaceae (13 spp.), Gramineae (11 spp.), Leguminosae (8 spp.), Solanaceae (6 spp.), Cruciferae (6 spp.), Convolvulaceae (6 spp.) and Malvaceae (4 spp.). The origin of these species is diverse, being attributed to Europe (33.6%), North America (26.2%), Asia (20.6%), Middle and South America (17.8%) and Africa and Oceania (0.9%).

History of vegetation, and the origins of the present biogeographic regions: results and discussion

From the evidence presented above, it is clear that, in Korea, 16 genera of evergreen arboreal fossils, all of which have a long-term presence in the peninsula, also have representatives in the existing flora of today. Their relationships to the biogeographic divisions of Korea, presented in Chapter 2, are given in Table 3.9. In this, some of the genera which appear to have reached the peninsula since the Upper Pleistocene, e.g. *Em-*

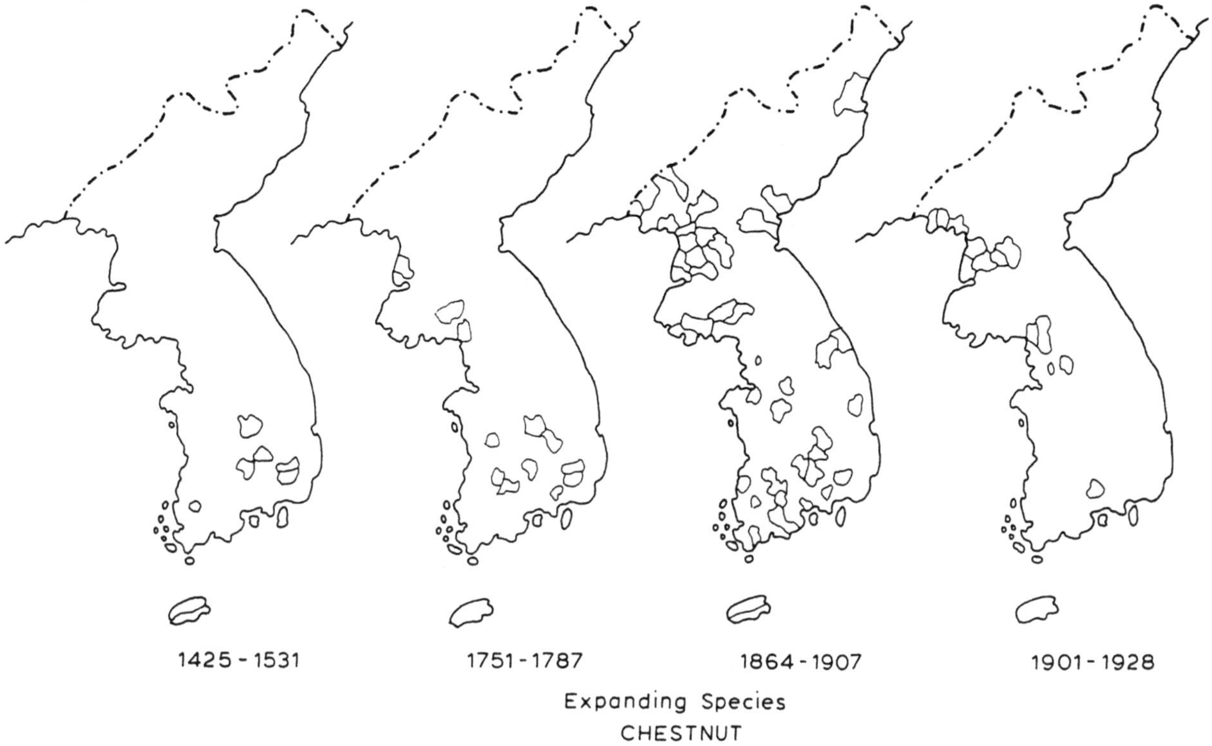

Fig. 3.9.(c) Temporal distributional changes of plants, 1425–1928-III.

petrum, *Andromeda*, *Ledum*, *Vaccinium* and others, also are given where appropriate. The significance of these latter genera to the history of Korean flora is to be further examined in subsequent chapters.

At this stage, a few general comments may be made about the history of these long-term taxa present within Korea.

Taxus occurs today in four regions, i.e. the Northern Alpine (I), the North-south Subalpine (II), the Midland Mountain (III) and the West-South-Eastern Insular and Associated Inland (VII) Regions. *Taxus* fossils were present in central Korea (Chummal) during the cold phases of the Middle Pleistocene, and also in approximately the same area (Chummal, Turubong) during similar phases of the Upper Pleistocene.

Abies grows in five regions today, i.e. the North-South Subalpine (II), the Midland Mountain (III), the Southern Mountain (IV), the Southern Insular (VI) and the West-South-Eastern Insular and Associated Inland (VII) Regions. Fossils of this genus were present in central Korea (Chummal) during the cold phases of the Middle Pleistocene, and in both northern (Hwadae, Haesang) and central Korea (Chummal, Turubong) during similar phases of the Upper Pleistocene. The genus also was present in central Korea (Younglang) during the distinctly cold period, from 17 000 to 15 000 years B.P.

Picea is found in two regions today, i.e. the North-South Subalpine (II) and the West-South-Eastern Insular and Associated Inland (VII) Regions. *Picea* fossils were present both in northern Korea (Yongdong) and southern Korea (Changgi) during the cooling climates of the Lower Miocene, in northern (Hwadae) and southern (Chummal) Korea during the Upper Pleistocene cold phases, and in central Korea (Younglang) from 17 000 to 15 000 years B.P.

Pinus occurs in all of the biogeographic regions of Korea at the present time, with nine species representatives. *Pinus* fossils were present in central Korea (Tongchon) during the Middle Miocene, and in the south (Yeonil) during the Upper Miocene. The genus was also located in

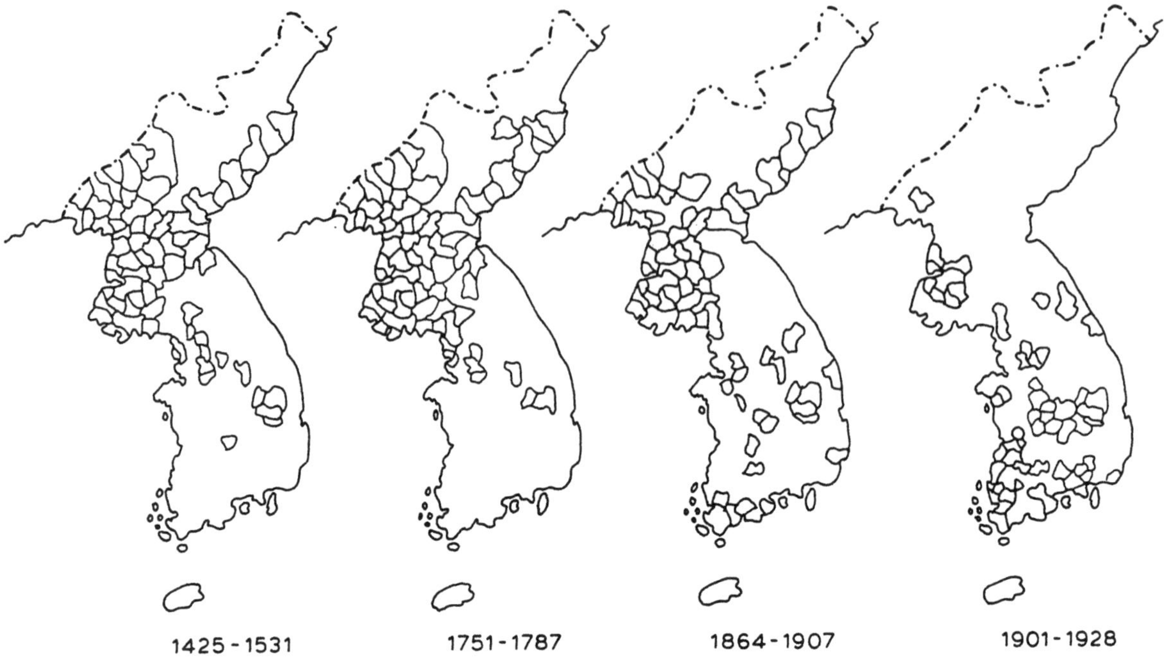

Fig. 3.9.(d) Temporal distributional changes of plants, 1425–1928-IV.

northern (Kumunmoru) and central (Chummal) Korea during the Middle Pleistocene. *Pinus* (Haploxylon) was present in northern Korea (Hwadae) and was common in northern (Haesang) and central (Chummal, Turubong) Korea during the Upper Pleistocene as well. *Pinus* (Haploxylon) was found in central Korea (Younglang) from 17 000 to 15 000 years B.P., and *Pinus* (Diploxylon) was common from 10 000 years B.P., at the same site.

Thuja grows in three regions today, i.e. the North-South Subalpine (II), the Midland Mountain (III) and the Mid-Western Insular (V) Regions. *Thuja* fossils were present in northern Korea (Kumunmoru) during both the Middle and Upper Pleistocene cold phases.

Juniperus is found in six regions presently, i.e. the Northern Alpine (I), the North-South Subalpine (II), the Midland Mountain (IV), the Mid-Western Insular (V), the Southern Insular (VI) and the West-South-Eastern Insular and Associated Inland (VII) Regions. *Juniperus* fossils, however, have only been reported from the north (Hwadae) thus far, where they date to the Upper Pleistocene cold phases.

Rhododendron grows in the Northern Alpine (I) and the North-South Subalpine (II) Regions currently. *Rhododendron* fossils were present in the south (Changgi) during the Lower Miocene, and in both the north (Hamjindong) and the south (Yeonil) during the Upper Miocene. The Ericaceae family further was present in central Korea (Turubong) during the Upper Pleistocene.

The evergreen broadleaved genus *Castanopsis* occurs in the Southern Insular (VI) and the West-South-Eastern Insular and Associated Inland Region (VII) today. Fossils of this genus were present in the south (Yeonil) during the Upper Miocene, and were also in the south (Bangojin, Kaheung) between 6700 and 1400 years B.P. It is indicative of a relatively warm climate, and the same is true for the remaining genera described in this section.

Cyclobalanopsis at the present time has a broadly similar distribution to *Castanopsis*, growing in the same two regions. Fossils of this genus

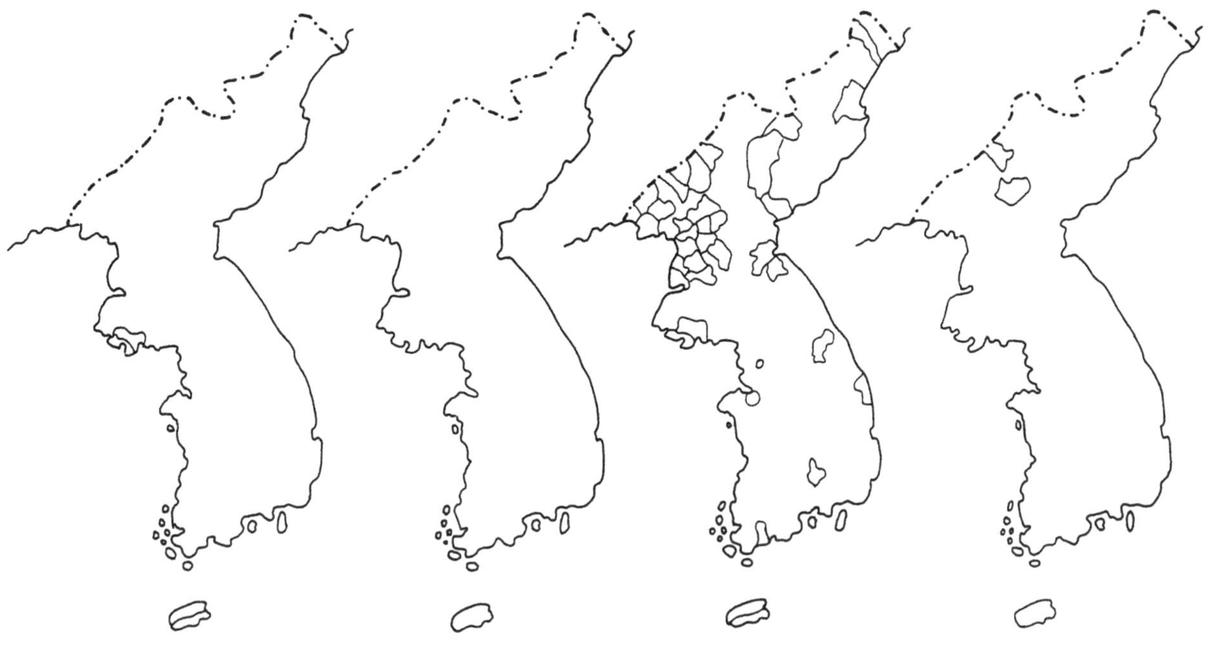

Fig. 3.9.(e) Temporal distributional changes of plants, 1425–1928-V.

were present in the north during the Middle Miocene (Kogeonweon) and during the Upper Miocene in the south (Yeonil). The genus was further recorded as being located in the south (Bangojin, Kaheung) from 6700 to 1400 years B.P.

Ficus too is found both in the Southern Insular (VI) and the West-South-Eastern Insular and Associated Inland (VII) Regions. *Ficus* fossils are known to have been present in the north (Pongsan) during the Oligocene, but as yet have not been recorded in the south.

Cinnamomum occurs in the Southern Insular Region (VI) today, and fossils of this genus were present both in the north (Jangpungri) and in the south (Yeonil) during the Upper Miocene.

Euonymus grows in the Southern Insular Region (VI) at present. *Euonymus* fossils were present in the north (Kogeonweon) during the Middle Miocene, but so far have not been discovered in southern regions.

Camellia is found in the Mid-Western Insular Region (V) today, and fossils of this genus were present in the south (Bangojin) from 4500 to 1400 years B.P.

Hedera currently occurs in three regions, i.e. the Southern Mountain (IV), the Mid-Western Insular (V) and the West-South-Eastern Insular and Associated Inland (VII) Regions. *Hedera* fossils were present in the north (Pongsan) during the Oligocene, but so far have not been detected elsewhere.

Ligustrum is found in the Southern Insular Region (VI) today, and fossils of this genus were present in central Korea (Turubong) during the Upper Miocene.

In general, one may say that the evidence indicates that thermophilous evergreen broadleaved plant genera in Korea can be traced back predominantly to the pre-Pleistocene period. Such is the case for *Ficus*, *Hedera* (Oligocene), *Cyclobalanopsis*, *Euonymus* (Middle Miocene), *Castanopsis*, *Cinnamomum* and *Ligustrum* (Upper Miocene). On the other hand, *Camellia* is known only as far back as the Holocene. Two genera, which contain both thermophilous and cryophilous species, were present prior to the Pleistocene, i.e. *Pinus* (Middle Miocene) and *Rhododendron* (Upper Miocene). *Picea* dates back to the Lower

Table 3.9. Flora of the biogeographic regions, and their fossil representatives in Korea.

Biogeographic Division	Present Flora	First Recorded	Micro- & Macrofossils	Site*
Northern Alpine Region (I)	*Taxus*, *Juniperus*, *Pinus* (Haploxylon), *Empetrum*, *Andromeda*, *Ledum*, *Rhododendron*, *Vaccinium*, etc.	Lower Miocene	*Rhododendron*	Changgi, S.
		Middle Miocene	*Pinus*	Tongcheon, C.
		Upper Miocene	*Rhododendron*	Hamjindong, N.
			Rhododendron, *Pinus*	Yeonil, S.
		Middle Pleistocene	*Pinus*	Kumunmoru, N.
			Taxus, *Pinus*	Chummal, C.
		Upper Pleistocene	*Pinus pumila*,	Hwadae, N.
			Juniperus	Hwadae, N.
			Pinus	Haesang, N.
			Taxus, *Pinus*	Chummal, C.
			Taxus, *Pinus*	Turubong, C.
			(Ericaceae)	Turubong, C.
North-South Subalpine Region (II)	*Taxus*, *Abies*, *Picea*, *Thuja*, *Pinus* (Haploxylon) (Diploxylon), *Biota*, *Juniperus*, *Vaccinium*, *Rhododendron*, etc.	Lower Miocene	*Picea*	Yongdong, N.
			Picea, *Rhododendron*	Changgi, S.
		Middle Miocene	*Pinus*	Tongcheon, C.
		Upper Miocene	*Picea*, *Rhododendron*	Hamjindong, N.
			Pinus, *Rhododendron*	Yeonil, S.
		Middle Miocene	*Pinus*, *Thuja*	Kumunmoru, N.
			Taxus, *Abies*, *Pinus*,	Chummal, C.
		Upper Pleistocene	*Pinus* (Haploxylon),	Hwadae, N.
			Juniperus, *Abies*,	Hwadae, N.
			Picea	Hwadae, N.
			Thuja	Kumunmoru, N.
			Abies, *Pinus*	Haesang, N.
			Taxus, *Abies*, *Picea*,	Chummal, C.
			Pinus (Cupressaceae)	Chummal, C.
			Taxus, *Abies*,	Turubong, C.
			Pinus	Turubong, C.
		17 000–15 000 years B.P.	*Picea*, *Abies*,	Younglang, C.
			Pinus (Haploxylon)	Younglang, C.
Midland Mountain Region (III)	*Taxus*, *Cephalotaxus*, *Abies*, *Pinus* (Diploxylon), *Biota*, *Thuja*, *Juniperus*, *Sabina*, *Hypear*, *Bifaria*, *Masakia*	Middle Pleistocene	*Thuja*	Kumunmoru, N.
			Taxus, *Abies*,	Chummal, C.
		Upper Pleistocene	*Juniperus*, *Abies*	Hwadae, N.
			Thuja	Kumunmoru, N.
			Abies	Haesang, N.
			Taxus, *Abies*,	Chummal, C.
			(Cupressaceae)	Chummal,C.
			Abies, *Taxus*	Turubong, C.
		17 000–15 000 years B.P.	*Abies*	Yonunglang, C.
Southern Mountain Region (IV)	*Cephalotaxus*, *Abies*, *Pinus* (Diploxylon) (Haploxylon), *Biota*, *Sabina*, *Buxus*, *Hedera*, etc.	Oligocene	*Hedera*	Pongsan, N.
		Middle Miocene	*Pinus*	Tongcheon, C.
		Upper Miocene	*Pinus*	Yeonil, S.
		Middle Pleistocene	*Pinus*	Kumunmoru, N.
			Pinus	Chummal, C.
		Upper Pleistocene	*Pinus* (Haploxylon),	Hwadae, N.
			Juniperus, *Abies*	Hwadae, N.
			Abies, *Pinus*	Haesang, N.
			Abies, *Pinus*,	Chummal, C.
			(Cupressaceae)	Chummal, C.
			Abies, *Pinus*	Turubong, C.
		17 000–15 000 years B.P.	*Abies*, *Pinus* (Haploxylon)	Younglang, C.
Mid-western Insular Region (V)	*Pinus* (Diploxylon), *Thuja*, *Juniperus*, *Hedera*, *Camellia*, etc.	Oligocene	*Hedera*	Pongsan, N.
		Middle Miocene	*Pinus*	Tongcheon, C.
		Upper Miocene	*Pinus*	Yeonil, S.
		Middle Pleistocene	*Pinus*, *Thuja*	Kumunmoru, N.
			Pinus	Chummal, C.
		Upper Pleistocene	*Pinus* (Haploxylon)	Hwadae, N.

Table 3.9. Continued.

Biogeographic Division	Present Flora	First Recorded	Micro- & Macrofossils	Site*
			Juniperus	Hwadae, N.
			Thuja	Kumunmoru, N.
			Pinus	Haesang, N.
			Pinus, (Cupressaceae)	Chummal, C.
			Pinus	Turubong, C.
		17 000 years B.P.	*Pinus*	Younglang, C.
		4500–1400 years B.P.		
Southern Insular Region (VI)	*Taxus*, *Abies*, *Pinus* (Diploxylon), *Juniperus*, *Castonopsis*, *Cyclobalanopsis*, *Euonymus*, *Ficus*, *Cinnanomum*, *Viburnum*, *Ligustrum*, etc.	Oligocene	*Ficus*, *Viburnum*	Pongsan, N.
		Lower Miocene	*Viburnum*	Changgi, S.
		Middle Miocene	*Cinnanomum*	Jangpungri, N.
			Cyclobalanopsis,	Kogeonweon, N.
			Euonymus	Kogeonweon, N.
		Upper Miocene	*Ligustrum*	Turubong, C.
			Cyclobalanopsis,	Yeonil, S.
			Castanopsis,	Yeonil, S.
			Cinnanomum	Yeonil, S.
		6700–1400 years B.P.	*Cyclobalanopsis*,	Bongojin, S.
			Castanopsis	Bangojin, S.
			Cyclobalanopsis,	Kaheung, S.
			Castanopsis	Kaheung, S.
West-south-eastern insular and associated inland region (VII)	*Taxus*, *Abies*, *Picea*, *Pinus* (Diploxylon, Haploxylon), *Juniperus*, *Castanopsis*, *Cyclobalanopsis*, *Ficus*, *Hedera*, etc.	Oligocene	*Hedera*, *Ficus*	Pongsan, N.
		Lower Miocene	*Picea*	Yongdong, N.
			Picea	Changgi, S.
		Middle Miocene	*Cyclobalanopsis*	Kogeonweon, N.
			Pinus	Tongcheon, C.
		Upper Miocene	*Picea*	Hamjindong, N.
			Pinus, *Castanopsis*,	Yeonil, S.
			Cyclobalanopsis	Yeonil,S.
		Middle Pleistocene	*Pinus*	Kumunmoru, N.
			Taxus, *Abies*, *Pinus*	Chummal, C.
		Upper Pleistocene	*Pinus*, *Abies*, *Picea*	Hwadae, N.
			Pinus, *Abies*	Haesang, N.
			Taxus, *Abies*	Chummal, C.
			Picea, *Pinus*	Chummal, C.
			Taxus, *Abies*, *Pinus*	Turubong, C.
		17 000–15 000 years B.P.	*Picea*, *Abies*,	Younglang, C.
			Pinus (Haploxylon)	Younglang, C.
		6700–4500 years B.P.	*Cyclobalanopsis*,	Bangojin, S.
			Castanopsis	Bangojin, S.
		4500–1400 years B.P.	*Cyclobalanopsis*,	
			Castanopis	Bangojin, S.
			Castanopsis,	
			Castanopsis	Kaheong, S.
North-South Disjunctive Region (VIII)	*Pinus*, *Juniperus*, *Picea*, *Sinoarundinaria*, *Sasa*	Lower Miocene	*Picea*	Youngdong, N.
		Middle Miocene	*Picea*	Changgi, S.
		Upper Miocene	*Pinus*	Tongcheon, C.
			Picea	Hamjindong, N.
		Middle Pleistocene	*Pinus*	Yeonil,S.
			Pinus	Kumunmoru, N.
		Upper Pleistocene	*Pinus*	Chummal, C.
			Pinus, *Juniperus*,	Hwadae, N.
			Picea	Hwadae, N.
			Pinus	Haesang, N.
			Picea, *Pinus*	Chummal, C.
		17 000–15 000 years B.P.	*Pinus*	Turubong, C.
			Picea, *Pinus*	Younglang, C.

(Compiled by authors)
* Abbreviations:
N: Northern Korea,

Miocene. However, other cryophilous evergreen coniferous plant genera, such as *Taxus*, *Abies*, *Thuja* (Middle Pleistocene) and *Juniperus* (Upper Pleistocene), can be traced back solely to the Pleistocene. As a broad conclusion, it may be said, from an examination of this evidence, that the thermophilous floral elements of the southern mainland and the islands have been present in the peninsula from pre-Pleistocene times, but the cryophilous floral elements of the north and of the highlands are of much more recent origin.

In respect of the Pleistocene itself, most of the evidence for vegetation change thus far comes from the last two glacial periods, the intervening interglacial and the Holocene. Tentative vegetation zones relating to the last glacial period (Fig. 3.6) undoubtedly show some general dissimilarities from present-day biogeographic ranges and regions (Fig. 2.2) and indeed, bearing in mind the associated changes in climate, this is only what one would expect. These general dissimilarities may be summarised as follows: first, all vegetation zones associated with this glacial period seem to have retreated northward and upward during the subsequent post-glacial phase; second, the three cryogenetic vegetation zones of the glacial period (i.e. local alpine glaciers or intensive periglacial activities in the northernmost high mountains above 2000 m a.s.l.) have now largely disappeared; third, the alpine tundra or montane-subarctic vegetation of the northernmost high mountains above 500–1000 m a.s.l. has moved upslope, and is now very restricted; fourth, the steppe vegetation (in northwestern Korea) has been replaced largely by northern coniferous species; and finally, the mixed coniferous and deciduous broadleaved vegetation in the Devensian of southern Korea has been substituted by a much more strongly thermophilous evergreen broadleaved and coniferous vegetation today, along with a range of deciduous vegetation as well. In short, the basic structure of the present-day biogeographic regions of the Korean peninsula has been established during the post-glacial period. A further examination of the consequences of this, particularly in respect of the arctic-alpine and alpine floristic elements, follows in the next two chapters.

CHAPTER 4

Arctic-alpine & alpine floristic elements

Definitions

The term 'alpine' originates from the Latin '*Alpinus*', referring to the Alpine mountain chain of Europe (The Alps). Over the years, it also has come to be applied, in vegetation studies, to the description and analysis of plants of any mountain (Spomer, 1962; Löve, 1970) high enough to have an 'arctic' climate. Many such plants will grow in arctic regions as well: hence, the linked treatment of the two groups by many authors.

'Arctic-alpine' and 'alpine'

It is perhaps relevant to note at this stage previous definitions of the term 'arctic-alpine'. These are as follows. First, the term has been used to refer to all plants and animals which occur within the circumpolar regions (Hultén, 1962). Secondly, it has also been used to describe species which grow on the northern flanks of the latitudinal timberline in the northern hemisphere, and above the altitude of the vertical timberline (Billings & Mooney, 1968). Thirdly, it has been associated with species which are common both to the arctic tundra, and to high mountains (Slavik, 1977; Walton 1984; Reader's Digest, 1985). Finally, it may describe species which grow generally in arctic or alpine regions, or in both (Prime, 1978). In this work, the term 'arctic-alpine' is used to indicate plants which grow in both the arctic and alpine regions.

More especially in respect of alpine plants, with which this chapter is particularly concerned, a string of definitions has also been used in the past. 'Alpine plants' *per se* have been described as being those species flourishing within the alpine zone (Arber, 1910), or those that can endure mountain climates above the timberline (Rydberg, 1913; Billings, 1964; Moggi, 1985; Polunin & Walters, 1985). Species which are mainly distributed above the timberline have also been placed into this category (Dahl, 1955). Species which occur regularly in alpine floras have also been described as belonging to this group, irrespective of altitude (Packer, 1974). In addition, some authorities have suggested that all plants that grow exclusively at high elevations should be regarded as being alpine (Mani, 1973, 1978). Finally, according to Packer (1974), some species which grow below the timberline should also be included in the category of alpine plants, if they are in other respects characteristic of alpine species. In this work, the term 'alpine' normally refers to plants which grow mainly above the timberline. If the local environment is favourable for the growth of alpine plants at lower altitudes (below the timberline), then these plants are also included in this category as well.

'The alpine zone'

Definitions of the term 'alpine zone' vary from discipline to discipline, and this has caused some confusion in the past. In biology, it has been referred to as the 'tundra' region beyond the upper limit of normal tree growth (Spomer, 1962); the place above the upper limit of continuous or closed forest (Shetler, 1963; Mani, 1978); the zone above the upper timberline (Butzer, 1964; Troll, 1972; Zimma, 1973); the ground above the potential altitudinal limit of tall shrub and forest (Birks, 1973); and finally, the area above the upper limit of trees and tall shrubs (Wardle, 1974; Grey-Wilson & Blamey, 1986). In geography, on the other hand, it has been described as referring to a zone which has a typical high mountain landform complex, with its lower limit defined as being that of actual active cryonival denudation processes (Troll, 1972, 1973); the zone above the lower limit of active surface solifluction (Heuberger, 1974); and the zone above an elevation of 3000 m or more (Heath & Williams, 1979). In this work, the 'alpine zone' has been taken to refer to all locations that occur above the climatic upper limit of tall shrubs and trees, in which cryopedogenic activities and/or cryonival denudation processes are frequent.

'The subalpine zone'

The precise definition of the 'subalpine zone' has also caused some difficulties in the past. It has been described as being the vertical zone which exists between continuous forest and the alpine treeline (Hustrich, 1953); the natural belt extending from the upper altitudinal timberline to closed forest at lower elevations (Löve, 1970); a transitional zone of open woodland and tall shrubs (Birks, 1973); the zone immediately below the treeline, including both dense forest and ecotonal krummholz (Wardle, 1974, 1981a); and finally, the transitional zone in which relatively continuous forest gives way to treeless alpine tundra (MacMahon & Anderson, 1982; Daly, 1984). In this work, the 'subalpine zone' has been taken to refer to the transitional zone between the upper limit of tall shrubs or trees, and the continuous montane forest belt.

The origins of arctic-alpine and alpine plants, and their plant communities

It is very difficult to determine the origins of present-day alpine floras, due to a lack of relevant macrofossils (Billings, 1974, 1978a). Most of the available information comes from microfossils and pollen data, and these vary in quality. Raup (1953) and Tolmachev (1965), however, have pointed out the advantages of comparing range maps of alpine species in the elucidation of origins of arctic-alpine plants. This is a point which is taken up further at a later stage herein (pp. 96–98).

From the available evidence, many research workers have considered what might have been the possible times of origin of alpine and arctic-alpine plants, and it is only fair to say that their conclusions have varied widely. For example, in the case of arctic-alpine species, Hare (1954) suggested that they evolved during successive coolings of the Pleistocene, from a former Miocene flora; Hadac (1963) believed that they evolved on mountain ranges in the Arctic, Eurasia and America during the Neogene (Miocene and Pliocene) and Pleistocene; and Tralau (1961, 1963, 1973) postulated a much more recent origin for them, considering that the oldest pollen record of their existence was known only from the Riss glaciation of central Europe. Other views are that they evolved from the late Miocene to early Pleistocene in the highlands of central Asia, as well as in the Rocky Mountains of North America (Hoffmann & Taber, 1967; Bliss, 1971, 1981; Yurtsev, 1972; Zwinger & Willard, 1972; Billings, 1974; Murray, 1981); that they evolved from a former Tertiary flora in alpine areas (Löve & Löve, 1974; Larsen, 1980); and that they evolved during the Pleistocene (Haber, 1986).

Many research workers have also discussed the origin of alpine plants. According to Charlesworth (1957), they evolved from plains floras in temperate latitudes of the northern hemisphere during the mid-Tertiary and Pleistocene; Packer (1974) postulated that they evolved in the mountains of central Asia; Löve & Löve (1975b) suggested that they evolved at all times in all climates where mountains were formed; and, finally, Speirs (1979) suggested that they have evolved from montane plants since the Miocene.

For Eurasia, one may say in general that arctic-alpine and alpine plants, and their plant communities, most likely evolved in the central Asian high mountains, and surrounding localities, mainly during the late Miocene and Pleistocene periods.

Migration

Early theories on the general nature of the migration of alpine and arctic-alpine plants during the Pleistocene have been reviewed by Rydberg (1914), Du Rietz (1940), Porsild (1951), Berg (1963) and Tolmachev (1965). Since then, two contradictory views on migration have dominated. One, put forward by Löve & Löve (1974), suggests a complete northward dispersal of arctic-alpine plants in the northern hemisphere from their areas of origin in alpine central Asia after each major glaciation. Alternatively, Hultén (1958), Tolmachev (1965) and Crawford (1989) have proposed that the present alpine floras should be regarded as being remnants of a formerly much wider circumpolar arctic-alpine plant community. Hultén (1958), in particular, postulated that most of the temperate and subtropical plants evolved and reached their circumpolar distribution during the Tertiary period. Later, during the Pleistocene cooling, many warmth-loving species were forced out from the northern regions, and the circumpolar community of cold-tolerant plants which remained, then subsequently became isolated into a series of fragments, some of which, especially on mountains, then evolved genetically into distinctly separate and different entities, which are now termed alpine communities.

Both of these views require the adoption of certain assumptions about the ability of alpine and arctic-alpine plants to migrate quickly, as climate ameliorates or deteriorates. But despite numerous investigations, still little is known about the migration processes of such species or the mechanisms of alpine and arctic-alpine plant survival during the Pleistocene glaciations. We also know very little about the patterns of survival of arctic-alpine plants immediately outside the areas of heavy glaciation, as well as their southernmost limits during the glacial phases, particularly in respect of glacial refugia. In this work, we examine the migration and survival patterns of arctic-alpine plants in East Asia, including the Korean peninsula, from which no extensive glaciation thus far has been reported (pp. 61–62).

In East Asia, evergreen broadleaved plants were dominant up to 50° N during the Eocene. Subsequently, the main vegetation constituents were deciduous broadleaved plants, throughout the Oligocene and the lower Miocene (Wang, 1961). During the late Tertiary (Middle and Upper Miocene, and Pliocene), the gradual fall in mean annual temperature, especially at these latitudes, brought about the beginnings of a southward migration of species. Both coniferous and broadleaved forests then shifted southward conspicuously during the early Pleistocene (Grichuk, 1984). At the same time, even the cold-tolerant species of the broad arctic-alpine plant communities moved down from the high mountains and to the south (see Fig. 4.1). The general north–south mountain range structure of East Asia facilitated these southward migrations. As a result, both the lowlands and mountain chains of East Asia are suggested as being one of the three principal worldwide southward migration routes of plants in the Pleistocene by Polunin (1960), the others being the north-south mountain ranges of North America, and the lowlands of Scandinavia and adjacent areas of western and central Europe. The southward migration of arctic-alpine plants during the glacial periods has been recognised by many workers in the case of East Asia, including Hultén (1937), Vaskovsky (1964), Powers (1973), Kong *et al.* (1976), CLIMAP (1981) and Kurentsova (1984). Their maximum southward extension is possibly associated with the major climatic minimum during the second half of the upper Pleistocene, which seems to correspond to the Riss glaciation (Buks, 1984). A massive disjunction of arctic-alpine plant ranges then took place in the upper (post-Riss) Pleistocene, due to the climatic amelioration which pushed them back northwards and upwards, and also expanded the range of thermophilous temperate and warm temperate plants to cover large areas of lowland, so isolating the mountain floras which were becoming established. It is, however, difficult to determine the exact periods of migration, even though the evidence in East Asia suggests a succession of alternating advances and retreats of the arctic-alpine plant communities (Hultén, 1928; Shikama, 1962; Murata, 1977). In the case of Korea, cryophilous evergreen coniferous genera can be traced back mainly to particular phases of the Pleistocene (see Table 3.9). Thus, *Taxus*, *Abies* and *Thuja* date back to the Middle Pleistocene, and *Juniperus*, *Pinus pumila*, *P. sibirica* and *P. koraiensis* to the Upper Pleistocene.

On the basis of the present-day latitudinal distribution of arctic-alpine and alpine plants in Korea (Fig. 2.2), their altitudinal ranges (Fig. 2.6), the fossil evidence (Table 3.9), the distributional patterns of circumpolar plants (Hultén, 1958, 1962, 1970) and the above-mentioned East Asian vegetation history, and using all the available evidence, the pattern of migration of arctic-

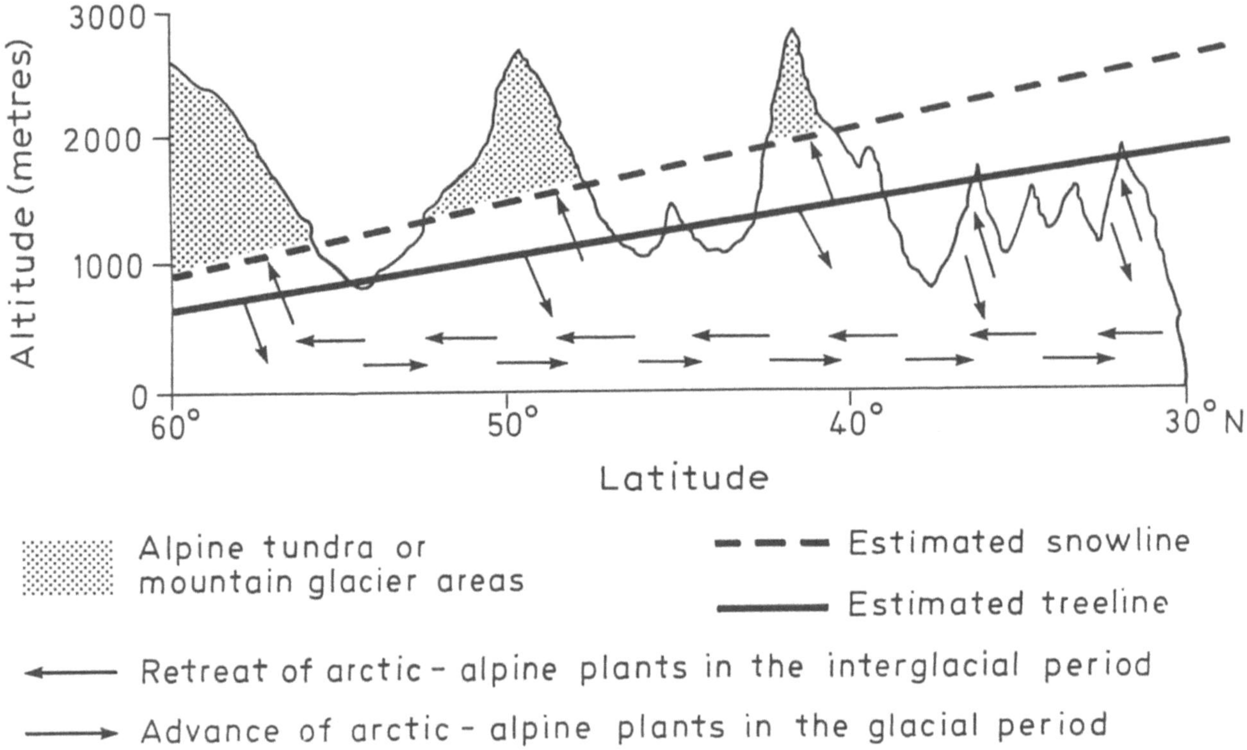

Fig. 4.1. Schematic diagram for the migration of arctic-alpine plants into Korea.

alpine plants may be reconstructed in general terms for the Korean peninsula, and its immediate vicinity, for the first time. This is presented in Fig. 4.1, which shows a schematic diagram for the migration of arctic-alpine plants into the Korean peninsula. In this, the estimated treeline and snowline are approximate though, according to Lautensach (1945), the upper timberline during the Pleistocene glacial phases probably lay at ca. 1300 m a.s.l. in northern Korea, i.e. 700 m lower than today. Further, more detailed examples of possible migrant routes for individual species are discussed in the next section.

As the climate deteriorated during the Pleistocene cold phases, most of the coniferous and broadleaved forests would have shifted downward from the mountains and southward from the north. At the same time, cold-tolerant arctic-alpine and alpine plants moved downward and southward as well. However, the climatic amelioration during the post-glacial period caused glaciers, and the areas of permafrost, to retreat towards the north and upslope. This appears to have resulted in the removal of both the permafrost and alpine glaciers from mountain lands in most of Korea. The temperate species of both lowlands and mountain lands then became more common than the arctic-alpines. As a result of the increased competition associated with this process, arctic-alpine plants retreated upwards towards mountain tops in the peninsula, and a fragmentation of the arctic-alpine plant distributional range took place as elsewhere in East Asia. At present, therefore, arctic-alpine plants are mainly confined to the alpine and subalpine zones of mountains in Korea, along with particular habitat-adapted species which grow below the alpine zone.

Refugia

After reviewing all the available information, Holland (1981) presented evidence for the existence of six different types of Pleistocene refugia, viz. nunataks, ice-free areas, refugia near the ice

Table 4.1. Suggested refugia of arctic-alpine plants during the Pleistocene glaciation.

Areas	Sources									
	Selander (1950)	Charlesworth (1957)	Hultén (1958)	Hadac (1963)	Billings (1964)	Hoffman & Taber (1967)	Kornas (1972)	Löve & Löve (1974)	Haber (1986)	Müller (1986)
Greenland	0									
Iceland	0									
Faroe Islands	0									
Scandinavia									0	
Scotland	0									
Spitzbergen	0			0						
Mediterranean		0								0
Armenia		0								0
Iran										0
Asia Minor		0								
Transcaucasia						0	0			
Central Europe								0		
Southern Europe							0			
Central Asia						0				0
Afghanistan		0								
Siberia	0		0	0		0			0	0
East Asia							0	0	0	
China		0			0					
Mongolia		0								
Japan		0			0					
Beringia			0						0	
Alaska-Yukon			0	0		0		0	0	0
Arctic North America	0	0						0	0	
Pacific North America										0
Atlantic North America				0	0		0			0
Mexico										0

(Compiled by the authors)

margin, refugia on the continental shelf, mainland refugia which are far from the ice margin, and dispersed mainland refugia. On a worldwide scale, many regions have been suggested as refugia for arctic-alpine plants during the Pleistocene glaciations (though, interestingly, Korea has never been one of them!) (Table 4.1). Evidence to the contrary is presented herein.

The importance of refugia in unglaciated areas close to the ice margin during the Pleistocene glaciations, for the survival of arctic-alpine plants, is especially noteworthy, and this has been emphasised by Larsen (1980). In the case of Europe, for example, nunataks and coastal areas played a significant role in this respect (Du Rietz, 1935; Matthews, 1937; Hultén, 1937; Böcher, 1972; Polunin, 1960; Gjaerevoll, 1963; Löve, 1963; Lindroth, 1970; Goudie, 1979; Dahl, 1987). Recently, Lindsey (1981) has introduced a dual concept of refugia type; one which is 'fixed', for plants with a low mobility, and the other which is 'shifting', according to environmental circumstances, for plants with a greater ability to disperse; but how far these ideas are generally applicable to refugia worldwide remains to be worked out.

A study of the Korean flora (see Appendices 2–5) suggests that 180 arboreal and 187 herbaceous alpine plants exist in the high mountains of Korea (Table 4.2).

In addition, 75 species of arctic-alpine plants are present (Hultén, 1958, 1962, 1970).

During the Pleistocene cold phases, the Korean peninsula undoubtedly was under very different environmental conditions to those of western Europe. For example, it is likely to have been much more intensively cold and dry in Korea than in western Europe, so much so that the climate in the high alpine zone of Korea was too cold and harsh for any arctic-alpine and alpine plants to survive. Therefore, arctic-alpine and alpine plants had to confine themselves to lower mountain and/or lowland areas during these phases, which effectively then became their *primary* refugia. Both in the case of Siberia and in Korea, the

Table 4.2. Estimated number of Korean alpine plants, including arctic-alpines growing solely in alpine areas (23 in number).

	Arboreal	Endemic Arboreal	Herbaceous	Endemic Herbaceous	Total
Alpine species	122	58	–	–	180
	–	–	140	47	187
Total	122	58	140	47	367

(Sources: Appendices 2–5.)

mountain ranges strike in a north-south direction. This orientation facilitated the southward migration of arctic-alpine and alpine plants, again emphasising the refugia role of the Korean lowlands. Then, as the climate subsequently ameliorated, arctic-alpine and alpine plants moved back upslope on to the mountain tops. The relatively long period of isolation of arctic-alpine and alpine plants on these scattered mountain tops then resulted in a relatively large number of endemic alpine plants being produced. In a sense, the mountain tops proved to be important *secondary* refugia within the peninsula, in consequence becoming significant especially for the emergence of endemics only after glaciation began to wane. Some of the resulting species variability has already been presented in the analysis of mountain top vegetation presented in Chapter 2, and its further significance is to be discussed later (pp. 100–103).

Species composition of alpine and arctic-alpine plant communities in Korea

General comments

For the purpose of the remainder of this chapter, bearing in mind the definitions of the two terms as indicated previously (p. 84), the term 'alpine plant' in the Korean context is interpreted as referring to plants which grow mainly above the timberline, but also includes such plants growing below the timberline where the local environment permits; the term 'arctic-alpine plant' refers to plants which grow both in the arctic, and in the alpine regions of Korea. For ease of marshalling the large amount of available information, alpine and arctic-alpine plants are examined separately in the following pages.

It was Wilson (1918, 1919) who, following a north-south expedition along the peninsula, listed for the first time several Korean 'alpine' species, including *Pinus pumila*, *P. koraiensis*, *Abies holophylla*, *Picea jezoensis*, *Empetrum nigrum*, *Vaccinium vitis-idaea*, *V. uliginosum*, *Ledum palustre*, *Arctous alpina*, *Juniperus chinensis* var. *sargentii*, *J. communis* var. *montana*, *Thuja*, dwarf *Salix*, *Dryas* and *Rhododendron*. Then, in 1926, Nakai reported *Taxus cuspidata*, *Abies koreana*, *Juniperus sargentii*, *Empetrum nigrum*, *Rhododendron mucronulatum*, *Diapensia lapponica*, *Prunus sachalinensis* and *Daphne kamtschatica* from Mt. Halla (1950 m a.s.l.) in Cheju Island, and also suggested that these could have been descendants of a former wider group of arctic-alpines which originally had a continuous range throughout the mainland. This view assumed that the floristic linkages across the Korean Straits (between the peninsula and Cheju Island) had been strongly maintained at some times in the past, presumably during low sea-level phases. More recently, Cha *et al.* (1975) suggested that the distribution of *Empetrum nigrum* var. *japonicum* and *Diapensia lapponica* subsp. *obovata*, which are found as arctic-alpines both on Mt. Halla and in the far north of Korea and eastern Siberia, but nowhere in between, also reflected the former existence of this land connection. Further discussion as to the validity of this assumption follows later (pp. 95–97). In the north, according to Takahasi (1944), there is a particularly high diversity of alpine plants on Mt. Paikdoo (2744 m), which he thinks is due to the locally high precipitation and humidity. Later, Küchler & Takenouchi (1948) also noted the presence of alpine meadows on Mt. Paikdoo. Recently, Xu *et al.* (1987) reported alpine tundra above 2000 m a.s.l. on the northern slope of Mt. Paikdoo. There have been, however, no other

Table 4.3. Species composition and physiognomy of Korean arboreal alpine plants.

Family name	Number of species			Percentages		Physiognomy				
	Alpine spp.	End. Alp. spp.	Endemic Alpine spp.	Alpine spp./ Total spp.	End. Alp. spp./ Alpine spp.	ES	ET	DS	DT	DV
Taxaceae	4	3	1	75.0	33.3	1	2	0	0	0
Abietaceae	14	12	7	85.7	58.3	0	9	0	3	0
Pinaceae	13	3	0	23.1	0	1	2	0	0	0
Cupressaceae	6	1	1	16.7	100.0	0	1	0	0	0
Juniperaceae	13	3	1	23.1	33.3	3	0	0	0	0
Salicaceae	55	25	11	45.5	44.0	0	0	17	7	1
Betulaceae	34	21	6	61.8	28.6	0	0	4	17	0
Fagaceae	50	1	0	2.0	0	0	0	0	1	0
Ulmaceae	31	1	0	3.2	0	0	0	1	0	0
Ranunculaceae*	93 (32)	8	6	8.6	75.0	0	0	0	0	8
Berberidaceae	11	5	2	45.5	40.0	0	0	5	0	0
Philadelphaceae	22	1	0	4.5	0	0	0	1	0	0
Spiraceae	38	6	3	15.8	50.0	0	0	6	0	0
Ribesiaceae	16	7	0	43.8	0	0	0	7	0	0
Malaceae	68	11	7	16.2	63.6	0	0	3	8	0
Rosaceae*	135 (64)	11	1	8.2	9.1	1	0	10	0	0
Amygdalaceae	42	2	0	4.8	0	0	0	0	2	0
Fabaceae*	141 (42)	1	1	0.7	100.0	0	0	0	1	0
Rhamnaceae	22	1	1	4.5	100.0	0	0	1	0	0
Araliaceae	23	2	1	8.7	50.0	0	0	2	0	0
Cornaceae	11	2	0	18.2	0	1	0	1	0	0
Diapensiaceae	1	1	0	100.0	0	1	0	0	0	0
Empetraceae	1	1	0	100.0	0	1	0	0	0	0
Ericaceae	47	28	1	69.6	3.6	20	0	7	1	0
Labiatae	105	1	0	0.9	0	0	0	1	0	0
Caprifoliaceae	83	21	8	25.3	38.1	1	0	20	0	0
Asteraceae	341 (4)	1	0	0.3	0	0	0	1	0	0
Total	1420	180	58	12.7%	32.2%	30	14	87	40	9

(Source: See Appendix 3).
*Families which include both arboreal and herbaceous alpine plants (Number of shrubs in parentheses).
Abbreviations: ES: Evergreen shrub; ET: Evergreen tree; DS: Deciduous shrub; DT: Deciduous tree; DV: Deciduous vine.

detailed studies on arctic-alpine and alpine plant communities in Korea to date.

Arboreal alpine plants

As indicated in the previous section (Table 4.2), the estimated number of Korean arboreal alpine plants is ca. 180 species in 62 families. These are listed in summary in Table 4.3, along with their physiognomy for the full list of species in Appendix 2.

From this it can be seen that the dominant families (omitting for the time being the Korean arboreal endemics) numerically are the Ericaceae (28 spp.), Salicaceae (25 spp.), Betulaceae (21 spp.), Caprifoliaceae (21 spp.) and Abietaceae (12 spp.) with other components, such as Rosaceae (11 spp.), Malaceae (11 spp.), Ranunculaceae (8 spp.), Ribesiaceae (7 spp.), Spiraceae (6 spp.), Berberidaceae (5 spp.), Taxaceae (3 spp.), Pinaceae (3 spp.), Juniperaceae (3 spp.), Amygdalaceae (2 spp.), Araliaceae (2 spp.), Cornaceae (2 spp.), Cupressaceae (1 sp.), Fagaceae (1 sp.), Ulmaceae (1 sp.), Philadelphaceae (1 sp.), Fabaceae (1 sp.), Rhamnaceae (1 sp.), Diapensiaceae (1 sp.), Empetraceae (1 sp.), Labiatae (1 sp.) and Asteraceae (1 sp.), also present. It should also be noted that some families, viz. the Ranunculaceae, Rosaceae, Asteraceae and Fabaceae, contain both herbaceous and arboreal plants.

Of the 58 Korean endemic arboreal alpine plant species, 11 come from the Salicaceae, 8 from the Caprifoliaceae, 7 from the Abietaceae and 7 more from the Malaceae, 6 from the Betulaceae, and 6 more from the Ranunculaceae, 3 from the Spiraceae, 2 from the Berberidaceae and one each from the Taxaceae, Cupressaceae, Juniperaceae, Rosaceae, Fabaceae, Rhamnaceae, Araliaceae and Ericaceae. Additionally, it should be noted that the single-species (in Korea) families of the Diapensiaceae and Empetraceae are characterised by having one arboreal alpine representative each in the peninsula.

The proportion of arboreal alpine plants out of the total number of arboreal species present in each family other than the Diapensiaceae and Empetraceae (above), is relatively high in the peninsula in the case of the Abietaceae (85.7%), Taxaceae (75.0%), Ericaceae (69.6%), Betulaceae (61.8%), Salicaceae (45.5%), Berberidaceae (45.5%), Ribesiaceae (43.8%), Caprifoliaceae (25.3%), Pinaceae (23.1%) and Juniperaceae (23.1%), and especially in comparison with the mean of 12.7% for all Korean plant species.

All the arboreal alpine species of the Cupressaceae, Fabaceae and Rhamnaceae families are endemic. A relatively high proportion of endemic arboreal alpine plants out of the total number of alpine plants also is to be found in the Ranunculaceae (75.0%), Malaceae (63.6%), Abietaceae (58.3%), Araliaceae (50.0%), Spiraceae (50.0%), Salicaceae (44.0%), Berberidaceae (40.0%), Caprifoliaceae (38.1%), Taxaceae (33.3%) and Juniperaceae (33.3%). The mean of this parameter for all arboreal Korean plant species stands at 32.2%.

As regards plant physiognomy, 87 species belong to the category of deciduous shrubs, 40 species to that of deciduous trees, 30 species are evergreen shrubs, 14 species belong to the evergreen tree group, and nine deciduous vines are found. The number of broadleaved arboreal alpine plants (158 spp.) greatly predominates over those belonging to the coniferous arboreal alpine plant (22 spp.) group.

Herbaceous alpine plants

Table 4.4 shows that, out of 187 Korean herbaceous alpine plants, the peninsula's herbaceous alpine plant community consists predominantly of the Asteraceae (21 spp.), Scrophulariaceae (15 spp.), Gramineae (12 spp.), Rosaceae (11 spp.), Fabaceae (10 spp.), Cyperaceae (10 spp.), Geraniaceae (8 spp.), Aspidiaceae (8 spp.), Polygonaceae (7 spp.), Orchidaceae (7 spp.), Alsinaceae (6 spp.), Ranunculaceae (6 spp.), Brassicaceae (6 spp.), Primulaceae (6 spp.), Saxifragaceae (5 spp.) and Melanthaceae (5 spp.). Others include the Violaceae (4 spp.), Campanulaceae (4 spp.), Juncaceae (4 spp.), Lycopodiaceae (4 spp.), Helleboraceae (3 spp.), Apiaceae (3 spp.), Gentianaceae (3 spp.), Pyrolaceae (2 spp.), Rubiaceae (2 spp.), Alliaceae (2 spp.), Selaginellaceae (25 spp.), Papaveraceae (1 sp.), Crassulaceae (1 sp.), Spiraceae (1 sp.), Euphorbiaceae (21 sp.), Hypericaceae (1 sp.), Lentibulariaceae (1 sp.), Dipsacaceae (1 sp.), Scheuchzeriaceae (1 sp.), Araceae (1 sp.), Liliaceae (1 sp.), and Pteridaceae (1 sp.). (See Appendix 3 for the full list of 187 species.)

Out of a total of 47 endemic herbaceous alpine species, the number is relatively large in the case of the Asteraceae (7 spp.), Scrophulariaceae (5 spp.), Ranunculaceae (4 spp.), Fabaceae (4 spp.), Alsinaceae (3 spp.), Brassicaceae (3 spp.) and Geraniaceae (3 spp.). Others include the Polygonaceae (2 spp.), Rosaceae (2 spp.), Apiaceae (2 spp.), Gramineae (2 spp.), Papaveraceae (1 sp.), Saxifragaceae (1 sp.), Spiraceae (1 sp.), Euphorbiaceae (1 sp.), Violaceae (1 sp.), Pyrolaceae (1 sp.), Rubiaceae (1 sp.), Dipsacaceae (1 sp.), Melanthaceae (1 sp.) and Lycopodiaceae (1 sp.). Further, the Korean endemic herbaceous montane-alpine genus *Hanabusaya* has two species (*H. asiatica* and *H. latisepala*), and these can be included also as endemic alpine plants.

As regards the percentages of alpine representation in each family the only species representative of the Scheuchzeriaceae is an alpine plant. Then, a relatively high proportion of herbaceous alpine plants also is found in the Geraniaceae (36.4%), Melanthaceae (33.3%), Lycopodiaceae (30.8%), Papaveraceae (20.0%), Primulaceae (20.0%), Dipsacaceae (20.0%), Selaginellaceae (20.0%), Pyrolaceae (16.7%) and Juncaceae (16.7%). The mean value for the representation of herbaceous alpines within the total of Korean species as a whole is 7.7%.

All the herbaceous alpine species representatives of the Papaveraceae, Spiraceae, Euphorbiaceae and Dipsacaceae are endemic plants.

Table 4.4. Species composition and physiognomy of Korean herbaceous alpine plants.

Family name	Number of species			Percentage		Physiognomy		
	Total spp.	Alpine spp.	Endemic Alpine spp.	Alpine spp./ Total spp.	End. Alp. spp./ Alpine spp.	PH	BN	AN
Polygonaceae	78	7	2	9.0	28.6	6	0	1
Alsinaceae	64	6	3	9.4	50.0	6	0	0
Ranunculaceae*	93	6	4	6.4	66.7	6	0	0
Helleboraceae	68	3	0	4.4	0	3	0	0
Papaveraceae	5	1	1	20.0	100.0	1	0	0
Brassicaceae	54	6	3	11.0	50.0	3	3	0
Crassulaceae	34	1	0	2.9	0	1	0	0
Saxifragaceae	40	5	1	12.5	20.0	5	0	0
Spiraceae	38	1	1	2.6	100.0	1	0	0
Rosaceae*	135	11	2	8.1	18.2	11	0	0
Fabaceae*	141	10	4	7.1	40.0	10	0	0
Geraniaceae	22	8	3	36.4	37.5	8	0	0
Euphorbiaceae	27	1	1	3.7	100.0	1	0	0
Hypericaceae	15	1	0	6.7	0	1	0	0
Violaceae	59	4	1	6.8	25.0	4	0	0
Apiaceae	84	3	2	3.6	66.7	3	0	0
Pyrolaceae	12	2	1	16.7	50.0	2	0	0
Primulaceae	30	6	0	20.0	0	6	0	0
Gentianaceae	28	3	0	10.7	0	3	0	0
Scrophulariaceae	108	25	5	13.9	33.3	10	0	5
Lentibulariaceae	8	1	0	12.5	0	1	0	0
Rubiaceae	68	2	1	2.9	50.0	2	0	0
Dipsacaceae	5	1	1	20.0	100.0	0	1	0
Campunulaceae	47	4	0	8.5	0	4	0	0
Asteraceae	341	21	7	6.2	33.3	21	0	0
Scheuchzeriaceae	1	1	0	100.0	0	1	0	0
Gramineae	238	12	2	5.0	16.7	12	0	0
Cyperaceae	217	10	0	4.6	0	10	0	0
Araceae	16	1	0	6.3	0	1	0	0
Juncaceae	24	4	0	16.7	0	4	0	0
Melanthaceae	15	5	1	33.3	20.0	5	0	0
Alliaceae	17	2	0	11.8	0	2	0	0
Liliaceae	20	1	0	5.0	0	1	0	0
Orchidaceae	71	7	0	9.9	0	7	0	0
Lycopodiaceae	13	4	1	30.8	25.0	4	0	0
Selaginellaceae	10	2	0	20.0	0	2	0	0
Pteridaceae	34	1	0	2.9	0	1	0	0
Aspidiaceae	140	8	0	5.7	0	8	0	0
Total	2420	187	47	7.7%	25.1%	177	4	6

(Sources: See Appendix 3).
*Families which include both arboreal and herbaceous alpine plants.
Abbreviations: PH: Perennial herb BN: Biennial AN: Annual.

Also a relatively high proportion of endemic herbaceous alpine plants is found in the Ranunculaceae (66.7%), Apiaceae (66.7%), Alsinaceae (50.0%), Brassicaceae (50.0%), Pyrolaceae (50.0%), Rubiaceae (50.0%), Fabaceae (40.0%), Geraniaceae (37.5%), Scrophulariaceae (33.3%), Asteraceae (33.3%), Polygonaceae (26.6%), Violaceae (25.0%) and Lycopodiaceae (25.0%). The mean value for endemic representation within the herbaceous alpine group in Korea as a whole is 25.1%.

On the basis of life form, the Korean herbaceous alpine plants can be divided into 177 perennials, six annuals and four biennials. Overall, the proportion of endemics in the arboreal alpine plant group (32.2%) is slightly higher than that of endemics in the herbaceous alpine plant group (25.1%).

Arboreal arctic-alpine plants

On a worldwide scale, Polunin (1951, 1955, 1959) and Speirs (1979) have estimated that there are around 900 species of arctic-alpine plants. However, other authorities place the figure as being much lower than this. Billings (1974) and Price (1981) proposed the existence of ca. 500 arctic-alpine plants, although Billings (1983) later revised this downwards to 450 species. On the other hand, Haber (1986) has suggested that only up to 200 species of arctic-alpine plants may be found.

According to Hultén (1958, 1962, 1970), there are 17 arboreal arctic-alpine and 58 herbaceous arctic-alpine species in the Korean peninsula (see Appendices 4 & 5). Ericaceae (9 spp.) and Rosaceae (5 spp.) are the major families of Korean arboreal arctic-alpine plants, with one species each from the Juniperaceae, Diapensiaceae, Empetraceae and Caprifoliaceae. Out of the 75 arctic-alpine species, however, only 15 arboreal and 8 herbaceous ones grow solely within the alpine areas, and others occur elsewhere either on mountains or in lowlands. The fifteen arboreal arctic-alpine plants which grow within the alpine zone alone are *Juniperus communis* var. *montana* (= ? *Sabina sargentii*), *Dryas octopetala* var. *asiatica*, *Rubus articus*, *Diapensia lapponica* subsp. *obovata*, *Empetrum nigrum* var. *japonicum*, *Chamaedaphne calyculata*, *Ledum palustre* var. *diversipilosum*, *L. palustre* subsp. *decumbens*, *Oxycoccus microcarpus*, *O. quadripetalus*, *Phyllodoce caerulea*, *Rhododendron lapponicum* (= ? *R. parvifolium*), *Vaccinium uliginosum* var. *alpinum*, *V. vitis-idaea* var. *minus* and *Linnaea borealis*. All of these species are shrubs, and 13 species out of the original 17 are evergreen plants. This situation will be discussed further in detail in Chapter 5.

Herbaceous arctic-alpine plants

Korean herbaceous arctic-alpine plants consist of representatives of the Gramineae (9 spp.), Cyperaceae (5 spp.), Pyrolaceae (4 spp.), Polygonaceae (2 spp.), Ranunculaceae (2 spp.), Labiatae (2 spp.), Lentibulariaceae (2 spp.), Rubiaceae (2 spp.), Juncaceae (2 spp.), Aspidiaceae (2 spp.) and one species each from the Chenopodiaceae, Alsinaceae, Nymphaceae, Helleboraceae, Brassicaceae, Rosaceae, Fabaceae, Elatinaceae, Violaceae, Droseraceae, Primulaceae, Boraginaceae, Typhaceae, Zosteraceae, Ruppiaceae, Scheuchzeriaceae, Juncaginaceae, Alliaceae, Orchidaceae, Lycopodiaceae, Equisetaceae, Botrychiaceae, Ophioglossaceae, Osmundaceae, Hymenophyllaceae and Polygodiaceae. Out of the 58 species, 8 grow solely within the alpine zone, namely *Oxyria digyna*, *Moneses uniflora*, *Pinguicula villosa*, *Scheuchzeria palustris*, *Poa glauca*, *Carex rotundata*, *Eriophorum alpina* and *Lycopodium complanatum*.

Unlike the arboreal arctic-alpine group, well over one half of the herbaceous arctic-alpines grow below the alpine zone, as well as growing within it. On the basis of life form, the Korean herbaceous arctic-alpine plants can be divided into eight annuals, viz. *Suaeda maritima*, *Rubus chamaemorus*, *Glaux maritima*, *Galeopsis bifida*, *Lamium amplexicaule*, *Elatine triandra*, *Bromus japonicus* and *Digitaria ischaemum*; two biennials (*Turritis glabra* and *Galium spurium*); and the other 48 are perennials.

Distributional patterns of arctic-alpine plants: East Asia and Korea

The number of arctic-alpine, alpine and endemic alpine species in Korea, and their distributional patterns, are presented in Table 4.5. In this, on the basis of scale, a 'continuous' or 'disjunctive' distribution refers to the spatial patterns of arctic-alpine and alpine plants in East Asia, and 'widespread' or 'restricted' distribution refers to the spatial patterns of alpine and endemic alpine plants in Korea. For these purposes, the boundary of East Asia is taken to be a line which connects, from the west, the Lena Delta, Lake Baikal, the Gobi Desert, Lanchow, the Tsinling Shan Mountain, the Hwaiyang Shan Mountain, Shanghai, Korea and Japan, to eastern Siberia in the east (Fig. 4.2). For these purposes, too, the Korean peninsula is also divided into four regions, namely, the North, Midland and South

Table 4.5. Distribution types of arctic-alpine, alpine and endemic alpine plants in Korea

	Continuous – widespread	Continuous – restricted	Disjunctive – widespread	Disjunctive – restricted	Endemic – widespread	Endemic – restricted	Total
Arctic-alpine	5 (1, 4)	12 (3, 9)	30 (3, 27)	28 (10, 18)	–	–	*75 (17, 58)
Alpine	–	–	87 (44, 43)	152 (64, 88)	–	–	239 (108, 131)
Endemic Alpine	–	–	–	–	24 (18, 6)	81 (40, 41)	105 (58, 47)
Total	5	12	117	180	24	81	**419

N.B. Numbers in parentheses represent the number of arboreal and herbaceous species, respectively.
*Out of the 75 arctic-alpine species, only 23 species grow solely in alpine areas.
**The total number of arctic-alpine, alpine and endemic alpine species in Korea is 367 species, exclusive of 52 arctic-alpine species which are found below the alpine areas of Korea.

regions, and Cheju Island. Within this framework, species which grow in one region alone are taken to belong to the 'restricted' type; and those occurring in two or more adjacent regions are taken to be 'widespread'.

Table 4.5 shows that four distributional types are to be found, from the 75 arctic-alpine species taken into consideration (17 arboreal and 58 herbaceous species), along with two distributional types from both the 239 alpine and 105 endemic alpine species. For the *arctic-alpines*, the range of types are as follows: (i) species with a continuous distribution in East Asia, and widespread in Korea (5 spp.); (ii) those with a continuous distribution in East Asia, but restricted in Korea (12 spp.); (iii) those with a disjunctive distribution in East Asia, but widespread in Korea (30 spp.); and (iv) those with a disjunctive distribution in East Asia, and restricted in Korea (28 spp.).

Species with a continuous distribution in East Asia and widespread in Korea

Five arctic-alpine species occur continuously in East Asia and are widespread in Korea. These include one arboreal and four herbaceous plants, and they display four distributional patterns in Korea.

Arboreal Species:
- (i) North and Cheju (1 sp.)
  Empetrum nigrum var. *japonicum*

Herbaceous Species:
- (i) Throughout the Peninsula (1 sp.)
  Equisetum hiemale
- (ii) North to Midland (2 spp.)
  Stellaria longiflora, *Calamagrostis canadensis* subsp. *langsdorffii*
- (iii) North and Cheju (1 sp.)
  Luzula rufescens

Species with a continuous distribution in East Asia, but restricted in Korea

Twelve arctic-alpine species grow continuously in East Asia, but display a restricted distribution in Korea. Three arboreal and 9 herbaceous plants belong to this group, and all species are confined to the north of Korea.

Arboreal Species:
- (i) North (3 spp.)
  Oxycoccus quadripetalus, *Vaccinium vitis-idaea* var. *alpinum*, *Phyllodoce caerulea*

Herbaceous Species:
- (ii) North (9 spp.)
  Anemone dichotoma, *Carex rotunda*, *Caltha palustris* var. *sibirica*, *Eriophorum alpina*, *Agrostis canina* subsp. *trinii*, *Allium schoenoprasum*, *Carex loliacea*, *C. pseudo-curaica*, *Calypso bulbsa*.

Species with a disjunctive distribution in East Asia, but widespread in Korea

Thirty arctic-alpine species are found disjunctively in East Asia, but show a widespread distributional pattern in Korea. This group contains 3 arboreal and 27 herbaceous plants, and displays four distributional patterns in Korea.

Arboreal Species:
- (i) Throughout the Peninsula (3 spp.)
  Juniperus communis var. *montana* (= ? *Sabina sargentii*), *Sorbus amurensis*, *S. commixta*

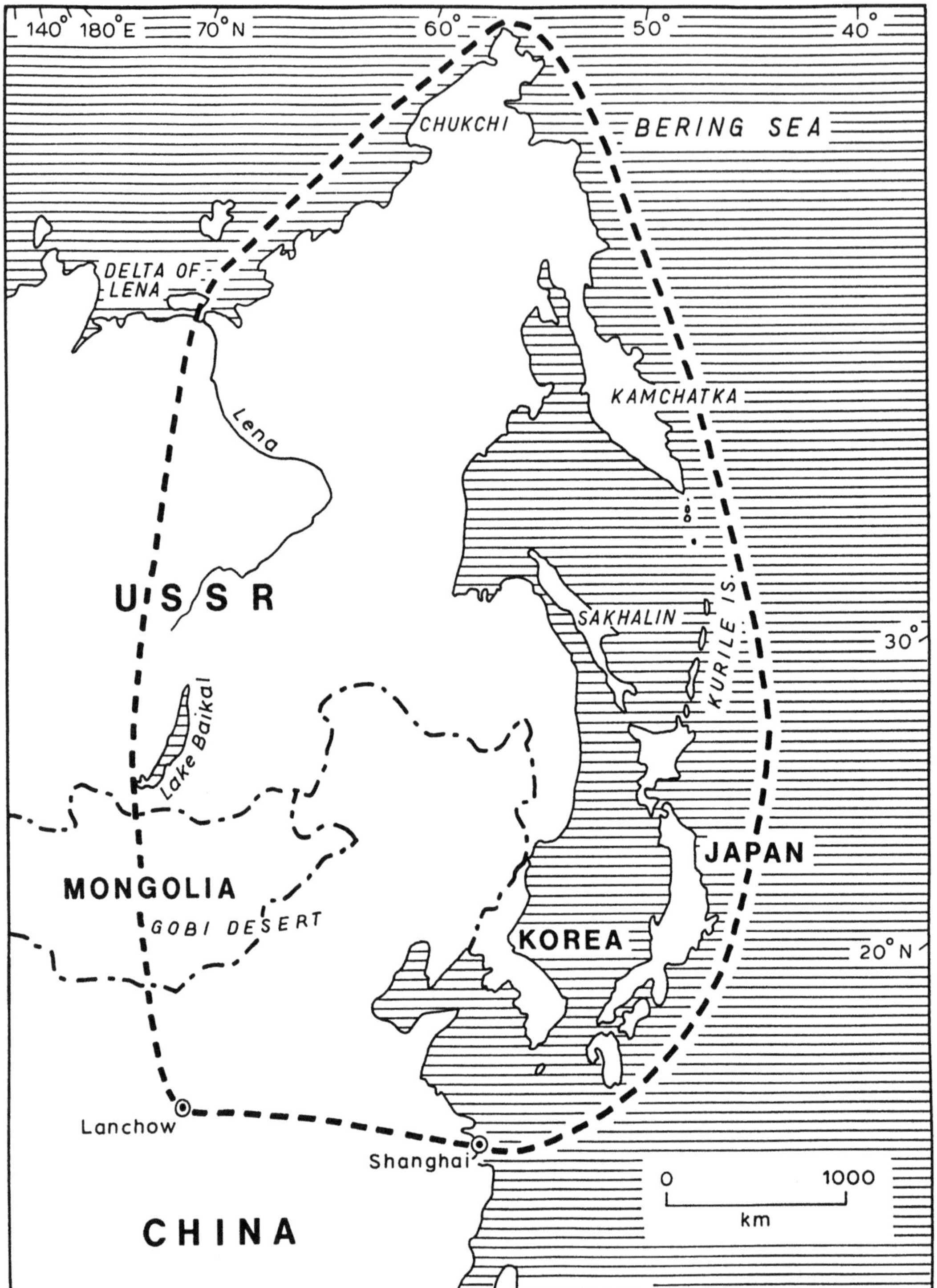

Fig. 4.2. The biogeographic boundary of East Asia.

Herbaceous Species:

(i) Throughout the Peninsula (15 spp.)
Turritis glabra, *Triglochin maritimum* var. *asiaticum*, *Lathyrus maritimus*, *Bromus japonicus*, *Viola selkirkii*, *Digitaria ischaemum*, *Pyrola minor*, *Lycopodium complanatum*, *Lamium amplexicuale*, *Osmunda cinnamomea* var. *asiatica*, *Galium aparine*, *G. spurium*, *Dryopteris filix-mas*, *Polystichum braunii*, *Zostera marina*

(ii) North to Midland (7 spp.)
Rumex maritimus, *Typha latifolia*, *Glaux maritima*, *Cinna latifolia*, *Galeopsis bifida*, *Polypodium vulgare* subsp. *virginianum*, *Utricularia intermedia*

(iii) Midland to South (5 spp.)
Suaeda maritima, *Ruppia maritima*, *Brasenia schreberi*, *Cynodon dactylon*, *Monotrapa hypopitys* subsp. *japonica*

Species with a disjunctive distribution in East Asia, and restricted in Korea

Twenty-eight arctic-alpine species occur disjunctively in East Asia, and are restricted in Korea. They include 10 arboreal and 18 herbaceous plants and display six distributional patterns in Korea.

Arboreal Species:

(i) North (9 spp.)
Dryas octopetala var. *asiatica*, *Oxycoccus microcarpus*, *Rubus articus*, *Rhododendron lapponicum* (= ? *R. parvifolium*), *Chamaedaphne calyculata*, *Vaccinium vitis-idaea* var. *minus*, *Ledum palustre* var. *diversipilosum*, *L. palustre* subsp. *decumbens*, *Linnaea borealis*.

(ii) Cheju (1 sp.)
Diapensia lapponica subsp. *obovata*

Herbaceous Species:

(i) North (14 spp.)
Oxyria digyna, *Hackelia deflexa*, *Rubus chamaemorus*, *Scheuchzeria palustris*, *Ranunculus hyperboreus*, *Poa glauca*, *P. palustris*, *P. trivalis*, *Moneses uniflora*, *Pyrola secunda*, *Drosera angelica*, *Carex rostrata*, *Pinguicula villosa*, *Luzulu wahlenbergii*

(ii) Midland (1 sp.)
Botrychium virginianum

(iii) South (1 sp.)
Elatine triandra

(iv) Cheju (2 spp.)
Ophioglossum vulgatum, *Hymenophyllum wrightii*

Distributional patterns and ranges of Korean arctic-alpine plants in East Asia

Out of the 17 arboreal arctic-alpine plants in Korea, four species, viz. *Ledum palustre* var. *diversipilosum*, *Oxycoccus quadripetulus*, *Phyllodoce caerulea* and *Vaccinium uliginosum* show a continuous distribution from the arctic to northern Korea, and at the same time a disjunctive distribution from northern Korea to Cheju island in the south. The remainder of the arboreal arctic-alpine plants, namely *Juniperus communis* var. *montana*, *Dryas octopetala* var. *asiatica*, *Rubus articus*, *Sorbus amurensis*, *S. commixta*, *Diapensia lapponica* subsp. *obovata*, *Chamaedaphne calyculata*, *Ledum palustre* subsp. *decumbens*, *Oxycoccus microcarpus*, *Rhododendron lapponicum*, *Vaccinium vitis-idaea* var. *minus* and *Linnaea borealis* show a disjunctive distribution from the arctic to Korea (Fig. 4.3).

Six arboreal arctic-alpine species grow in the Korean peninsula at the world's southernmost limits of their distribution. They include *Phyllodoce caerulea*, *Ledum palustre* var. *diversipilosum*, *L. palustre* subsp. *decumbens*, *Rhododendron parvifolium*. *Diapensia lapponica* subsp. *obovata* and *Empetrum nigrum* var. *japonicum*. The survival of many arctic-alpine plants, including the six which are at the southernmost limits of their distribution worldwide, again suggests that the Korean peninsula has served as a major glacial refugium during the Pleistocene glaciation.

Similar general distributional patterns can be seen in the case of the herbaceous arctic-alpine plants. Thus, there are 13 herbaceous arctic-alpine plants which grow continuously from the arctic to Korea, namely *Stellaria longiflora*, *Anemone dichotoma*, *Caltha palustre* var. *sibirica*, *Agrostis canina* subsp. *trinii*, *Calamagrostis canadensis* subsp. *langsdorffii*. *Carex loliacea*, *C. pseudo-curaica*, *C. rotundata*, *Eriophorum alpina*, *Luzula rufescens*, *Allium schoenoprosum*, *Calypso bulbsa* and *Equisetum hiemale*. On the

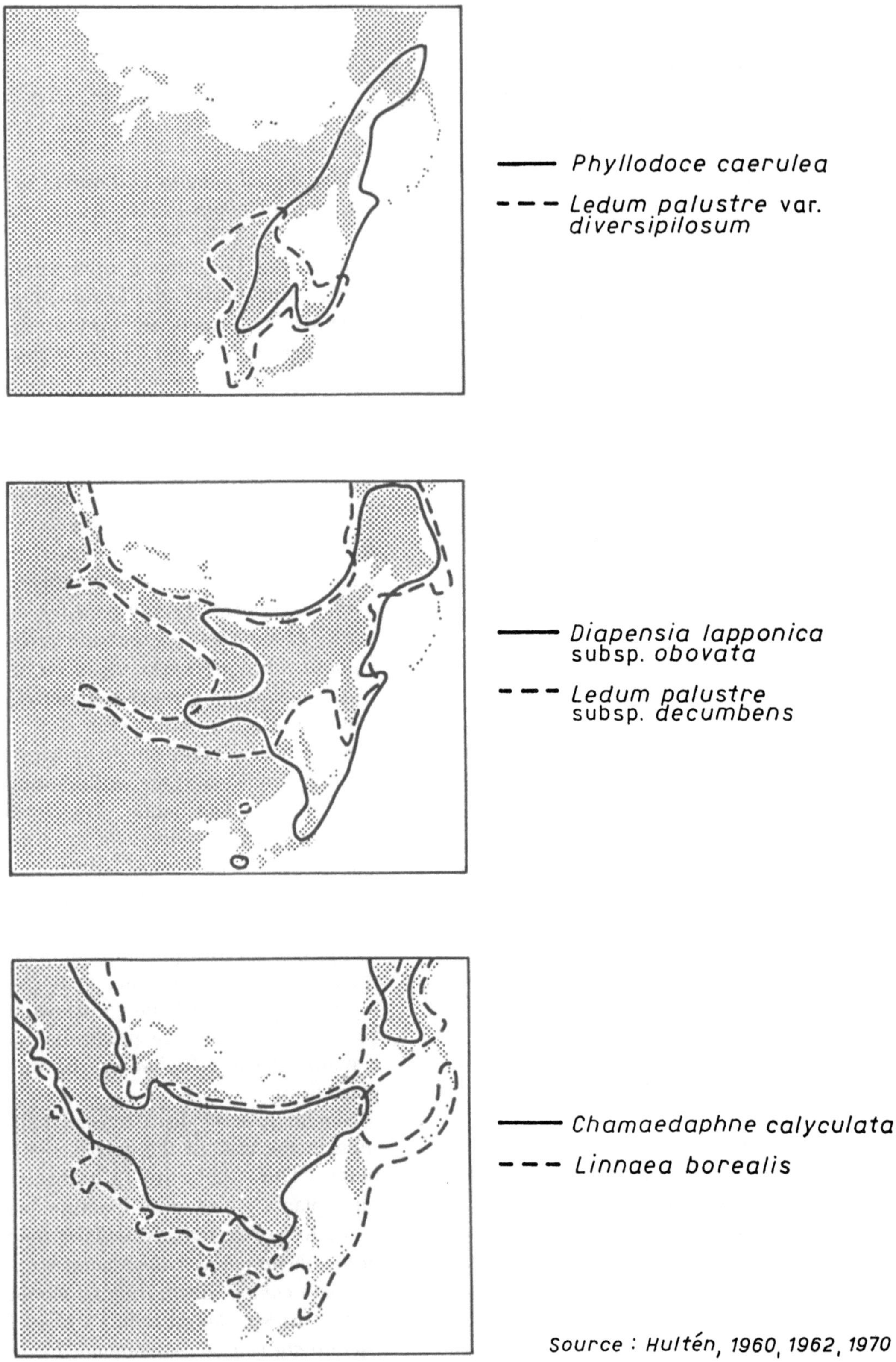

Fig. 4.3. Continuous-disjunctive distributions of arctic-alpine plants from the circumpolar area to Korea.

other hand, eight annual herbs, viz. *Suaeda maritima*, *Rubus chamaemorus*, *Glaux maritima*, *Galeopsis bifida*, *Lamium amplexicaule*, *Elatine triandra*, *Bromus japonicus* and *Galium spuricum*, as well as another 35 perennial herbs (see Appendix 5), show a characteristic disjunctive distributional pattern. Only 13 perennial herbs show a continuous distributional pattern and this may result from their ability both to withstand the severe climatic conditions of the Pleistocene glacial phases, and to respond quickly to the subsequent climatic ameliorations by spreading widely on to open ground. Such species, with a high genetic plasticity, are long-term survivors and are expected to grow over a wide area. Other species which are genetically less plastic, cannot adapt easily to conditions outside of their range, and are less competitive there, even though they can grow very well in their own climate.

Overall, up to 76.5% (13 spp.) of the arboreal and 77.6% (45 spp.) of the herbaceous arctic-alpine plants show a disjunctive distributional pattern in Korea. Many hypotheses are available to explain the worldwide disjunctive distribution of cold-loving arctic-alpine plants; e.g. first, that they occur in response to general environmental variations over time, and especially climatic changes during the Pleistocene glaciation; secondly, that the expansion of forest tree communities on low land subsequent to the end of the Pleistocene has had the effect of restricting formerly continuous arctic-alpine groups into disjunctive areas on mountain lands; thirdly, and associated with the second hypothesis, that there has been an ever greater restriction of available habitats for arctic-alpines following this climatic amelioration; and finally, that the creation of new biogeographical barriers by the expansion of new forest has, in itself, caused non-mountain, as well as mountain disjunction (Hultén, 1955; Löve, 1967; Goudie, 1979; Pielou, 1979; Tivy, 1982; Bremer, 1983).

It is likely that the disjunctive distribution of arctic-alpine plants in Korea is due primarily to the southward expansion of species towards the Korean peninsula from the arctic regions into refugia, as the Pleistocene glacial phases approached, and then their subsequent isolation upslope in mountain areas as the post-Devensian climatic amelioration followed. It could also be due to their continued preservation in isolation in some mountain refugia during the Devensian itself, and a subsequent persistence in the same areas (See Fig. 4.1).

On the basis of analysis of the present distributional pattern of numerous arctic-alpine plants by Hultén (1958, 1962, 1970), it is possible to suggest that the main southward migration route of arctic-alpine plants from the arctic to Korea was through the maritime territory of the USSR, and especially along the Sikhote-Alin mountain ranges, which connect the arctic of Siberia and the alpine regions of northern Korea (see Fig. 4.3). However, the existence of several solely East Asian arctic-alpine species, subspecies and varieties in Korea also may suggest the possibility of some species evolution in East Asia itself, possibly since the post-Devensian. This may be the case for *Empetrum nigrum* var. *japonica*, *Sorbus amurensis*, *S. commixta*, *Triglochin maritimum* var. *asiaticum*, *Osmunda cinnamomea* var. *asiatica*, *Monotropa hypopitys* subsp. *japonica*, *Dryas octopetala* var. *asiatica* and *Hymenophyllum wrightii*.

Distributional patterns of alpine plants in Korea

The occurrence of such a large number of arboreal (180 spp.) and herbaceous (187 spp.) alpine plants, which form up to 10% of the total Korean flora, can be explained by the following hypotheses. First, there has been no catastrophic environmental event in the peninsula during the Pleistocene, such as that deriving from the existence of a continental ice-sheet, and consequently a long history of the flora is assured. However, it was very cold in the peninsula during the glacial phases (see p. 62). This clearly restricted the distribution of plants during these glacial phases. Secondly, the unique geographic setting (a north-south linkage of mountain ranges in East Asia and in the peninsula) has enabled smooth migrations of arctic-alpine and alpine plants from north to south to take place, with the onset of the Pleistocene cold phases. Thirdly, the existence of 79 high mountains (over 2000 m a.s.l. in height) in northern Korea, and ca. 264 mountains of 2000 to 1500 m a.s.l. in height, mainly in the north, but also in central and southern Korea, ensured that isolated space was available for these species to colonise during the post-Devensian warming. Finally, the different climatic regimes present in the peninsula today, namely a continental one in the north, an intermediate one in the midlands and a slightly oceanic one in the south, have created a range of diverse habitats, further enhanced by local differences in soil condition, and topography, all of which has encouraged the long-term survival of these species at high altitudes in recent times.

Though 180 arboreal and 187 herbaceous alpine plants grow in Korea, only 239 alpine species (108 arboreal and 131 herbaceous alpine species), i.e. excluding the 23 arctic-alpines growing solely in alpine areas are discussed here (see Table 4.5). Two broad distributional types are noticed, i.e.: (i) a disjunctive distribution in East Asia, but widespread in Korea (87 spp.), and (ii) a disjunctive distribution in East Asia, but restricted in Korea (152 spp.).

Species with a disjunctive distribution in East Asia, but widespread in Korea

Eighty-seven alpine species grow disjunctively in East Asia, but are widespread in Korea. These include 44 arboreal and 43 herbaceous alpine species; and they display 11 distributional patterns in Korea.

Arboreal Species:

(i) Throughout the Peninsula (23 spp.)
Taxus cuspidata, *Philadelphus pekinensis*, *Abies holophylla*, *A. nephrolepis*, *Sorbus amurensis* var. *rufea*, *S. commixta*, *Picea jezoensis*, *Eleutherococcus senticosus*, *Pinus koraiensis*, *Rhododendron dauricum*, *R. fauriae* form. *rufescens*, *Salix hultenii*, *Betula chinensis*, *B. costata*, *B. ermanii* var. *acutifolia*, *B. schmidtii*, *Vaccinium uliginosum*, *Thymus quinquecostatus* var. *ibukiensis*, *Lonicera caerulea* subsp. *edulis*, *L. chrysantha*, *L. sachalinensis*, *Clematis koreana*, *Berberis amurensis*,

(ii) North to Midland (19 spp.)
Taxus cuspidata var. *latifolia*, *Sorbaria stellipilla* var. *typica*, *Pinus pumila*, *Ribes distans* var. *typicum*, *Salix floderusii*, *S. floderusii* var. *glabra*, *S. hallaisanensis* var. *longiflora*, *Rosa koreana*, *R. maretii*, *R. pimpinellifolia*, *Alnus maximowiczii*, *Arctous ruber*, *Betula ermanii* var. *ganjuensis*, *B. ermanii* var. *incisa*, *B. patyphylla* var. *latifolia*, *Rhododendron aureum*, *Lonicera caerulea* var. *emphyllocalyx*, *L. caerulea* form. *ovata*, *Berberis amurensis* var. *latifolia*.

(iii) Midland to South (1 sp.)
Betula ermanii var. *genuina*

(iv) North and Cheju (1 sp.)
Salix myrtilloides var. *manshurica*

Herbaceous Species:

(i) Throughout the Peninsula (9 spp.)
Bilderdykia vulgaris var. *yezoensis*, *Ligularia fischeri*, *Sanguisorba hakusanensis*, *Senecio flanmeus*, *Geranium dahuricum*, *Veratrum versicolor* form. *albidum*, *Cnidium tachiroei*, *Athyrium rupestre*, *Pedicularis verticillata*

(ii) North to Midland (20 spp.)
Anemone narcissiflora, *Agrostis hiemalis*, *Astragalus davuricus*, *A. membranaceus*, *Eriophorum japonicum*, *E. polystachyon*, *E. vaginatum*, *Geranium eriostemon* var. *typicum*, *Trientalis europaea* var. *arctica*, *T. europaea* var. *eurasiatica*, *Luzula oligantha*, *Allium senescens*, *Gentiana jamesii*, *G. makinoii*, *Platanthera maximowicziana*, *P. sachalinensis*, *Pedicularis mandshurica*, *Selaginella rupestris* var. *shakatanensis*, *Leontopodium japonicum*, *Athyrium melandepis*

(iii) Midland to South (1 sp.)
Allium taquetii

(iv) North and South (2 spp.)
Rhodiola tachiroei, *Herminium monorchis*

(v) North and Cheju (7 spp.)
Ranunuculus borealis, *Hieracium japonicum*, *Pedicularis spicata*, *Saussurea manshurica*, *Adenophora liliifolia*, *A. triphylla* var. *canenscens*, *Senecio nemorensis* subsp. *fuchsii*.

(vi) Midland and Cheju (1 sp.)
Lastrea quelpaertensis

(vii) South and Cheju (3 spp.)
Potentilla matsumurae, *Ligularia stenocephala*, *Hypericum vaniotii*

Photo 4.1. Landscape near the summit of Mt. Paikdoo (2744 m a.s.l.), the highest summit of the Korean peninsula, on the border between North Korea and China, Kaema Upland. *Larix olgensis* is the dominant tree at the timber-line and at the forest limit on the Korean part of Mt. Paikdoo. It occurs as trees and krummholz at the timber-line. The altitudinal range of the timber-line is approximately 1950 to 2100 m a.s.l. (Photograph by M. Šrûtek.)

Species with a disjunctive distribution in East Asia, and restricted in Korea

One hundred and fifty-two alpine species occur disjunctively in East Asia and show a restricted distribution in Korea. They include 64 arboreal and 88 herbaceous alpine species; and they display six distributional patterns.

Arboreal Species:

(i) North (56 spp.)

Larix olgensis, *Cretaegus maximowiczii*, *C. pinnatifolia* var. *major*, *Pinus tabulaeformis*, *Sabina davurica*, *Sinomalus sambucifolia*, *Populus maximowiczii* var. *barbinervis*, *Dasiphora fruticosa*, *D. fruticosa* var. *coreana*, *D. fruticosa* var. *glabra*, *Salix berberifolia* var. *brayi*, *S. berberifolia* var. *genuina*, *S. integra*, *S. pentandra* var. *intermedia*, *S. pseudolinearis*, *S. roridaeformis*, *S. sibirica* var. *brachypoda*, *S. siuzevii*, *Rosa suavis* var. *polyphylla*, *Rubus arcticus* var. *tentipetala*, *Chamaepericlimenum canadense*, *Cornus alba*, *Andromeda polifolia*, *A. polifolia* form. *acerosa*, *Alnus hirsuta* var. *japonica*, *A. tinctoria* var. *typica*, *A. vermicularis*, *Ledum palustre* var. *latifolium*, *L. palustre* var.

minus, *L. palustre* var. *nipponicum*, *L. palustre* var. *yesoense*, *Betula chinensis* var. *angusticarpa*, *B. cyclophylla*, *B. gmelini*, *B. platyphylla* var. *mandshurica*, *Rhododendron parvifolium* var. *alpinum*, *Therorhodion redawskianum*, *Vaccinium vitis-idaea* var. *genuinum*, *Quercus mongolia* var. *tomentosa*, *Lonicera caerulea* var. *longibracteata* form. *normalis*, *L. ruprechtiana*, *L. sachalinensis* var. *barbinervis*, *L. sachalinensis* var. *stenophylla*, *L. subhispida*, *Ulmus mandshurica*, *Clematis ochotensis*, *Berberis poiretii* var. *angustifolia*, *Spiraea pseudocrenata*, *Ribes bureiense*, *R. diacantha*, *R. glabellum*, *R. horridum*, *R. ussuriensis*, *Sambucus latipinna* var. *rhodoclotus*, *Artemisia triroba*.

(ii) Midland (2 spp.)
Picea schrenkiana, *Lonicera caerulea* var. *venulosa*

(iii) South (4 spp.)
Rhododendron parvifolium form. *albiflorum*, *R. trinerve*, *R. tschonoskii*, *Lonicera tatarinowii* var. *leptantha*

(iv) Cheju (2 spp.)
Prunus buergeriana, *Hugeria japonica*

Herbaceous Species:

(i) North (83 spp.)
Bilderdykia ochotensis, *B. pacifica*, *Asperula platyglium* var. *alpina*, *Adenophora divaricata*, *A. lamarkii*, *Pleuropteropyrum ochreatum*, *Dianthus repens*, *D. superbus* var. *speciosus*, *Achillea paektussanense*, *Antennaria dioica*, *Minuartia laricina*, *Artemisia laciniata*, *A. sinensis*, *A. stelleriana*, *Trollius hondoensis*, *T. japonicus*, *T. macropetalus*, *Erigeron thunbergii* subsp. *glabratus*, *Arabis lyrata* var. *kamtschatica*, *Hieracium umbellatum* subsp. *umbellatum* var. *coronopilfolium*, *Cardamine bellidifolia*, *Draba incana* var. *multicaulis*, *Deschampia caespitosa*, *Bergenia pacifica*, *Hierochloe alpina*, *Saxifraga cernua*, *S. laciniata*, *S. octopetala* var. *albiflora*, *Lasiagrostis mongolica*, *Phleum alpinum*, *Poa kanboensis*, *Parageum calithifolium* var. *nipponicum*, *Trisetum spicatum*, *Potentilla bifurea* var. *glabrata*, *P. nivea* var. *vulgaris*, *P. nivea* form. *alpina*, *Carex capillaris*, *C. subumbellatum* var. *koreana*, *Eriophorum angustifolium*, *E. manshiricum*, *Sanguisorba alpina*, *S. obutusa*, *S. parcifolia*, *Kobresia myosuroides*, *Symplocarpus buchenensis*, *Astragalus uliginosus*, *Oxytropis anertii*, *O. koreana*, *O. stroloilacea*, *Juncus brachyspathus*, *J. triglumis*, *Luzula sudetica* var. *nipponica*, *Geranium eriostemon* var. *megalanthum*, *G. eriostemon* var. *orientale*, *G. maximowiczii*, *Tofieldia nutans* var. *rubescens*, *Zygadenus sibirica*, *Lloydia serotina*, *Viola biflora*, *V. cressa*, *V. silvestriformis*, *Epipactis aphyllum*, *Orchis bracteata*, *Poneorchis parciflora* var. *joo-iwokiana*, *Gentiana algida*, *Lycopodium selgo*, *L. sitchense*, *Melampyrum arcuatum*, *M. roseum* var. *alpinum*, *Selaginella selaginoides*, *Pedicularis apodochila*, *P. glandiflora*, *P. sudetica*, *Cryptogramma crispa*, *Athyrium crenatum* var. *glabrum*, *Woodsia glabella*, *W. ilvensis*, *W. saitonana*, *Veronica stelleri* var. *longistyla*

(ii) Cheju (5 spp.)
Primula modesta var. *genuina*, *P. modesta* var. *fauriae*, *Agrostis flaccida*, *Veratrum maximowiczii* var. *coreanum*, *Euphrasia multifolia*

Distributional patterns of endemic alpine plants in Korea

Two general distributional types are found from the 105 endemic alpine plants of Korea (see Table 4.5). They are: (i) endemic species, but with a widespread distribution in Korea (24 spp.), and (ii) endemic species, and a restricted distribution in Korea (81 spp.). The latter are clearly the more numerous.

Endemic species but widespread distribution in Korea

Twenty-four Korean endemic alpine species grow widely in the peninsula. They include 18 arboreal and six herbaceous endemic alpine species; and they display seven distributional types.

Arboreal Species:

(i) Throughout the Peninsula (5 spp.)
Salix hallaisanensis, *Oplopanax elatum*, *Clematis chiisanensis*, *Viburnum wrightii* var. *stipellatum*, *Sorbus amurensis* var. *rufea*

Photo 4.2. Some species of the herb layer at the timber-line of Mt. Paikdoo: *Oxytropis anertii*, *Saussurea alpicola* (silver leaves). (Photograph by M. Šrûtek.)

(ii) North to Midland (11 spp.)
Larix olgensis var. *koreana*, *Berberis koreana*, *Thuja koraiensis*, *Sorbaria stellipilla* var. *glandulosa*, *Salix metaformosa*, *Sinomalus komrovii*, *Alnus fruticosa* var. *mandshurica* form. *normalis*, *Sorbus chinophylla*, *Rubus idaeus* var. *concolor*, *Clematis mandshurica* var. *lancifolia*, *Viburnum koreana*

(iii) South and Cheju (2 spp.)
Abies koreana, *Betula ermanii* var. *parvifolia*

Herbaceous Species:

(i) North to Midland (1 sp.)
Chrysanthemum zawadskii

(ii) North to South (1 sp.)
Cardamine valida

(iii) North and Cheju (3 spp.)
Sieboldia coreana, *Scabiosa japonica* var. *alpina*, *Veronica subincanovelutina*

(iv) Midland and Cheju (1 sp.)
Leontopodium coreanum

Endemic species and restricted distribution in Korea

Eighty-one Korean endemic alpine species occur in this category. They include 40 endemic arboreal alpine and 41 herbaceous alpine species; and they fall into 7 distributional types.

Arboreal Species:

(i) North (32 spp.)
Larix olgensis form. *viridis*, *Betula fusenensis*, *B. microphylla* var. *coreana*, *B. paisanensis*, *Picea intercedens*, *P. koraiensis*, *P. pungsanensis*, *P. tonaiensis*, *Clematis nobilis*, *C. nobilis* var. *plea*, *C. subtriternata*, *C. subtriternata* var. *tenuifolia*, *Juniperus sibirica*, *Salix bicarpa*, *S. hultenii* var. *elongata*, *S. kangensis*, *S. orthostemma*, *S. rotundifolia*, *S. rotundifolia* var. *lanata*, *S. sericeo-cinerea*, *S. sericeo- cinerea* var. *lanata*, *Spiraea sylvestris*, *Crataegus pinnatifida* form. *betulifolia*, *C. pinnatifida* var. *partita*, *Sinomalus komarovii* var.

Photo 4.3. Lower montane forest (ca. 1350 to 1650 m a.s.l.) at the Korean part of Mt. Paikdoo, Kaema Upland, North Korea, *Larix olgensis* and *Betula platyphylla* are dominant trees, *Picea koraiensis* is frequent, *P. jezoensis* is rare in the canopy. *Ledum palustre* var. *angustrum* is a frequent species of the lower shrub layer. (Photograph by M. Šrůtek.)

major, *S. komarovii* var. *pilosa*, *Ledum palustre* var. *maximum*, *Lonicera monantha*, *L. nigra* var. *barbinervis*, *L. sachalinensis* var. *stenophylla*, *Alnus vermicularis*, *Sambacus velutina*

(ii) Midland (2 spp.)
Taxus caespitosa, *Pentactina rupicola*

(iii) South (1 sp.)
Lonicera cerasoides

(iv) Cheju (5 spp.)
Salix blinii, *Maackia fauriei*, *Berberis amurensis* var. *quelpaertensis*, *Rhamnus taquetii*, *Prunus densiflora*

Herbaceous Species:

(i) North (27 spp.)
Persicaria lyrata, *Viola koraiensis*, *Phleum coreanum*, *Homopterya nakaiana*, *Dianthus morii*, *Euphrasia coreanalpina*, *E. mucronulata*, *Minuartia arctica* var. *major*, *Pulsatilla nivalis*, *Pedicularis lunaris*, *P. nigrescens*, *Ranunculus acris* var. *elatus*, *Papaver coreanum*, *Artemisia angustissima*, *Cardamine resedifolia* var. *morii*, *Leontopodium leoiolepis*, *Saxifraga furumii*, *Ligularia jamesii*, *Sanguisorba argutipens*, *Saussurea rectinervis*, *Astragalus adsurgens* var. *alpinus*, *A. membranceus* var. *manshuricus*, *A. setsureianus*, *Poa deschampsioides*, *Lycopodium alpinum*, *Geranium eriostemon* var. *glabrescens*, *G. eriostemon* var. *hypoleueus*

(ii) Midland (2 spp.)
Peucedanum coreanum, *Calamagrostis langsdorffii* var. *ramosa*

(iii) Cheju (11 spp.)
Silene fasciculata, *Geranium skikokianum* var. *quelpaertense*, *Ranunculus crucilobus*, *R. crucilobus* var. *chrysotrichus*, *Galarhoeus fauriei*, *Arabis serrata* var. *hallaisanensis*, *Galium pusillum*, *Aruncus aethusifolius*, *Anaphalis sinica* subsp. *morii*, *Astragalus membranaceus* var. *alpinus*, *Tofieldia fauriei*

Overall, 32.2% (58 spp.) of the total arboreal alpine plants of Korea belong to the endemic arboreal alpine plant group, and 25.1% (47 spp.) of the total herbaceous alpine plants of Korea belong to the endemic herbaceous alpine plant group. In the case of Europe, Favarger (1972) has suggested that historical factors have played an important part in mountain endemism, probably because genetic drift has occurred on these isolated populations (as also reported by Böcher, 1972). According to Flenley (1984), the previous 10 000 to 100 000 year time scale is particularly significant for Quaternary biological and evolutionary events, with all their important biogeographical effects. Watts (1984) has also suggested that many new species may have emerged only within the Post-Devensian (Holocene) period of change on land especially over the last 10 000 to 12 000 years. However, more detailed data are still required to clarify the timing of the evolution of the endemic alpine plants in Korea.

Results and discussion

For Eurasia, one may say that arctic-alpine and alpine plants most likely evolved in the central Asian high mountains and surrounding localities, mainly during the late Miocene and Pleistocene epochs, from which area they dispersed widely elsewhere.

The estimated number of Korean arboreal alpine plants is ca. 180 species in 62 families and that of its herbaceous alpine plants is ca. 187 species in 38 families. Seventy-five species (17 arboreal and 58 herbaceous species) of arctic-alpines are also present. Among these groups, 105 species are endemic to Korea.

The most characteristic feature of arctic-alpine, alpine and endemic alpine plants in Korea is their disjunctive distribution pattern, either overall, within East Asia, or more locally, within the Korean peninsula. This disjunctive distribution of arctic-alpine plants in Korea is likely to be due to the southward expansion of plants toward the Korean peninsula from the arctic region, as recent phases of the Pleistocene glaciations approached, and then their subsequent isolation caused by upslope movement in mountain areas as the post-Devensian climatic amelioration followed.

Despite the presence of so many arctic-alpine, alpine and endemic alpine plants in Korea, there has been no report to date to suggest that the Korean peninsula contained glacial refugia on a large scale. However, we suggest that such refugia undoubtedly existed there, on both a primary and secondary level. Although the cold-phase Pleistocene climates in the present alpine zones of Korea seem to have been too cold and harsh for arctic-alpine plants to survive, and these plants then had to confine themselves to lower montane and/or lowland areas, possibly including coastal areas, it is likely that these latter accordingly, at these times, effectively became *primary* refugia. The north–south orientation of mountain chains facilitated the southward migration of arctic-alpine plants from Siberia to Korea. The main southward migration route of arctic-alpine plants from the arctic to Korea was probably mainly through the maritime territory of Russia, especially via the Sikhote-Alin mountain ranges, which connect arctic Siberia and the alpine regions of northern Korea. Then, as the climate subsequently ameliorated, arctic-alpine plants moved upslope on to the mountain tops and northwards, and the range of thermophilous temperate and warm temperate plants was then expanded in the wake of this movement to cover large areas of lowland, and the lower mountains. As a result, arctic-alpine plants retreated further upwards on to mountain tops, as well as on to montane and further coastal areas in the Korean peninsula, and a fragmentation of arctic-alpine plant ranges took place. Today, therefore, arctic-alpine plants are confined mainly to the alpine and subalpine zones of the Korean mountains. The relatively long period of post-Devensian isolation of arctic-alpine and alpine plants on these scattered sites (mainly mountain tops), the earlier environmental stresses to which they had to adapt within their primary refugia (e.g. in coastal lowlands), and the extreme degree of environmental diversity, together have resulted in a relatively large number of endemic alpine plants being produced. The mountain tops accordingly have proved to be important *secondary* refugia within the peninsula. The importance of these primary and secondary refugia within the general environmental framework of Korea (especially within the predominant north–south mountain chains) to the biogeography of East Asia is generally emphasised by: first, the presence of 75 arctic-alpine plants within these refugia; and secondly, by the fact that six of these (e.g. *Phyllodoce caerulea*, *Ledum palustre* var. *diversipilosum*, *L. palustre* subsp. *decumbens*, *Rhododendron parvifolium*, *Diapensia lapponica* subsp. *obovata* and *Empe-*

trum nigrum var. *japonicum*) lie at the southernmost global limits of their range.

Among the 75 arctic-alpine plants, four types of distribution are found. These include: (i) a continuous distribution in East Asia and a widespread one in Korea (5 spp.); (ii) a continuous distribution in East Asia, but a restricted one in Korea (12 spp.), (iii) a disjunctive distribution in East Asia, but a widespread one in Korea (30 spp.); and (iv) a disjunctive distribution in East Asia, and a restricted one in Korea (28 spp.).

The 239 alpine plants produced two distributional types: (i) a disjunctive distribution in East Asia, but one which is widespread in Korea (87 spp.), and (ii) a disjunctive distribution in East Asia, and restricted in Korea (152 spp.). A distinctive feature is the existence of 139 alpine species (58.2% of the total alpine species of Korea) exclusively in the north of Korea.

As to the 105 endemic alpine plants in Korea, two distributional types are found: one with a widespread distribution (24 spp.), and a larger group with a restricted distribution (81 spp.). 32.2% (58 spp.) of the arboreal alpine plants and 25.1% (47%) of the herbaceous alpine plants of Korea belong to the endemic alpine plant group, and many of these seem to have emerged as individual species probably within the post-Devensian period.

Overall, the present-day occurrence of such a large number of arboreal (180 spp.) and herbaceous (187 spp.) alpine plants, which form up to 10% of the total number of Korean plants, is likely to have been due to first, the absence of any prolonged phases of ice accumulation during the Pleistocene within the peninsula; secondly, the unique geographic setting (a north–south linkage of mountain ranges in East Asia and in the peninsula), which has enabled smooth migrations of arctic-alpine and alpine plants from north to south to take place, with the onset of Pleistocene cold phases; thirdly, the existence of suitable habitats, e.g. the montane, lowland and coastal areas, which served as primary refugia for migrating species during the Pleistocene glaciation; fourthly the existence of ca. 79 high mountains (over 2 000 m a.s.l. in height) in northern Korea, and ca. 264 mountains of 2000 to 1500 m a.s.l. in height in the peninsula as a whole, ensuring that isolated space was available for these species to colonise, often as secondary refugia during the immediate post-Devensian warming phase; and finally the presence of many different climatic regimes within the peninsula which have created a range of diverse habitats, all of which in turn ultimately have encouraged the long-term survival of these species.

CHAPTER 5

Evergreen arboreal arctic-alpine & alpine plants & their environments

In this chapter, the relationships between the distribution of evergreen arboreal arctic-alpine and alpine plants and their environments are discussed. The number of evergreen arboreal arctic-alpine and alpine plants selected for examination has been reduced to 41, so as to yield a better understanding of the survival mechanisms of these species in Korea's alpine environment. Results from this research will be helpful in the scrutiny of the patterns, processes and mechanisms for the survival of cryophilous plants in alpine areas generally, and it will also be useful to compare the ecological amplitudes of arctic-alpine and alpine species here with those from other parts of the world. In order to achieve this, the species associates, physiognomy, distributional patterns and ranges of the 41 relevant species, and their relationships with the environment, are detailed.

Species composition

The 41 selected evergreen arctic-alpine and alpine plants can be placed into 18 genera and 10 families, as presented in Table 5.1.

They are dominated by two families, viz. the Ericaceae (18 spp.) and the Abietaceae (9 spp.). Another 8 families, viz. the Taxaceae (3 spp.), the Pinaceae (3 spp.), the Juniperaceae (3 spp.), the Cupressaceae (1 sp.), the Rosaceae (1 sp.), the Diapensiaceae (1 sp.), the Empetraceae (1 sp.) and the Caprifoliaceae (1 sp.) are also present. The coexistence of 19 coniferous plants (46.3% of the total evergreen arboreal arctic-alpine and alpine plants in Korea) and 22 broadleaved plants (53.7% of these) in alpine communities in the peninsula has inevitably yielded a complicated pattern of species physiognomy, of 13 trees, 2 small trees and 26 shrubs. This characteristic mixture of trees, small trees and shrubs often makes it difficult to define the boundary of alpine communities in East Asia.

Table 5.1 indicates that the 41 evergreen arboreal plants in Korea consist of two broad distributional types; an alpine type (28 spp.) and an arctic-alpine type (13 spp.). Almost all of the coniferous plants (except for *Sabina sargentii*) and 10 broadleaved plants (*Andromeda polifolia*, *A. polifolia* form. *acerosa*, *Ledum palustre* var. *latifolium*, *L. palustre* var. *maximum*, *L. palustre* var. *minus*, *L. paulustre* var. *yesoense*, *Rhododendron aureum*, *R. dauricum*, *R. fauriae* form. *rufescens* and *Vaccinium vitis-idaea* var. *genuinum*) belong to the alpine group. One coniferous species (*Sabina sargentii*), and another 12 broadleaved plants are arctic-alpine in type. Endemic evergreen arboreal alpine plants in Korea consist of 7 coniferous plants (*Taxus caespitosa*, *Abies koreana*, *Picea intercedens*, *P. koraiensis*, *P. pungsanensis*, *P. tonaiensis*, *Thuja koraiensis* and one broadleaved plant (*Ledum palustre* var. *maximum*)

Distribution

The results of Chapter 2 (Biogeographic Divisions) show that out of the 41 evergreen arboreal arctic-alpine and alpine plants, 20 species (or varieties) are found in the Northern Alpine Region (I). These are *Taxus cuspidata* var. *latifolia*, *Pinus pumila*, *Juniperus sibirica*, *Empetrum nigrum* var. *japonicum*, *Andromeda polifolia*, *A. polifolia* form. *acerosa*, *Chamaedaphne calyculata*, *Ledum palustre* var. *latifolium*, *L. palustre* var. *maximum*, *L. palustre* var. *minus*, *L. palustre* var. *procumbens*, *L. palustre* var. *yesoense*, *Oxy-*

Table 5.1. Selected evergreen arboreal arctic-alpine and alpine plants in Korea.

Family Name	Species name	Physiognomy	Distribution	Endemism
Taxaceae	*Taxus caespitosa*	S	A	Endemic
	T. cuspidata	T	A	
	T. cuspidata var. *latifolia*	T	A	
Abietaceae	*Abies holophylla*	T	A	
	A. koreana	T	A	Endemic
	A. nephrolepis	T	A	
	Picea intercedens	T	A	Endemic
	P. jezoensis	T	A	
	P. koraiensis	T	A	Endemic
	P. pungsanensis	T	A	Endemic
	P. schrenkiana	T	A	
	P. tonaiensis	T	A	
Pinaceae	*Pinus koraiensis*	T	A	
	P. pumila	ST	A	
	P. tabulaeformis	T	A	
Cupressaceae	*Thuja koraiensis*	ST	A	Endemic
Juniperaceae	*Juniperus sibirica*	S	A	? Endemic
	Sabina dahurica	S	A	
	S. sargentii	S	A-A	
Rosaceae	*Dryas octopetala* var. *asiatica*	S	A-A	
Diapensiaceae	*Diapensia lapponica* var. *obovata*	S	A-A	
Empetraceae	*Empetrum nigrum* var. *japonicum*	S	A-A	
Ericaceae	*Andromeda polifolia*	S	A	
	A. polifolia form. *acerosa*	S	A	
	Chamaedaphne calyculata	S	A-A	
	Ledum palustre var. *diversipilosum*	S	A-A	
	L. palustre var. *latifolium*	S	A	
	L. palustre var. *maximum*	S	A	Endemic
	L, palustre var. *minus*	S	A	
	L. palustre subsp. *decumbens* (=var. *procumbens*)	S	A-A	
	L. paluscre var. *yesoense*	S	A	
	Oxycoccus microcarpus	S	A-A	
	O. quadripetalus	S	A-A	
	Phyllodoce caerulea	S	A-A	
	Rhododendron aureum	S	A	
	R. dauricum	S	A	
	R. fauriae form. *rufescens*	S	A	
	R. parvifolium	S	A-A	
	Vaccinium vitis-idaea var. *genuinum*	S	A	
	V. vitis-idaea var *minus*	S	A-A	
Caprifoliaceae	*Linnaea borealis*	S	A-A	

(Sources: Lee & Ahn, 1963 and Hultén, 1958, 1962, 1970.)
Abbreviations: T, Tree; ST, Small Tree; S, Shrub; A, Alpine; A-A, Arctic-Alpine.

coccus microcarpus, *O. quadripetalus*, *Phyllodoce caerulea*, *Rhododendron aureum*, *R. fauriae* form. *rufescens*, *R. parvifolium*, *Vaccinium vitis-idaea* var. *genuinum* and *Linnaea borealis*. Eleven species (or varieties) grow in the North-South Subalpine Region (II), namely *Taxus cuspidata*, *T. cuspidata* var. *latifolia*, *Abies nephrolepis*, *Picea jezoensis*, *Pinus koraiensis*, *P. pumila*, *Thuja koraiensis*, *Sabina sargentii*, *Rhododendron aureum*, *R. fauriae* form. *rufescens* and *Vaccinium vitis-idaea* var. *minus*. Three species occur in the Midland Mountain Region (III) (*Taxus cuspidata*, *Abies nephrolepis* and *Thuja koraiensis*). Four species (*Abies holophylla*, *A. koreana*, *Pinus koraiensis* and *Sabina sargentii*) grow in the Southern Mountain Region (IV), and one species (*Thuja koraiensis*) is present in the Mid-Western Insular Region (V). Nine species (or varieties) are found in the Southern Insular Region (VI), namely *Taxus cuspidata*, *Abies holophylla*, *A. koreana*, *Sabina sargentii*, *Diapensia lapponica* subsp. *obovata*, *Empetrum nigrum* var. *japonicum*, *Rhododendron dauricum*, *R. fauriae* form. *rufescens* and *Vaccinium vitis-idaea* var. *minus*. Six species (*Taxus cuspidata*, *Abies koreana*, *Picea jezoensis*, *Pinus koraiensis*, *Sabina sargentii* and *Rhododendron fauriae* form. *rufescens*) grow in the West-South-Eastern Insular and Associated Inland Region (VII), and two species (*Picea pungsanensis* and *Pinus tabulaeformis*) are located in the North-South Disjunctive Region (VIII). Overall, 13 species (or varieties) occur in more than one biogeographic region, and these include *Taxus cuspidata*, *Abies holophylla*, *A. koreana*, *A. nephrolepis*, *Picea jezoensis*, *Pinus koraiensis*, *P. pumila*, *Thuja koraiensis*, *Sabina sargentii*, *Empetrum nigrum* var. *japonicum*, *Rhododendron aureum*, *R. fauriae* form. *rufescens* and *Vaccinium vitis-idaea* var. *minus*.

Overall, one may say that the 41 evergreen arboreal arctic-alpine and alpine plants of Korea display seven major patterns of distribution. In general, these are (i) species present throughout the peninsula; (ii) species present from the north to the midlands; (iii) species present in the north and in Cheju Island, but not in between; (iv) species present in the mainland south and on Cheju Island alone; (v) species restricted to the north; (vi) species restricted to the midlands; and (vii) species restricted to Cheju Island. These distributional patterns are examined in greater detail below.

Species present throughout the peninsula

The eight species which display this type of distributional pattern are *Taxus cuspidata*, *Abies holophylla*, *A. nephrolepis*, *Picea jezoensis*, *Pinus koraiensis*, *Sabina sargentii*, *Rhododendron dauricum* and *R. fauriae* form. *rufescens*. No endemic species occurs within this group. The overall ranges of each of these display a wide variability, and require further examination. Data for horizontal and vertical ranges of each species come from various sources, including Nakai (1915–1939), Uyeki (1926), Chung (1943), Chung & Lee (1965), Ann & Lee (1965), Ann *et al.* (1982), Lee (1982a) and local floral reports (Table 2.1).

Taxus cuspidata grows in Russia (East Siberia, Ussuri, Amur, Sakhalin, Kuriles), China, Japan (Hokkaido, Honshu) and Korea (Wilson, 1916; Nakai, 1935; Kitagawa, 1939, 1979; Dallimore & Jackson, 1954; Den Ouden & Boom, 1965; Ohwi, 1965; Knystautas, 1987). In Korea itself, the species ranges from 40° N to 33° N, and it is present on Mts. Pirae (750–1400 m a.s.l.), Seungjuk (700–1500), Myohyang (1000–1900), Haram (900–1500), Chuae (1300–1500), Keumkang (800–1650), Sorak (1500–1650), Whaak (1000–1400), Myungji, Kaebang, Odae, Yongmoon (800–1500), Taeji (1000–1200), Songin (300–900), Daesung, Taebaik (1200–1500), Sobaik, Sokri, Chiri (1200–1900) and Halla (1000–1900). The average lower and upper limits of its vertical range in Korea are 920 and 1507 m, respectively. The standard deviations of the lower and upper limits of its vertical range are 254 and 275 m respectively (data from 14 sites). The location of mountains in Korea is presented in Fig. 5.1.

Abies holophylla grows in Russia (East Siberia, Ussuri, Amur), China (Manchuria) and Korea (Hu, 1934; Nakai, 1935; Kitagawa, 1939, 1979; Dallimore & Jackson, 1954; Suslov, 1961; Wang, 1961; Den Ouden & Boom, 1965; Liu, 1971; Noda, 1971; Takhtajan, 1986), but not in Japan where *A. sachalinensis* and *A. veitchii* replace it (Shimizu, 1983). In Korea, it ranges from 42°30′ N to 33° N, and it is present on Mts. Songjin (600–), Chayoo (800–), Paikdoo, Mantap (900–1200), Chilbo (300–400), Hoochi (700–), Pirae (400–1000), Seungjuk (700–1250), Keumpae (700–950), Peenanduk (450–750), Myohyang (200–1200), Sasoo (400–200), Haram (600–1050), Chuae (500–1200), Keumkang (400–1100), Changsoo (300–), Myulak (200–), Sorak (300–850), Whaak (500–1400), Daeryong, Sori, Pukhan, Odae (600–1200), Kaebang, Yongmoon (100–1100), Taeji (750–1200), Chiak (600–1200), Daesung, Taebaik (750–1100), Sobaik, Ilwol (800–900), Sokri (650–950), Chuwang, Kaeryong (200–), Palgong (500–) Dukyoo (450–980), Kaya, Kaji (200–700), Naejang (400–), Paikyang (100–300), Moodeung (500–), and Halla (1500–). The average lower and upper limits of vertical range are 493 and 1008 m, respectively. The stan-

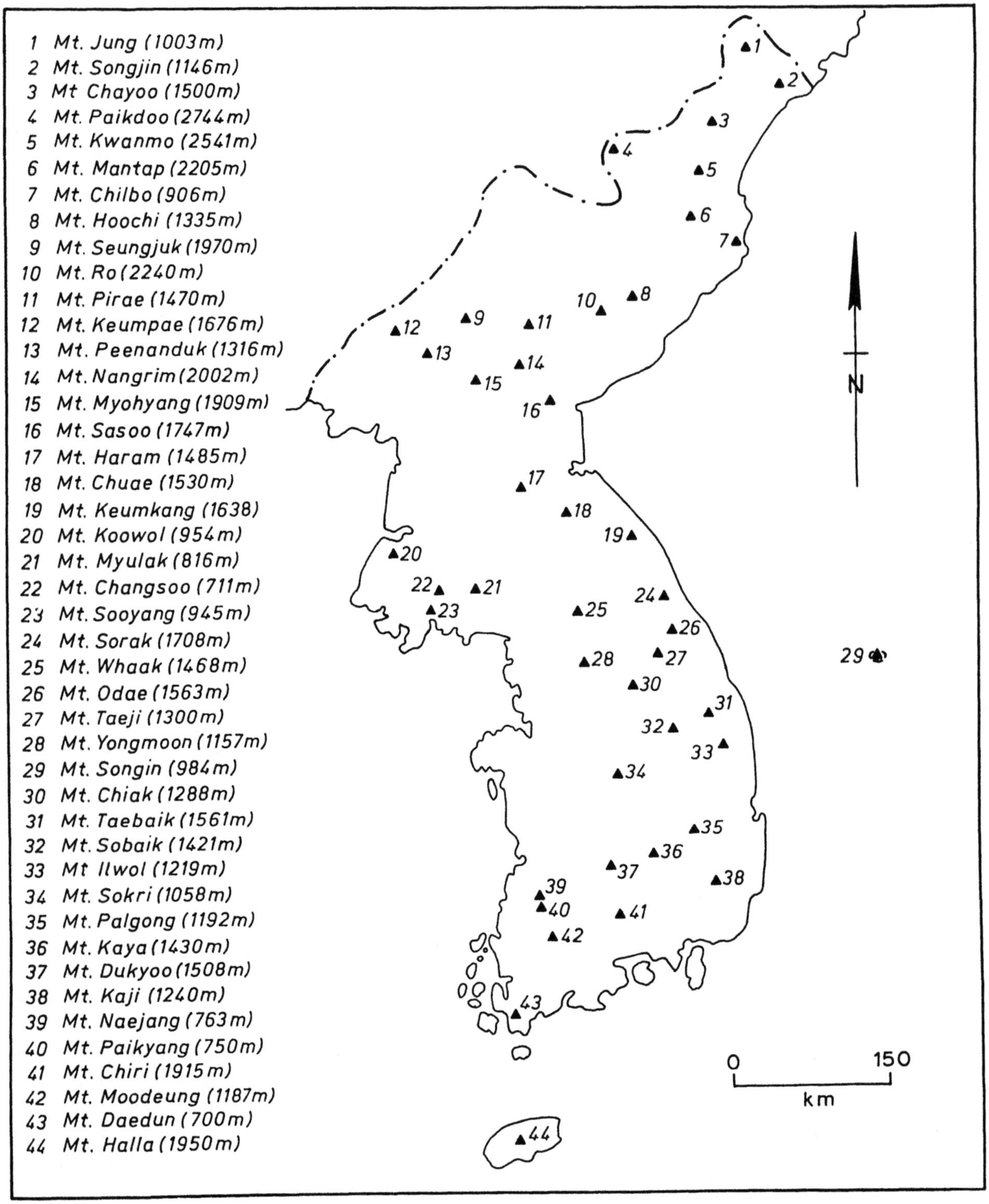

Figure 5.1. The location of Korean mountains (for range analysis).

dard deviations of lower and upper limits of vertical range are 274 and 262 m, respectively (data from 23 sites).

Abies nephrolepis occurs in Russia (East Siberia, Zeya, Kamchatka, Okhotsk, Amur, Ussuri), Mongolia, China (Manchuria), and Korea (Hu, 1934; Nakai, 1935; Kitagawa, 1939, 1979; Takahasi, 1944; Dallimore & Jackson, 1954; Suslov, 1961; Wang, 1961; Den Ouden & Boom, 1965; Liu, 1971; Gorchakovsky & Shiyatov, 1978; Takhtajan, 1986), but it is again absent from Japan. In Korea, it grows from 42° N to 35° N, and exists on Mts. Chayoo (800–1400), Paikdoo, Mantap (900–2200), Ro (1200–1900), Hoochi (700–1350), Pirae (700–1400), Seungjuk (700–1600), Nangrim (1300–1800), Keumpae (1200–1600), Peenanduk (1000–1250), Myohyang (700–1900), Sasoo (900–1750), Haram (1000–1400), Chuae (900–1450), Keumkang (780–), Sorak (700–1550), Whaak (1100–), Myungji, Odae (600–1500), Yongmoon (800–1100), Chiak (1000–1300), Daesung, Taebaik (1100–1500), Sokri, Chuwang, Dukyoo (1050–1500) and Chiri (1200–). The average lower and upper limits of its vertical range are 918 and 1507 m, respectively. The standard deviations of lower and upper limits of vertical range are 192 and 257 m respectively (data from 19 sites). *Abies nephrolepis* displays a typical pattern of distribution, and altitudinal range, for this type of species pattern (Fig. 5.2).

Picea jezoensis grows in Russia (East Siberia, Okhotsk, Kamchatka, Yakutia, Trans-Baikal, Ussuri, Amur, Sakhalin, Kuriles), China (Manchuria), Japan (Hokkaido) and Korea (Wilson, 1916, 1918; Hultén, 1927; Hu, 1934; Nakai, 1935; Kitagawa, 1939, 1979; Suslov, 1961; Wang, 1961; Den Ouden & Boom, 1965; Ohwi, 1965; Noda, 1971; Gorchakovsky & Shiyatov, 1978; Buks, 1984; Takhtajan, 1986), In Korea, it occurs from 42° N to 35° N, and is found on Mts. Chayoo (500–1400), Paikdoo, Mantap (1450–2300), Ro (1200–1700), Hoochi (700–1350), Pirae (800–1450), Seungjuk (700–1600), Nangrim (800–1900), Sasoo (1200–1750), Haram (1000–1400), Keumkang (800–1650), Chiak, Dukyoo (1400–) and Chiri (1200–1900). The average lower and upper limits of its vertical range are 989 and 1646 m, respectively. The standard deviations of the lower and upper limits of its vertical range are 273 and 263 m, respectively (data from 14 sites).

Pinus koraiensis grows in Russia (Southeast Siberia, Amur, Ussuri), China (Manchuria), Japan (Honshu) and Korea (Wilson, 1916, 1918; Hu, 1934; Kitagawa, 1939, 1979; Takahasi, 1944; Dallimore & Jackson. 1954; Suslov. 1961; Academy of Sciences of the USSR, 1964; Den Ouden & Boom, 1965; Ohwi, 1965; Critchfield & Little, 1966; Mirov, 1967; Noda, 1971; Takhtajan, 1986). In Korea, it occurs from 42° N to 35° N, and is present on Mts. Chayoo (800–1200), Paikdoo (700–), Mantap (1000–1600), Chilbo (400–), Ro (1200–), Hoochi (1000–1350), Pirae (300–1450), Seungjuk (600–1500), Nangrim (1300–1700), Keumpae (1200–1600), Peenanduk (700–1200), Myohyang (300–500), Sasoo (400–1750), Haram (700–300), Chuae (700–1450), Keumkang (100–1650), Koowol (200–600), Changsoo (–300), Sooyang (300–950), Sorak (400–1500), Whaak (300–1450), Myungji, Samak, Daeryong, Kari, Sori, Odae (1000–), Kaebang, Yongmoon (350–800), Taeji (800–1200), Chiak (900–1810), Taebaik (1100–1300), Sobaik, Ilwol (800–900), Sokri (650–1000), Kaeryong (200a), Palgong (500–1900), Hyangro, Kaya, Choijung, Biseul, Dukyoo (500–1500), Kaji (600–1250), Keumjung, Naejang, Paikyang, Moodeung and Chiri (1200–1900). The average lower and upper limits of its vertical range are 632 and 1308 m, respectively. The standard deviations of the lower and upper limits of its vertical range are 534 and 367 m, respectively (data from 28 sites).

Sabina sargentii grows in Russia (Sakhalin), Japan (Hokkaido, Honshu) and Korea (Wilson, 1916; Nakai, 1935). In Korea, it is present from 40°30′ N to 33° N. and is found on Mts. Seungjuk (1600–), Nangrim (2300–). Peenanduk, Myohyang (1600–1900), Sasoo (1600–1750), Chuae (1400–), Keumkang (1000–1650), Sorak (700–800), Kaya, Dukyoo (140–), Chiri (1400–1900) and Halla (1900–1950). The average lower and upper limits of its vertical range are 1367 and 1658 m, respectively. The standard deviations of the lower and upper limits of its vertical range are 421 and 446 respectively (data from six sites).

Rhododendron dauricum is found in Russia (East Siberia, Amur, Ussuri), Mongolia, China (Manchuria), Japan (Hokkaido) and Korea

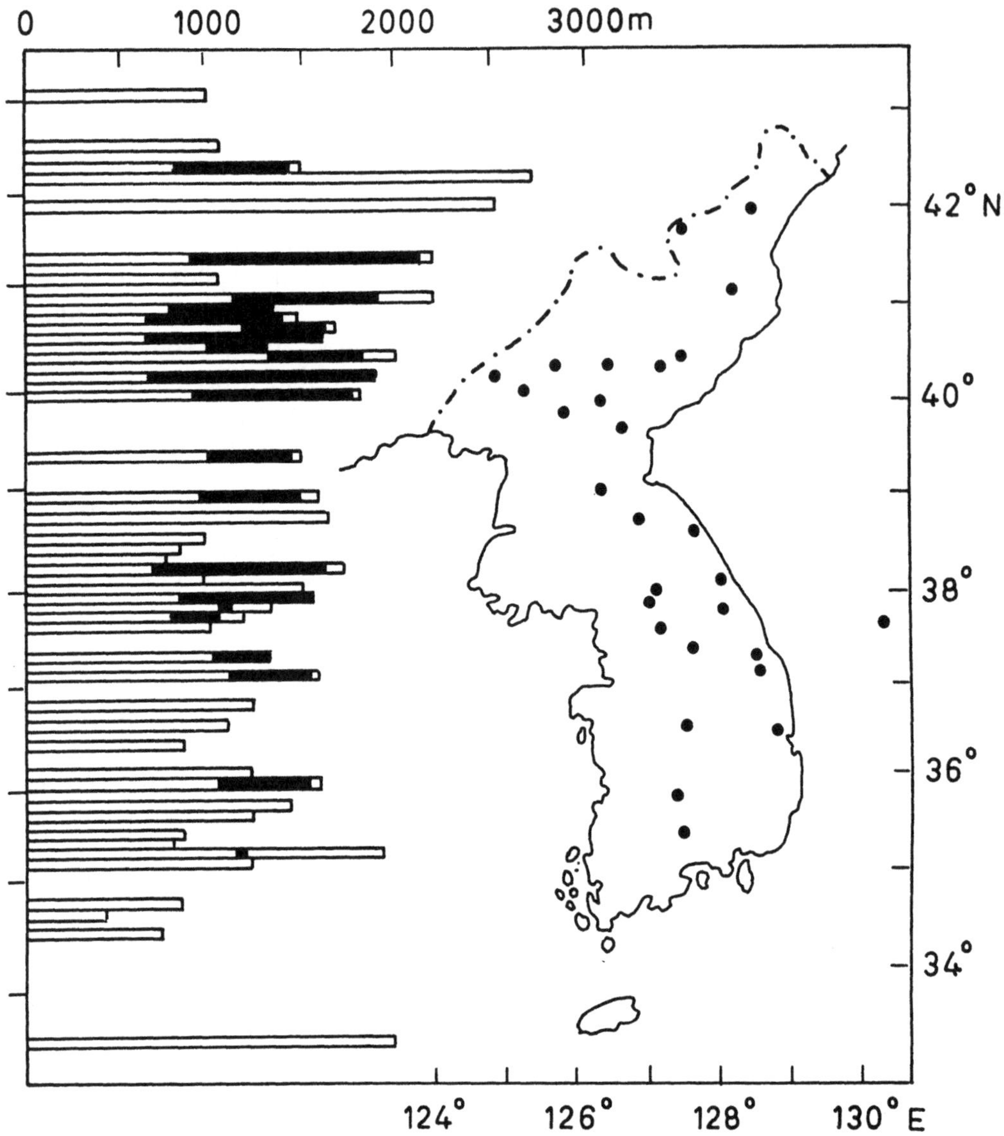

Figure 5.2. Horizontal and vertical ranges of *Abies nephrolepis* in Korea.

(Nakai, 1935; Kitagawa, 1939, 1979; Takahasi, 1944; Ohwi, 1965). In Korea, it occurs from 42°30′ N to 33° N, and it grows on Mts. Jung (300–1000), Songjin (600–1100), Mantap (800–2000), Ro, Hoochi (850–1350), Pirae, Seungjuk (800–900), Keumpae (1250–400), Myohyang (1600–1900), Keumkang, Sobaik, Pohyun and Halla (1900–1950). The average lower and upper limits of its vertical range are 1013 and 1450 m, respectively. The standard deviations of the lower and upper limits of its vertical range are 498 and 418 m (data from eight sites). Its relative *Rhododendron fauriae* form. *rufescens* grows from 41°30′ N to 33° N, and is present on many of the other mountains as well, including Mts. Mantap, Hoochi, Keumpae, Nangrim (1400–1500), Sasoo (800–1700), Haram (1100–1200), Keumkang (1000–1680). Sorak (700–1700), Kari, Odae, Taeji, Chiak, Taebaik (1300–1500), Dukyoo, Chiri and Halla. The average lower and upper limits of vertical range are 1050 and 1547 m, respectively. The standard deviations of the lower and upper limits of vertical range are 250 and 178 m (data from six sites).

Photo 5.1. Montane forest (ca. 1750 to 1800 m a.s.l.) on SE slopes of Mt. Paikdoo, Kaema Upland, North Korea. *Larix olgensis* is a dominant tree, *Picea jezoensis* and *P. koraiensis* are very frequent. *Abies nephrolepis* is rare in the canopy. (Photograph by M. Šrůtek.)

Species present from the north to the midlands

Five species or varieties fall into this category, and include *Taxus cuspidata* var. *latifolia*, *Pinus pumila*, *Thuja koraiensis*, *Rhododendron aureum* and *Vaccinium vitis-idaea* var. *minus*.

Taxus cuspidata var. *latifolia* grows in Russia (Ussuri, Sakhalin), China (Manchuria), Japan and Korea (Hu, 1934; Kitagawa, 1939, 1979). In Korea, it occurs from 41° N to 38° N, and is present on Mts. Kwanmo, Seungjuk, Pirae, Myohyang in the north, and Mt. Songin in the remote Ulreung Island in the East Sea. No vertical distributional data are available for this species.

Pinus pumila grows in Russia (Siberia, west of Bering Sea, Kamchatka, Lake Baikal area, Ussuri, Amur, Kuriles), China (Manchuria), Japan (Hokkaido, Honshu) and Korea (Wilson, 1916, 1918; Hultén, 1927; Hu, 1934; Nakai, 1935; Kitagawa, 1939, 1979; Dallimore & Jackson, 1954; Polunin, 1959; Suslov, 1961; Wang, 1961; Ohwi, 1965; Critchfield & Little, 1967; Noda, 1971; Gorchakovsky & Shiyatov, 1978; MacMahon & Anderson, 1982; Buks, 1984; Okitsu & Ito, 1984; Takhtajan, 1986; Knystautas, 1987). In Korea, it occurs from 41° N to 38° N, and is present on Mts. Mantap (2000–2200), Ro (2000–), Pirae (1350–), Seungjuk (1500–1600), Myohyang (1600–1900), Sasoo (1400–1750), Haram (1450–), Keumkang (900–1680) and Sorak (1500–1700). The average lower and upper limits of its vertical range in Korea are 1485 and 1805 m, respectively. The standard deviations of the lower and upper limits of its vertical range are 222 and 199 m, respectively (data from six sites).

Thuja koraiensis covers a small amount of territory, and it is a Korean endemic plant. It is present from 41° N to 37° N, and is located on Mts. Chilbo (1200–1700), Seungjuk (1000–1600), Nangrim (1900–2300), Ro, Keumpae (1400–1500), Peenanduk (700–900), Myohyang (1000–1800), Sasoo (400–1750), Haram (1300–1400), Chuae (1400–1500), Keumkang (800–1600), Sorak (700–1600), Whaak (1100–1400) and Taebaik (1100–1400). The average lower and upper limits of its vertical range are 1077 and 1573 m, respectively. The standard deviations of the lower and upper limits of its vertical range are 370 and 302 m respectively (data from 13 sites). *Thuja koraiensis* displays a typical distributional pattern of species present from the north to the midlands (Fig. 5.3).

Rhododendron aureum is found in Russia (Altai, Siberia, Kamchatka, Okhotsk, Amur, Ussuri, Kuriles, Sakhalin), Mongolia, China (Manchuria), Japan (Hokkaido, Henshu) and Korea (Wilson, 1923; Nakai, 1935; Kitagawa, 1939, 1979; Takahasi, 1944; Suslov, 1961; Ohwi, 1965, Knystautas, 1987). In Korea, it occurs from 42° N to 37° N, and is found on Mts. Paikdoo, Mantap (1800–2200), Ro (1700–2000), Pirae (1300–), Nangrim (1700–2300), Myohyang

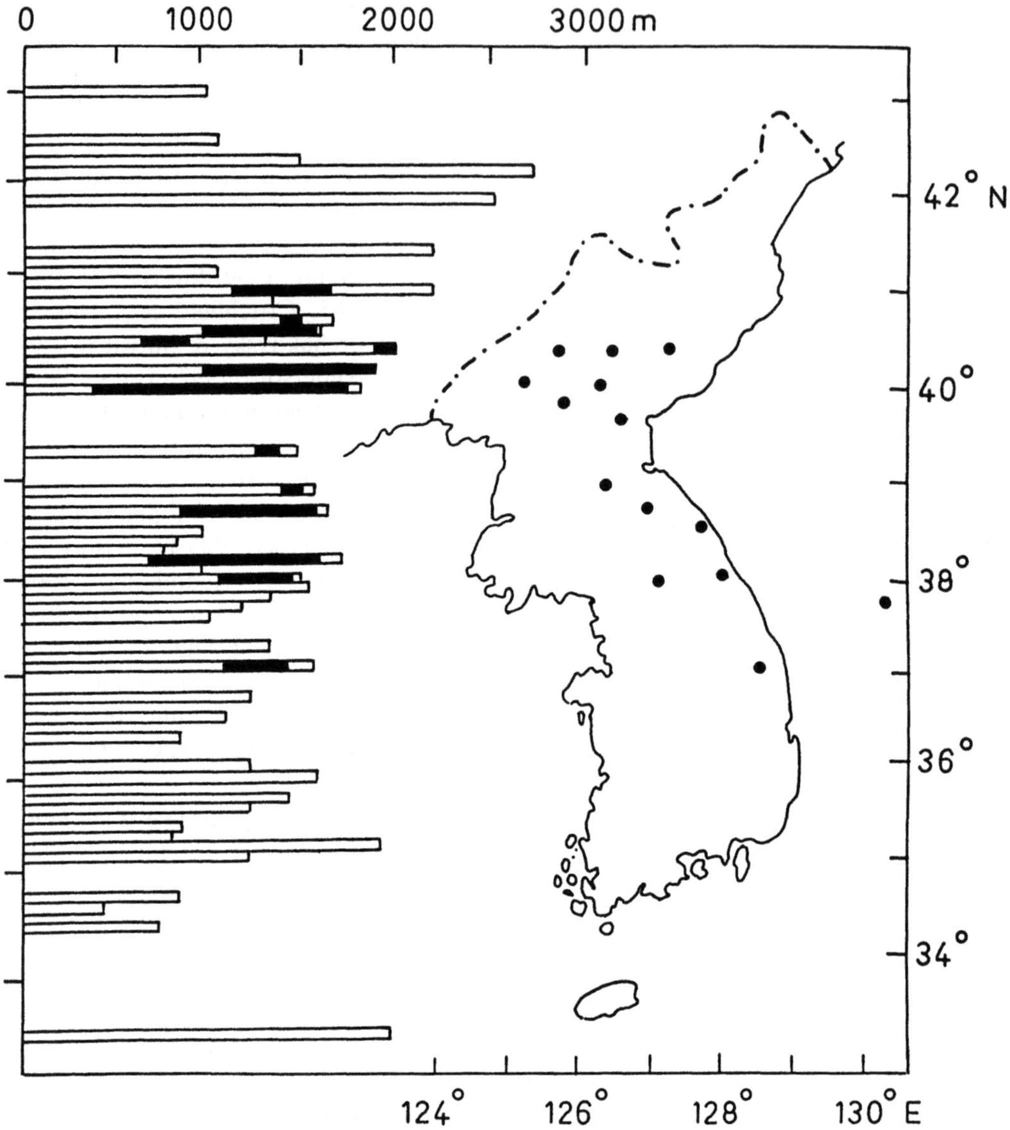

Figure 5.3. Horizontal and vertical ranges of *Thuja koraiensis* in Korea.

(1600–1700), Sasoo (1600–1800), Chuae (1400–1550) and Sorak (800–1600). The average lower and upper limits of its vertical range are 1586 and 1893 m, respectively. The standard deviations of the lower and upper limits of its vertical range are 180 and 281 m, respectively (data from seven sites).

Vaccinium vitis-idaea var. *minus* grows exceptionally widely, being found in Scandinavia, Alaska, North America, Russia (Siberia, Chukchi, Kamchatka, Sakhalin, Kuriles), Mongolia, China (Manchuria), Japan (Hokkaido, Honshu) and Korea (Hultén, 1927, 1950, 1968, 1970; Nakai, 1935; Kitagawa, 1939, 1979; Takahasi, 1944; Ritchie, 1955; Polunin, 1959, 1969; Wang, 1961; Ohwi, 1965; Tikhomirov, 1966; Huxley, 1967; Noda, 1971; Ingwersen, 1976; Bliss, 1979; Moggi, 1985; Grey-Wilson & Blamey, 1986). In Korea, however, it is restricted, being present from 42° N to 39°30′ N, and being located on Mts Chayoo (1400–), Paikdoo, Mantap (1900–2200), Chilbo (1000), Ro (1700–1200), Hoochi (1000–1300), Nangrim (1900–2200), Keumpae (1200–1600), Sasoo (1800–1860) and Keumkang (1000–

1680). The average lower and upper limits of its vertical range are 1500 and 1835 m, respectively. The standard deviations of the lower and upper limits of its vertical range are 374 and 307 m respectively (data from seven sites).

Species present in the north and in Cheju Island, but not in between

Only the species *Empetrum nigrum* var. *japonicum* is present in the north and Cheju Island, but not in between. The species grows in Russia (Kamchatka, Soviet Far East, Sakhalin), Japan (Hokkaido, Honshu), and Korea (Fernald, 1925; Good, 1927; Hultén, 1927, 1950, 1968, 1970; Nakai, 1935; Kitagawa, 1939, 1979; Takahasi, 1944; Ohwi, 1965; Tikhomirov, 1966; Huxley, 1967; Eshbaugh, 1968; Bell & Tallis, 1973; Claphan *et al.*, 1981; Shimizu, 1982). In Korea, it has been recorded only from Mts. Paikdoo and Hoochi in the north, and from Mt. Halla (1700–1900) in Cheju Island (Fig. 5.4).

Species present in the mainland south and on Cheju Island alone

Only the species *Abies koreana* is present in the mainland south and on Cheju Island alone. This is a Korean endemic species, occurring on Mts. Kaya, Dukyoo (1400–), Kaji (1000–), Chiri (1200–1900) and Moodung (500–) on the mainland, and on Mt. Halla (1200–1950) in Cheju Island (Fig. 5.5).

Species restricted to the north of Korea

Twenty-three species or varieties are restricted to the north of Korea. Five of these are Korean endemic species, namely *Picea intercedens*, *P. koraiensis*, *P. pungsanensis*, *P. tonaiensis*, and *Ledum palustre* var. *maximum*. Of these, *Picea intercedens* and *P. tonaiensis* grow at Moosan, at around 42° N, and *Picea pungsanensis* occurs at Pungsan, at around 40°30′ N. *Picea koraiensis* grows from 42° N to 40° N, and it is present on Mts. Paikdoo, Mantap (450–750), Hoochi (800–1350), Keumpae (1200–1600) and Pirae (Fig. 5.6). The average lower and upper limits of its vertical range (three sites only) are 817 and 1233 m respectively.

Ledum palustre var. *maximum* is extremely restricted in its distribution, growing on Mt. Paikdoo only, at around 42° N.

Of the more widely dispersed species, *Pinus tabulaeformis* grows in Russia (Ussuri), Mongolia, China (mid-western, northern, Manchuria) and Korea (Hu, 1934; Nakai, 1935; Kitagawa, 1939, 1979; Takahasi, 1944; Dallimore & Jackson, 1954; Den Ouden & Boom, 1965; Critchfield & Little, 1966; Richardson, 1966; Mirov, 1967). Within the Korean peninsula it occurs on Maengsan, at around 39°30′ N.

Juniperus sibirica occurs from 42° N to 41° N in Korea, including Mts. Chayoo (800–), Paikdoo (1600–) and Mantap (2300). Details of its range outside Korea are not presently forthcoming.

Sabina davurica grows in Russia (Yakutia, Dahuria, Ussuri, Amur), Mongolia, China (Manchuria) and Korea (Kitagawa, 1939, 1979; Den Ouden & Boom, 1965), being found from 42° N to 40° N, on Moosan, Kapsan and Dancheon, in the latter country.

Dryas octopetala has an extremely broad distribution, and is found in the Arctic, northern Europe, the European high mountains, the Himalayas, North America, Russia (Siberia, Chukchi, Soviet Far East, Sakhalia), Mongolia, northern China, Japan (Hokkaido, Honshu) and Korea (Schröter, 1900; Hegi, 1930; Takahasi, 1944; Hultén, 1950, 1959, 1970; Polunin, 1959; Suslov, 1961; Ohwi, 1965; Elkington, 1971; Bliss, 1979; Shimizu, 1982; Moggi, 1985; Landolt & Aeschimann, 1986). *Dryas octopetala* var. *asiatica* is more restricted in its distribution, occurring only in Japan and Korea. In Korea, it is found on Mts. Paikdoo, Ro (2200–) and Nangrim (1300–).

Andromeda polifolia also grows very widely in the Arctic, northern Europe, the southern European high mountains, northern Asia, North America, Russia (Chukchi, Kamchatka, Soviet Far East), Japan and Korea (Hultén, 1927, 1950, 1968; Nakai, 1935; Takahasi, 1944; Polunin, 1959; Ohwi, 1965; Tikhomirov, 1966; Huxley, 1967; Ingwersen, 1978; Grey-Wilson & Blamey, 1986). In Korea, its distribution is limited from 42° N to 41° N, and it is found on Moosan and Kilju. *Andromeda polifolia* var. *acerosa* also grows in the same areas.

Chamaedaphne calyculata, another broadly-ranging species in this group, grows in the Arctic,

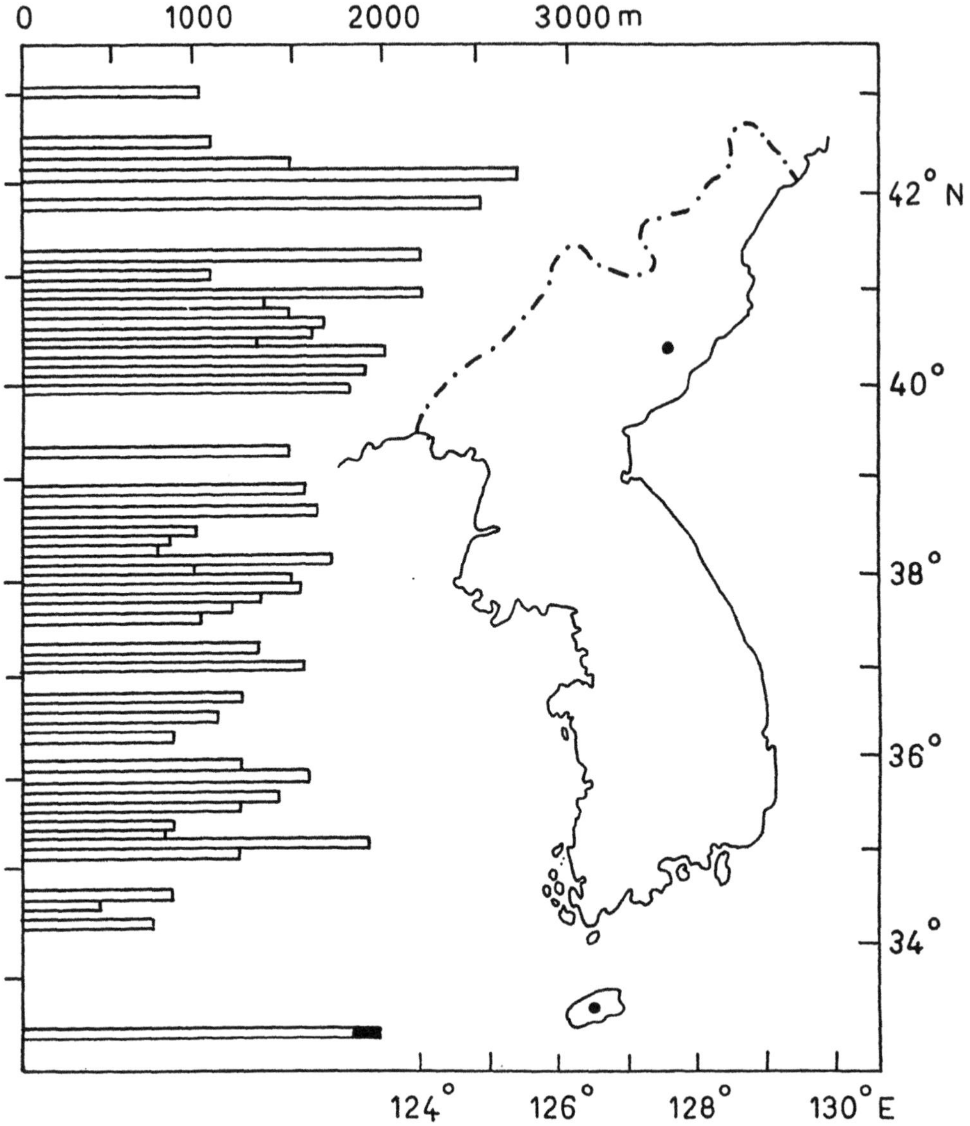

Figure 5.4. Horizontal and vertical ranges of *Empetrum nigrum* var. *japonicum* in Korea.

northwestern Europe, northern Asia, North America, Russia (Chukchi, Okhotsk, Kamchatka, Ussuri, Amur, Sakhalin), China (Manchuria), Japan (Hokkaido) and Korea (Kitagawa, 1939, 1979; Takahasi, 1944; Hultén, 1950, 1968, 1970; Polunin, 1959; Ohwi, 1965; Fitter *et al.*, 1985). In Korea, it is located from 42° N to 41° N, again on Moosan and Kilju.

Ledum palustre var. *latifolia* again grows widely, in northern Europe, western Greenland, North America, Russia (Siberia, Sakhalin) and Korea (Polunin, 1959; Clapham *et al.*, 1981). Within the peninsula, it occurs from 42° N to 40° N, and it is present on Moosan, Mts. Paikdoo, Kwanmo, Mantap, Chilbo and Hoochi.

Ledum palustre var. *diversipilosum* shows a similarly broad distributional range to *L. palustre* var. *latifolia*. In Korea, it grows from 42° N to 40° N, and it is present on Mts. Paikdoo, Kwanmo, Mantap (1100–1600), Chilbo (1000–), Hoochi (800–1100), Keumpae (1200–1600) and Ro (1700–2000). The average lower and upper limits of its vertical range (four sites only) are 1200 and 1575 m respectively.

Photo 5.2. Korean endemic species *Abies koreana* on Mt. Chiri (1915 m). (Photograph by J.M. Kim.)

Ledum palustre var. *minus* grows in Russia (Dahuria), China (Manchuria) and Korea (Kitagawa, 1939, 1979), in which in the latter territory it is found only on Mt. Paikdoo (1600–2300).

Ledum palustre subsp. *decumbens* occurs in North America, Russia (Siberia, Kamchatka, Soviet Far East, Sikhote-Alin, Kuriles), Japan (Hokkaido) and Korea (Nakai, 1935; Ohwi, 1965; Shimizu, 1982). In Korea, it is found only on Mt. Paikdoo (1800–2540 m). In Korea, *Ledum palustre* var. *yesoense* occurs only on Mt. Kwanmo.

Oxycoccus microcarpus and *O. quadripetalus* are other widely-present species representatives of this group, growing in north-western Europe, the southern European high mountains, Greenland, Arctic Eurasia, North America, Russia (Siberia, Kamchatka, Soviet Far East, Sakhalin, Ku-

Photo 5.3. Abies koreana.

riles), Japan (Hokkaido, Honshu) and Korea (Kitagawa, 1939, 1979; Camp, 1944; Hultén, 1950, 1968, 1970; Ohwi, 1965; Huxley, 1967; Clapham *et al.*, 1981; Shimuzu, 1982). In Korea, they occur from 42° N to 41° N, and are present on Moosan and Kilju.

Phyllodoce caerulea grows very widely in the Arctic, northern Europe, the southern European high mountains, North America, northern Eurasia, Russia (Siberia, Chukchi, Kamchatka, Soviet Far East, Sakhalin), Japan (Hokkaido) and Korea (Hultén, 1927, 1950, 1958, 1968; Nakai, 1935; Kitagawa, 1939, 1979; Takahasi, 1944; Polunin, 1959; Ohwi, 1965; Tikhomirov, 1966; Huxley, 1967; Bliss, 1979; Shimizu, 1982) In Korea, it occurs from 42° N to 40°30′ N, and it is located on Mts Chayoo,Paikdoo (1540–1800), Kwanmo and Hoochi.

Rhododendron parvifolium is found in northern Europe, Russia (Altai, Siberia, Verkhoyansk, Chukchi, Okhotsk, Amur, Sakhalin), Alaska, China (Manchuria), Japan (Hokkaido) and Korea (Wilson, 1923; Kitawaga, 1939, 1979; Ohwi, 1965; Tikhomirov, 1966; Hultén, 1968; Takhtajan, 1986). In Korea, it occurs on Mts. Paikdoo, Hoochi (800–1100) and Keumpae.

In Korea, *Vaccinium vitis-idaea* var. *genuinum* grows on Kilju, at around 40° N.

Linnaea borealis grows in the southern European high mountains, northern Eurasia, Russia (Chukchi, Kamchatka, Soviet Far East, Siberia Sakhalin, Kuriles), China (Manchuria), Japan (Hokkaido, Honshu) and Korea (Hagerup, 1914; Hultén, 1927, 1950, 1968, 1970; Nakai, 1935; Kitagawa, 1939, 1979; Takahasi, 1944; Polunin, 1959; Ohwi, 1965; Huxley, 1967; Noda, 1971; Grey-Wilson & Blamey, 1986). In Korea, it is present on Mts. Paikdoo, Kwanmo, Mantap and Ro.

Species restricted to the midlands of Korea

Two species grow solely in the midlands of Korea, namely *Taxus caespitosa* and *Picea schrenkiana*.

Taxus caespitosa is a Korean endemic species, which occurs from 39° N to 38° N, and it is found solely on Mts. Keumkang and Sorak (Fig. 5.7).

In Korea, *Picea schrenkiana* occurs on Mt.

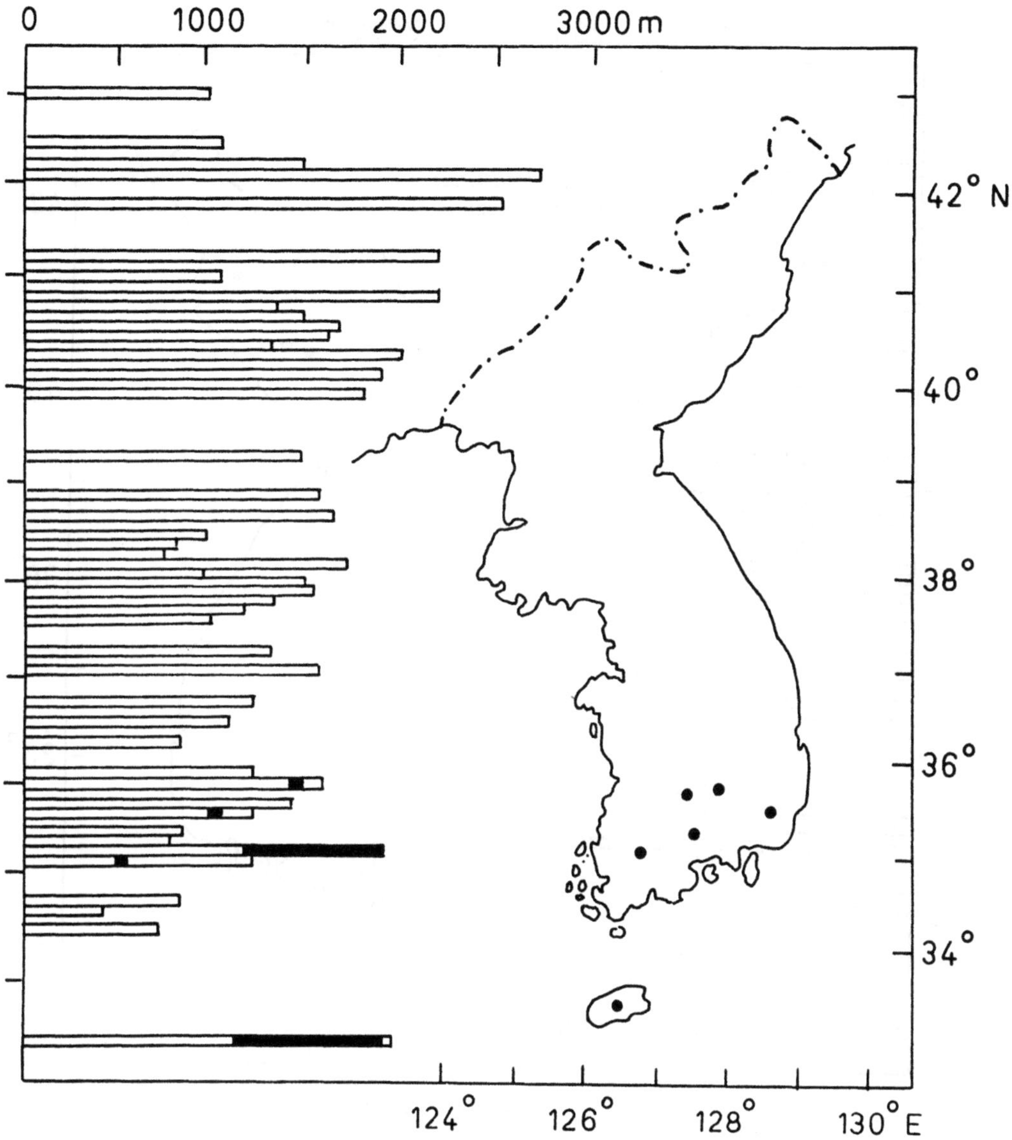

Figure 5.5. Horizontal and vertical ranges of *Abies koreana* in Korea.

Odae alone, at 37°40′ N. No information is presently forthcoming as to its distribution outside of Korea.

Species restricted to Cheju Island

Though *Diapensia lapponica* grows widely worldwide in Alaska, Russia (Chukchi, Kamchatka, Sakhalin), Japan (Hokkaido, Honshu) and Korea (Hultén, 1927, 1958, 1968; Nakai, 1935; Baldwin, 1939; Li, 1952; Polunin, 1959; Ohwi, 1965; Tikhomirov, 1966; Ingwersen, 1978; Day & Scott, 1984; Heywood, 1985), in Korea, a subspecies, *Diapensia lapponica* subsp. *obovata*, occurs solely on Mt. Halla (1800–1950), in Cheju Island (Fig. 5.8). This subspecies is also present on mountains in Japan (Ohwi, 1965; Shimizu, 1982).

Overall, out of the 41 evergreen arboreal arctic-alpine and alpine species selected in Korea, 23 species are restricted to the north, 8 species are present throughout the peninsula, 5 species are present from the north to the midlands, 2 species

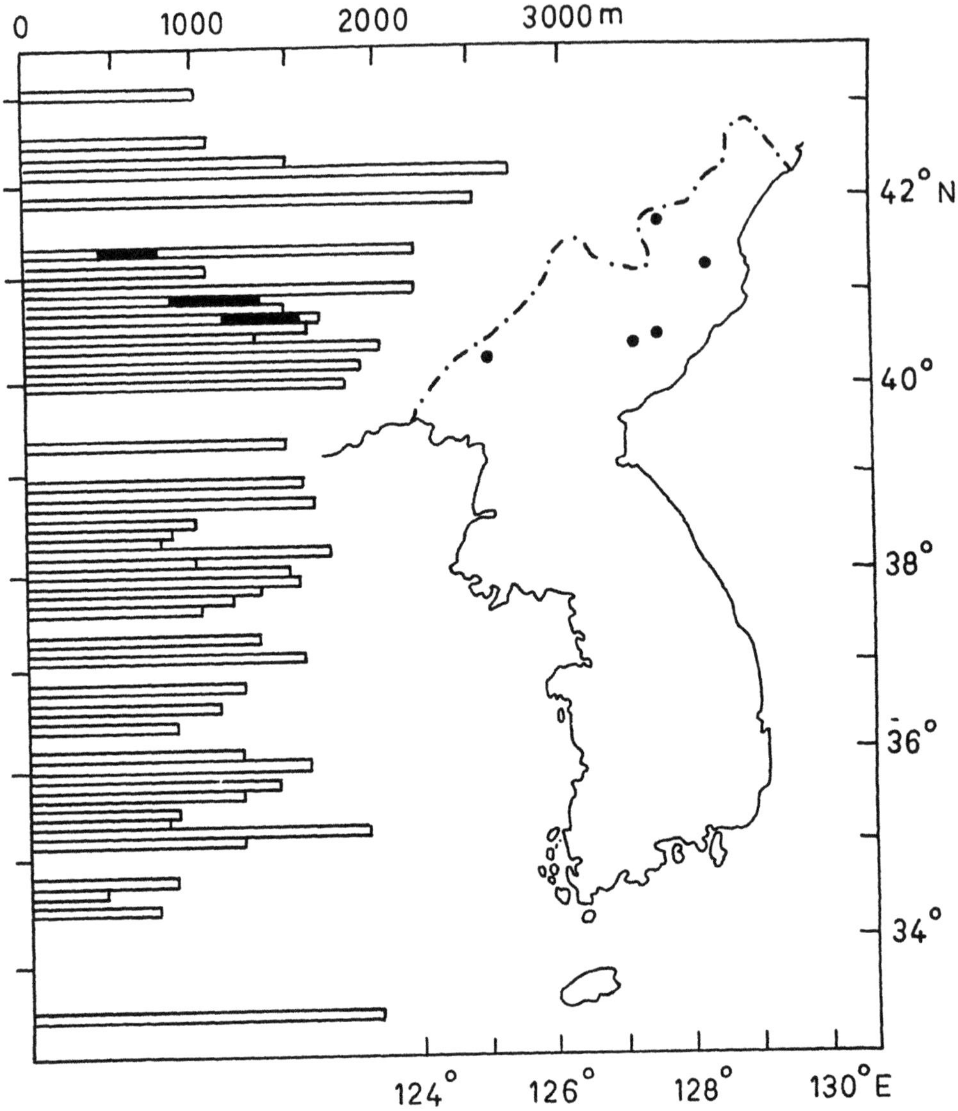

Figure 5.6. Horizontal and vertical ranges of *Picea koraiensis* in Korea.

are restricted to the midlands, 1 species is present in the north and in Cheju Island (but not in between), 1 species is present in the midland south and Cheju Island, and 1 species is restricted to Cheju Island. Thirty-eight species (92.7% of the total evergreen arboreal arctic-alpine and alpine plants in Korea considered in this chapter) grow in the north of the Korean peninsula. It is likely that this is a response partly to the frequent floristic exchanges of these floras with those of the Arctic and other parts of East Asia during alternate Pleistocene warm and cold phases, but it is also undoubtedly due to the availability of a relatively large number of suitable sites in the north of Korea.

In the situation described above, in which one considers the distribution and vertical ranges of Korean arctic-alpine and alpine plants, the maximum vertical range of which may, in many cases, be constrained by the heights of the mountains themselves, or by the scarcity of information relating to the upper limits of the species, it may make more sense to consider the lower limit of their altitudinal range as an aid to illustrating their biogeography.

The mean values of lower limits for 23 ever-

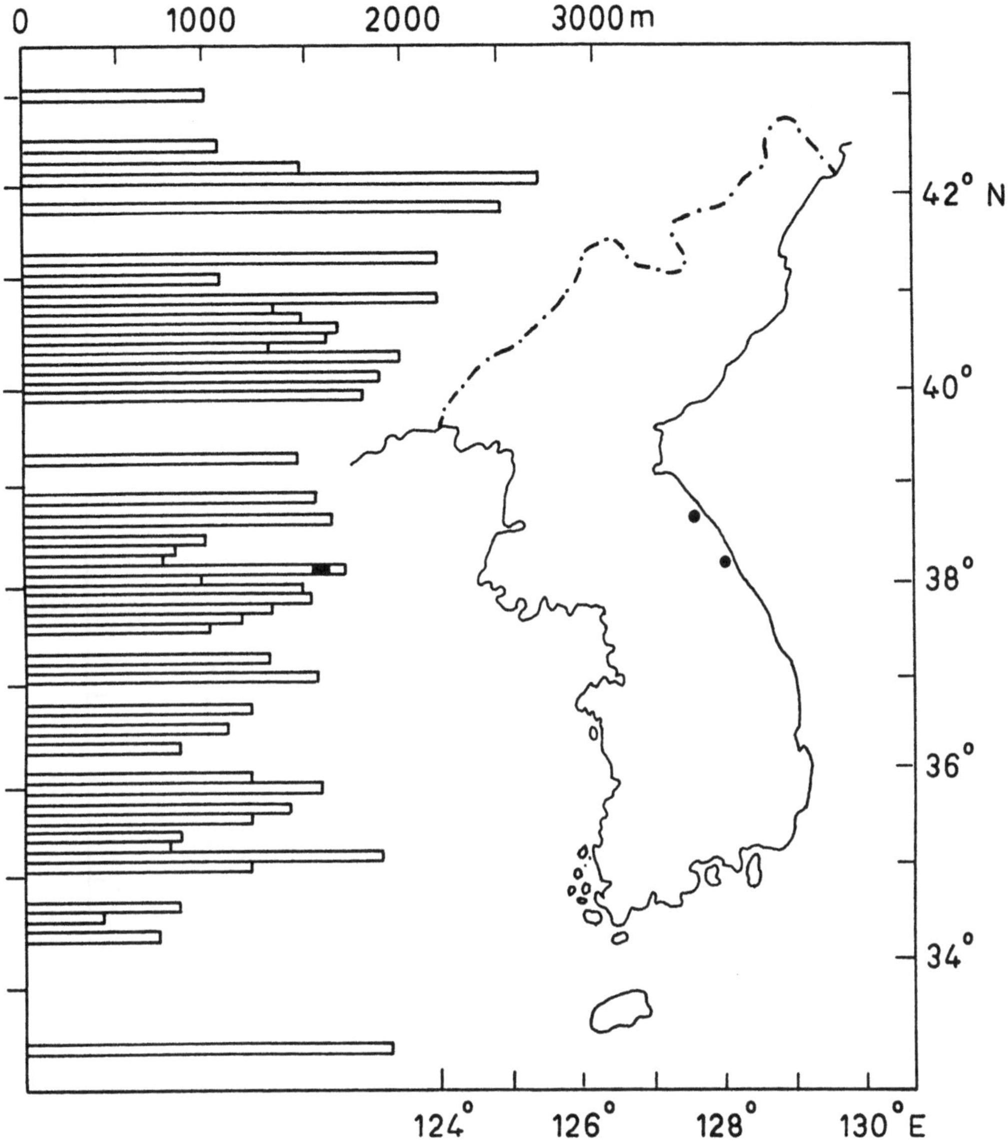

Figure 5.7. Horizontal and vertical ranges of *Taxus caespitosa* in Korea.

green arctic-alpine and alpine plants in Korea (Table 5.2) display four distinct vertical distributional patterns. First, those at 1500 to 1800 m a.s.l. or above are dominated by evergreen broadleaved plants, such as *Ledum palustre* subsp. *decumbens*, *Diapensia lapponica* subsp. *obovata*, *Dryas octopetala* var. *asiatica*, *Empetrum nigrum* var. *japonicum*, *Ledum palustre* var. *minus*, *Rhododendron aureum*, *Phyllodoce caerulea* and *Vaccinium vitis-idaea* var. *minus*. Second, those at 1000 to 1500 m a.s.l. consist of a mixture of evergreen coniferous plants and evergreen broadleaved plants, namely *Pinus pumila*, *Sabina sargentii*, *Ledum palustre* var. *diversipilosum*, *Rhododendron fauriae* form. *rufescens* and *R. dauricum*. Third, those at ca. 1000 m a.s.l. contain two endemic evergreen coniferous plants, including *Thuja koraiensis* at 1077 m a.s.l., and *Abies koreana* at 1060 m a.s.l. Fourth, those at 500 to 1000 m a.s.l. are dominated by evergreen coniferous plants, namely *Picea jezoensis*, *Taxus cuspidata*, *Abies nephrolepis*, *Picea koraiensis*, *Pinus koraiensis* and *Abies holophylla*. Overall, evergreen broadleaved plants dominate the highest parts of many mountains in Korea, and a mixture of evergreen broadleaved plants and

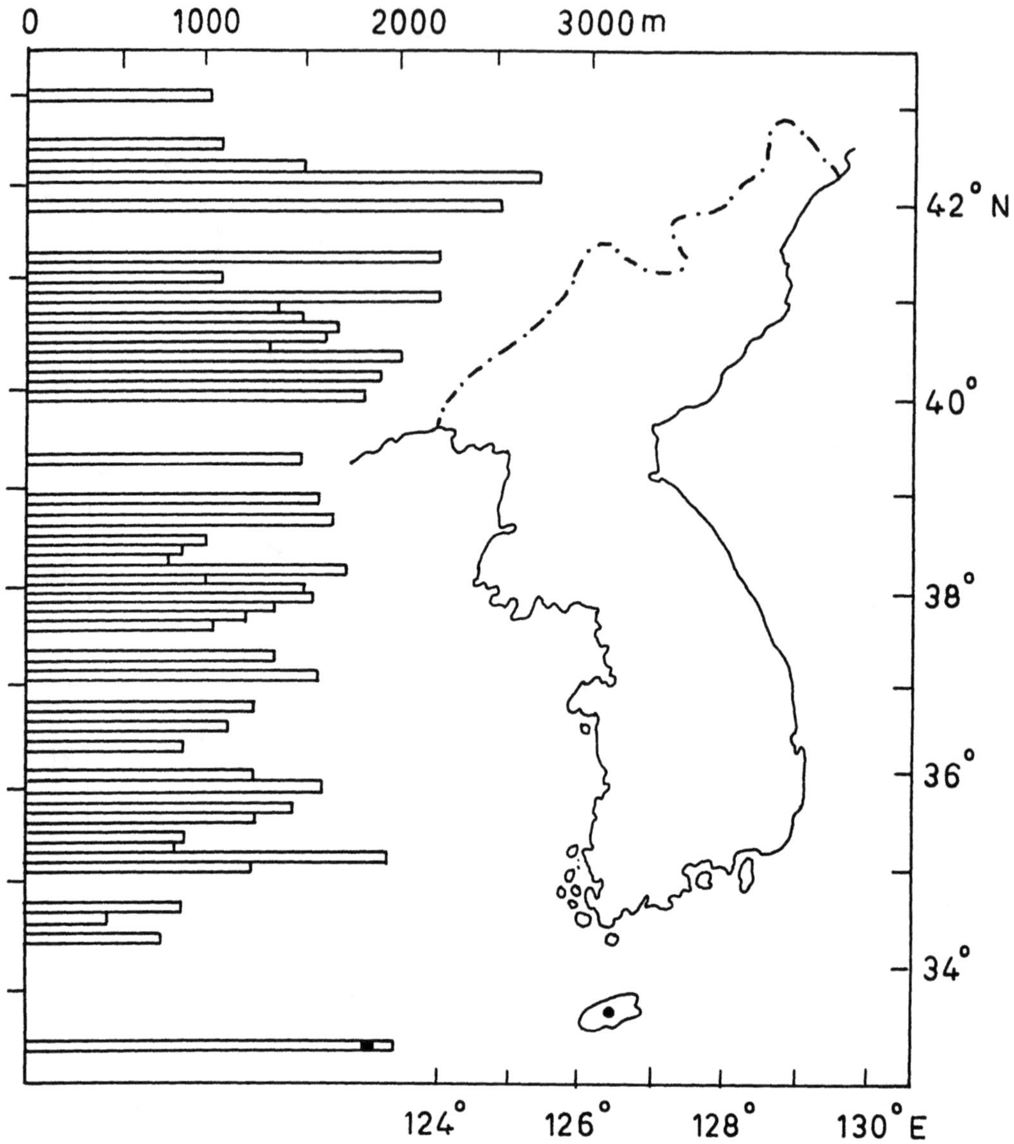

Figure 5.8. Horizontal and vertical ranges of *Diapensia lapponica* subsp. *obovata* in Korea.

evergreen coniferous plants is more common below this. Endemic evergreen coniferous plants appear to exist in the transitional zone between this latter zone and that of the evergreen coniferous plants below; and the arctic-alpine and alpine evergreen coniferous plants which grow on the lower parts of these mountains form a mixed forest with other more warmth-loving deciduous plants. A summary of this situation is given in Table 5.3.

The biogeographical implications of these plant ranges are interesting. First, while many evergreen arboreal broadleaved arctic-alpine and alpine plants are common elsewhere (except for *Diapensia lapponica* subsp. *obovata* and *Dryas octopetala* var. *asiatica*), they are very restricted in Korea, being located mainly in the north (except for *Diapensia lapponica* subsp. *obovata* on Cheju Island) at the heights of 1500 to 1800 m a.s.l. On a worldwide distribution, most of these species occur at the southernmost limit of their range in Korea, and they are now endangered because of the competition from downslope plants, and also from environmental changes,

Table 5.2. Mean values of vertical ranges for selected evergreen arctic-alpine and alpine plants in Korea.

Species name	Physiognomy	Lower limit (m)	Upper limit (m)	No. sites	Distribution
Ledum palustre subsp. *decumbens*	E.B.P.	1800	2540	1	N
Diapensia lapponica subsp. *obovata*	E.B.P.	1800	1950	1	S
Dryas octopetala var. *asiatica*	E.B.P.	1750	–	2	N
Empetrum nigrum var. *japonicum*	E.B.P.	1700	1900	1	N–C
Ledum palustre var. *minus*	E.B.P.	1600	2300	1	N
Rhododendron aureum	E.B.P.	1586	1893	7	N–M
*Juniperus sibirica?**	E.C.P.	1567	–	3	N
Phyllodoce caerulea	E.B.P.	1540	1800	1	N
Vaccinium vitis-idaea var. *minus*	E.B.P.	1500	1835	7	N–M
Pinus pumila	E.C.P.	1485	1805	6	N–M
Sabina sargentii	E.C.P.	1367	1658	6	N–M–S–C
Ledum palustre var. *diversipilosum*	E.B.P.	1200	1575	4	N
*Thuja koraiensis**	E.C.P.	1077	1573	13	N–M
*Abies koreana**	E.C.P.	1060	1900	5	S–C
Khododendron fauriae form. *rufescens*	E.B.P.	1050	1547	6	N–M–S–C
R. dauricum	E.B.P.	1013	1450	8	N–M–S–C
Picea jezoensis	E.C.P.	989	1646	14	N–M–S
Taxus cuspidata	E.C.P.	920	1507	14	N–M–S–C
Abies nephrolepis	E.C.P.	918	1507	19	N–M–S
*Picea koraiensis**	E.C.P.	817	1233	3	N
Rhododendron parvifolium	E.B.P.	800	1100	1	N
Pinus koraiensis	E.C.P.	632	1308	28	N–M–S
Abies holophylla	E.C.P.	493	1008	23	N–M–S–C

Abbreviations: E.B.P., Evergreen Broadleaved Plant; E.C.P., Evergreen Coniferous Plant; N, North; M, Midland; S, South; C, Cheju.
* Endemic species.

Photo 5.4. Rhododendron aureum (light coloured) is an important species of the alpine flora of the Korean part of Mt. Paikdoo. It is frequent along the periodic streams. (Photograph by M. Šrůtek.)

Table 5.3. Vertical groupings and distributional ranges of evergreen arboreal arctic-alpine and alpine species of Korea with known lower vertical limits.

Vertical grouping	Mean lower limit of spp.	Species names	General range	Ranges in Korea	
	1800 m	*Ledum palustre* subsp. *decumbens*	Wide	N	1 mt.
	1800 m	*Diapensia lapponica* subsp. *obovata*	Narrow	C	1 mt.
	1750 m	*Dryas octopetala* var. *asiatica*	Narrow	N	3 mts.
1500–1800 m	1700 m	*Empetrum nigrum* var. *japonicum*	East Asia	N&C	3 mts.
	1600 m	*Ledum palustre* var. *minus*	East Asia	N	1 mt.
	1586 m	*Rhododendron aureum*	East Asia	N–M	9 mts.
	1540 m	*Phyllodoce caerulea*	Wide	N	4 mts.
	1500 m	*Vaccinium vitis-idaea* var. *minus*	Wide	N–M	10 mts.
	1485 m	*Pinus pumila*	East Asia	N–M	9 mts.
1000–1500 m	1367 m	*Sabina sargentii*	East Asia	N–C	12 mts.
	1200 m	*Ledum palustre* var. *diversipilosum*	Wide	N	7 mts.
	1050 m	*Rhododendron fauriae* form. *rufescens*	East Asia	N–C	16 mts.
	1013 m	*Rhododendron dauricum*	East Asia	N–C	13 mts.
ca. 1000 m	1077 m	*Thuja koraiensis*	Endemic	N–M	14 mts.
	1060 m	*Abies koreana*	Endemic	S–C	6 mts.
	989 m	*Picea jezoensis*	East Asia	N–S	14 mts.
	920 m	*Taxus cuspidata*	East Asia	N–C	20 mts.
500–1000 m	918 m	*Abies nephrolepis*	East Asia	N–S	27 mts.
	817 m	*Picea koraiensis*	Endemic	N	5 mts.
	632 m	*Pinus koraiensis*	East Asia	N–S	48 mts.
	493 m	*Abies holophylla*	East Asia	N–C	42 mts.

Abbreviations: N, North of Korea; M, Midland of Korea; S, South of Korea; C, Cheju Island, Korea; Mts., Mountains.

such as the climatic amelioration associated with the greenhouse effect. Second, slightly more thermophilous evergreen coniferous and broadleaved plants again are common, mainly in East Asia, but at the same time are more widely distributed in Korea, from the north to the south at above 1000 to 1500 m a.s.l.; they are also less endangered as compared with the evergreen broadleaved arctic-alpine and alpine plants. Third, Korean endemic evergreen coniferous plants are found at ca. 1000 m a.s.l.; and this might be the result of evolution or hybridisation at the top of the evergreen coniferous zone. A similar pattern has been noticed for the horizontal transitional zone of biogeographical divisions in Chapter 2 (see p. 42). Fourth, relatively thermophilous evergreen coniferous plants are common in East Asia, and also widely distributed in Korea above 500 to 1000 m a.s.l., where they are often mixed with deciduous broadleaved plants. The appearance of the Korean endemic coniferous plant *Picea koraiensis* at 817 m a.s.l. in the north is interesting, but nothing is yet known about its local climatic and environmental situation. These several points also have implications for the relationship between the vertical distribution of evergreen arboreal arctic-alpine and alpine plants and climate, especially on Mt. Sorak, which is to be further considered in Chapter 6.

Physiognomy: general comments

Physiognomy, a study of the growth forms of plants, is particularly important in any examination of their relationships with their habitat, vegetation structure and physiological processes. According to Meusel *et al.* (1977) and Box (1981b), the analysis of physiognomy is a means of interpreting the ways in which vegetation changes along environmental gradients, and also provides important information as to the mechanisms of environmental limitation for them. The most useful parameters are morphology, life form, leaf form and phenology.

In respect of morphology, there are a number of significant general differences between plants which are strictly 'arctic' and those which are 'alpine' in their distribution. Some of these are listed below. First, there is a decrease in the average number of inflorescent branches, from south to north, and an increase of rhizome production in northern populations (Mooney & Billings, 1961). Second, a development of underground reproductive structures and an appearance of leaf hairs is common in alpine plants (Wehrmeister & Bonde, 1977); and third, a compact and mossy form of growth, which forms as an adjustment to the extreme drought and cold climates of alpine plant communities often is not common in the arctic region (Darlington, 1978; Foster, 1982). There are also, however, a few morphological similarities between 'arctic' and 'alpine' tundra plant communities, such as first, the customary lack of both trees and large bushes, and the tendency of small bushes to form a hemispherical cushion form, in which a large number of branches congregate together (Löve & Löve, 1975a); second, often a large number of dwarf shrubs prevail, which show mainly rosette, cushion and matted forms, and grow close to the ground (Prime, 1978); and third, there is an abundance of dwarf evergreen shrubs, small perennial herbs, bryophytes and lichens in arctic and alpine areas (Grime, 1979). In general, the physiognomic characteristics of arboreal alpine plants *per se* include: first, roots which are shallow, but long and well developed; second, branches which are often short and prostrate, with suppressed internodes, all forming a compact vegetal layer, which sometimes result in krummholz and mat forms; third, leaves which customarily are small, narrow and/or needle-like, often with a rigid, thick, coriaceous texture, and at times revolute edges, and abundant epidermal hairs on their surfaces; and finally, flowers which are frequently bright and large in order to encourage rapid pollination in a short growing season. More detailed information on the life form, leaf form and flower pattern of arctic-alpine and alpine plants in Korea is given in Table 5.4.

Evergreenness

Evergreenness has many advantages for growth in an harsh alpine environment. First, evergreen leaves are physiologically more active than deciduous leaves over a longer time during the year, and so are better able to store food and nutrients over a longer period (Bliss, 1962b; Monk, 1966); second, the life span of those evergreen leaves is normally longer than a year, which aids year-round photosynthesis; third, a late break in bud dormancy in evergreen plants generally avoids damage to them by end-of-season frosts (Billings & Mooney, 1968; Zwinger & Willard, 1972; Wielgolaski & Karenlampi, 1975; Wielgolaski, 1986); fourth, their stems and roots are perennial, and consequently there is no need for them annually to produce these, using vital nutrients and energy, in one short growing season (Zwinger & Willard, 1972; McCown, 1975; Raven & Walters, 1984; Haber, 1986); and fifth, they tend to have a low leaf conductance, which is immensely advantageous to growth and survival, as compared to the seasonality of deciduous alpines (Oberbauer & Miller, 1981). All the above-mentioned characteristics together are likely to be responsible for the prevalence of evergreen arboreal arctic-alpine and alpine species on impoverished mountain sites and at timberline.

Life form

Of the 41 selected evergreen arboreal arctic-alpine and alpine plants in Korea, 19 species belong to the conifers and taxads, viz. *Taxus* (3 spp.), *Abies* (3 spp.), *Picea* (6 spp.), *Pinus* (3 spp.), *Thuja* (1 sp.), *Juniperus* (1 sp.) and *Sabina* (2 spp.). In general, evergreen conifers are more efficient in their physiological processes than deciduous hardwoods in the temperate zone in several respects. First, they assimilate carbon dioxide throughout the year, in favourable environments (Bourdeau, 1959); second, their needlelike leaves reduce rates of evapotranspiration (Zwinger & Willard, 1972; Golte, 1976; Waring & Franklin, 1979); third, they are normally well-adapted to withstand both moderate and severe freezing temperatures in winter, and often too a scarcity of water during the summer (Box, 1981b; Bell & Woodcock, 1983; Eyre, 1984; Polunin & Walters,

Table 5.4. Physiognomy of selected evergreen arctic-alpine and alpine plants in Korea.

Species name	Life form	Height	Leaf size	Leaf shape
Taxus caespitosa	Shrub (decumbent)	–	–	Linear
T. cuspidata	Tree	17 m	1.5–2 cm	Linear
T. cuspidata var. *latifolia*	Tree	17 m	1.5–2 cm	Linear
Abies holophylla	Tree	40 m	4 cm	Linear
A. koreana	Tree	18 m	4–9 cm	Linear
A. nephrolepis	Tree	25 m	4 cm	Linear
Picea intercedens	Tree	–	–	Linear
P. jezoensis	Tree	40 m	1–2 cm	Linear
P. koraiensis	Tree	25 m	1.2–2 cm	Acicular
P. pungsanensis	Tree	20 m	1.2–2.5 cm	–
P. schrenkiana	Tree	–	–	Linear
P. tonaiensis	Tree	–	–	–
Pinus koraiensis	Tree	30 m	7–12 cm	Acicular
P pumila	Small tree (decumbent)	4–5 m	3–6 cm	Acicular
P. tabulaeformis	Tree	25 m	10–17 cm	Acicular
Thuja koraiensis	Small tree	1–2 m	1.8–2.4 cm	Rhomboid
Juniperus sibirica	Shrub (decumbent)	–	–	Linear
Sabina davurica	Shrub (decumbent)	–	0.6–0.8 cm	Acicular
S. sargentii	Shrub (procumbent)	–	0.3–0.5 cm	
Dryas octopetala var. *asiatica*	Shrub (decumbent)	–	–	–
Diapensia lapponica subsp. *obovata*	Small shrub (procumbent)	–	0.7–1.5 cm	Obovate
Empetrum nigrum var. *japonicum*	Small shrub (decumbent)	0.1–0.2 m	0.5–0.6 cm	Linear
Andromeda polifolia	Small shrub	0.1–0.3 m	1.5–3 cm	Linear
A. polifolia form. *acerosa*	Small shrub (decumbent)	0.1–0.3 m	1.5–3 cm	Linear
Chamaedaphne calyculata	Shrub	0.3–1 m	1–5 cm	Oblong
Ledum palustre var. *diversipilosum*	Small shrub	0.1–0.7 m	2–7 m	Oblong
L.p. var *latifolium*	Shrub	0.6 m	–	Lanceolate
L.p. var *maximum*	Shrub	–	7 cm	Lanceolate
L.p. var. *minus*	Shrub	–	–	Lanceolate
L.p. subsp. *decumbens*	Shrub (decumbent)	–	1.8 cm	Linear
L.p. var. *yesoense*	Shrub	–	–	Lanceolate
Oxycoccus microcarpus	Small shrub	–	0.3–0.6 cm	Ovate
O. quadripetalus	Small shrub (procumbent))	0.2 m	–	Oblong
Phyllodoce caerulea	Small shrub (decumbent)	0 1–0 25 m	0 4–1.0 cm	Linear
Rhododendron aureum	Shrub	1 m	3–8 cm	Elliptical
R. dauricum	Shrub	1–2 m	1–5 cm	Elliptical
R fauriae form. *rufescens*	Shrub	4 m	8–20 cm	Elliptical
R. parvifolium	Shrub	1–1.5 m	0.5–2 cm	Elliptical
Vaccinium vitis-idaea var. *genuinum*	Small shrub	0.2–0.3 m	1–3 cm	Elliptical
V v-i var. *minus*	Small shrub (procumbent)	0.2–0.3 m	1–3 cm	Ovate
Linnaea borealis	Small shrub (procumbent)	–	0.7–1.5 cm	Orbicular

1985; Riley & Young, 1987; Melzack & Watts, 1982a,b).

On the basis of height, the life forms of the 41 Korean evergreen arboreal arctic-alpine and alpine plants may be divided into two groups, after the schema worked out by Box (1981a). First, 15 species belong to a group of trees which are taller than 5 metres in height; and second, 26 species belong to shrubs which are smaller than 5 metres (Table 5.4). Among the 15 trees in the first group, 13 species are conifers, namely *Taxus cuspidata*, *T. cuspidata* var. *latifolia*, *Abies holophylla*, *A. koreana*, *A. nephrolepis*, *Picea intercedens*, *P. jezoensis*, *P. koraiensis*, *P. pungsanensis*, *P. schrenkiana*, *P. tonaiensis*, *Pinus koraiensis* and *P. tabulaeformis*. The remaining two conifers, *Pinus pumila* and *Thuja koraiensis*, display a decumbent small tree form. The other four species of conifer provide a decumbent or procumbent shrubby growth, e.g. *Taxus caespi-*

Table 5.4. Continued.

Leaf texture	Leaf trichome	Flowering period	Flower colour	Maturing seaaon	Fruit shape
–	–	April-May	–	Sept.	Nut
–	–	April	Dark pink	Aug.–Sept.	Drupe
–	–	April	Dark pink	Sept.	Drupe
–	–	April	Yellowish brown	Oct.	Cone
–	–	June	Dark purple	Sept.–Oct.	Cone
–	–	May	Purple	Sept.	Cone
–	–	–	–		Cone
–	–	June	Yellowish brown	Sept.	Cone
–	–	May	–	Oct.	Cone
–	–	June	–	Oct.	Cone
–	–	–	–	–	Cone
–	–	–	–	–	Cone
–	–	May	Dark green	Following Oct.	Cone
	–	June–July	Purple	Following Sept.	Cone
–	–	May	Purple	Following Sept.	Cone
–	–	May	Yellow	Sept.	Cone
–	–	May	–	Following Oct.	Cone
–	–	April	–	Oct.	Cone
–	–	May	–	Following Oct.	Cone
Subcoriaceous	Hairs	–	White	July–Aug.	Drupe
Coriaceous	No	July	White	Autumn	Capsule
Thick & revolute	Hairs	May	Purple	Aug.–Sept.	Berry
Coriaceous & revolute	No	May–June	White	Autumn	Capsule
Coriaceous & revolute	No	May–June	White	Autumn	Capsule
Coriaceous	Hairs	April–June	White	Autumn	Capsule
Revolute	Hairs	May–June	White	Sept.–Oct.	Capsule
–	Hairs	May	White	Sept.–Oct.	Capsule
–	Hairs	May	White	Sept.–Oct.	Capsule
–	Hairs	Summer	White	Autumn	Capsule
Revolute	Hairs	May	White	Autumn	Capsule
Revolute	Hairs	July	Light pink	Autumn	Berry
Revolute	No	July	Light pink	Autumn	Berry
Revolute	No	July–Aug.	Pink purple	Sept.	Capsule
Coriaceous & revolute	No	May–June	Yellow	Sept.	Capsule
Coriaceous	Scales	April	Light pink	Oct.	Capsule
Coriaceous & revolute	Hairs	July	White	Sept.	Capsule
Coriaceous	Hairs & scales	May–June	Pink	Sept.–Oct.	Capsule
Revolute	No	July	White	Sept.	Berry
Coriaceous & revolute	No	May–June	White	Aug.–Sept.	Berry
–	Hairs	July	White	Oct.	Subglobuse

tosa, *Juniperus sibirica*, *Sabina davurica* and *S. sargentii*.

Six different types of shrub form can be seen in the above four evergreen arboreal coniferous species, and the remaining 22 evergreen broad-leaved arboreal alpine plants. First, a 'standard' shrub form is found in 9 species (*Chamaedaphne calyculata*, *Ledum palustre* var. *latifolium*, *L. palustre* var. *maximum*, *L. palustre var.* minus, *L. palustre* var. *yesoense*, *Rhododendron aureum*, *R. dauricum*, *R. fauriae* form. *rufescens* and *R. parvifolium*). Second, a 'decumbent' shrub form is adopted by 5 species (*Taxus caespitosa*, *Juniperus sibirica*, *Sabina sargentii*, *Dryas octopetala* var. asiatica and *Ledum palustre* subsp. *decumbens*). Third, a 'small' shrub form of less than 1 metre in height is found in 4 species (*Andromeda polifolia*, *Ledum palustre* var. *diversipilosum*, *Oxycoccus microcarpus* and *Vaccinium vitis-idaea* var. *genuinum*). Fourth, a 'procumbent small' shrub form includes 4 species (*Diapensia lapponica* subsp. *obovata*, *Oxycoccus quadripet-*

Photo 5.5. Dead trunks of *Abies koreana* and *Rhododendron* community in spring on Mt. Halla (1950 m.)

alus, *Vaccinium vitis-idaea* var. *minus* and *Linnaea borealis*). Fifth, a 'decumbent small' shrub form is found in 3 species (*Empetrum nigrum* var. *japonicum*, *Andromeda polifolia* form. *acerosa* and *Phyllodoce caerulea*). Sixth, *Sabina sargentii* alone belongs to a 'procumbent' shrub form.

The effect of dwarfing and a prostrate growth habit, such as that displayed by many of the alpine plants listed above, is important in a harsh alpine zone. First, it enables species to resist dessication through the reduction of exposed shoots and inflorescences, and mechanical abrasion by wind, snow, ice crystals and sand (Bright, 1928; Porsild, 1951; Warren Wilson, 1959; Shetler, 1963; Savile, 1972; Fitter & Hay, 1981; Jones, 1983; Goudie, 1985); second, it provides an adaptation to low air and soil temperatures, using the benefit of relatively stable and high temperatures during the growing season near the ground (Bliss, 1956; Parkhurst & Loucks, 1972; Mani & Giddings, 1980); third, it ensures that the plant is likely to remain below a winter-snow cover, at which stage it becomes dormant and is further protected against extreme cold and dessication (Cornwell, 1970; Box, 1981b); and fourth, it enables the plant to utilise the relatively high growing season temperatures close to the ground, whenever conditions allow it to do so (Spomer, 1964; Stoutjesdijk, 1970; Wardle, 1971; Chabot & Billings, 1972; Speirs, 1979; Dahl, 1986).

Leaf size and form

In this research, out of the 41 species, 31 species of evergreen arboreal arctic-alpine and alpine plants have been studied for leaf size. Twenty species show leaf-lengths of 1 to 5 centimetres, 5 species leaf-lengths of less than 1 centimetre, namely *Sabina davurica*, *S. sargentii*, *Empetrum nigrum* var. *japonicum*, *Oxycoccus microcarpus* and *Phyllodoce caerulea*, and the remainder have leaves of greater than 5 centimetres, *viz*. *Abies koreana*, *Pinus koraiensis*, *P. tabulaeformis* and *Ledum palustre* var. *maximum*. Leaves of these 31 species accordingly are those of a predominantly small size.

There is, however, considerable variation in the outline of the studied leaves. Fifteen species are linear, five species are acicular and elliptical, four are lanceolate, three species are oblong, two species are ovate and rhomboid, and one species has rhomboid, obovate and orbicular leaves (Table 5.4). These variations of leaf shape may indicate diverse morphological adaptations to these difficult environments.

Seventeen species appear to have adapted their leaf texture against the severe climate and the harsh environment. Four species have coriaceous (leathery) leaves, namely *Diapensia lapponica* subsp. *obovata*, *Chamaedaphne calyculata*, *Rhododendron dauricum* and *Dryas octopetala* var. *asiatica*. However, in western Europe, *Dryas octopetala* is often deciduous. Six species have revolute leaves: *Ledum palustre* var. *diversipilosum*, *L. palustre* var. *yesoense*, *Oxycoccus microcarpus*, *O. quadripetalus*, *Phyllodoce caerulea* and *Vaccinium vitis-idaea* var. *genuinum*. Six species have coriaceous and revolute leaves, viz. *Empetrum nigrum* var. *japonicum*, *Andromeda polifolia*, *A. polifolia* form. *acerosa*, *Rhododendron aureum*, *R. fauriae* form. *rufescens* and *Vaccinium vitis-idaea* var. *minus*. Revolute leaves are regarded as an adaptive form against cold and dry climates, and also a lack of adequate mineral nutrients (Specht, 1979; Street & Opik, 1984).

Thirteen species have hairs on their leaves, viz. *Dryas octopetala* var. *asiatica*, *Empetrum nigrum* var. *japonicum*, *Chamaedaphne calyculata*, *Ledum palustre* var. *latifolium*, *L. palustre* var. *diversipilosum*, *L. palustre* var. *maximum*, *L. palustre* var. *minus*, *L. palustre* subsp. *decumbens*, *L. palustre* var. *yesoense*, *Oxycoccus microcarpus*, *Rhododendron fauriae* form. *rufescens*, *R. parvifolium* and *Linnaea borealis*. Two species (*Rhododendron dauricum* and *R. parvifolium*) have scales on their buds. The presence of hairs on leaves is thought to have several beneficial effects. First, they may protect pores and/or stomata, and help to minimise evapotranspiration during hot seasons, and to reduce the absorption of water during excessively wet periods (Arber, 1910; Mani, 1978; Mani & Giddings, 1980; Box, 1981b; Moggi, 1985; Sage, 1986); second, they are especially important in alpine and subalpine zones in that they may reflect a potential over-abundance of ultraviolet light energy (Billings & Morris, 1951; Bliss, 1956); and third, they can act as a heat trap, establishing a thermal cushion which will help the plant to heat sufficiently to develop and bloom, even when air temperatures are still near freezing (Krog, 1955; Zwinger & Willard, 1972; Dahl, 1986; Sage, 1986). According to Arber (1910), the function of scales is to protect leaf pores, thus minimising the rate of evapotranspiration.

Flower pattern

Of the 22 evergreen arboreal broadleaved arctic-alpine and alpine plants, 12 species bloom in the spring (March–May) and 10 species in the summer (June–August). Fifteen species display a white flower, six pink and purple flowers, two brown and yellow flowers and one greenish flowers. The dominant flower colour among the evergreen arboreal broadleaved arctic-alpine and alpine plants of Korea is white (15 spp.) (Table 5.4). The dominance of bright flower colours, viz white (15 spp.) and pink (6 spp.) is advantageous for rapid pollination, and also for the further enhanced reflection of potentially harmful ultraviolet (Mani & Giddings, 1980). The autumn is

Table 5.5. Climatic regimes of arctic and alpine areas.

Climate/area	Arctic	Alpine
Daylight	High seasonality	Low or moderate seasonality
Radiation	Low ultraviolet	Higher ultraviolet
Temperature	Low, with high seasonality	Low, with lower seasonality
Atmospheric pressure	'Normal'	Lower than 'normal'
Oxygen tension	'Normal'	Lower than 'normal'
Precipitation	Adequate to low	High variability, according to altitude
Snow cover	Deep to thin; long durability	High variability, according to altitude and topography; lower durability
Mean windspeed	High	High

dominated by maturing of fruits (29 spp.), the shapes of which vary considerably. Sixteen species have a cone nut, 15 species have a capsule, six species a berry, three species a drupe and one species a non-cone nut (Table 5.4).

Alpine environments

There are a number of general climatic differences between arctic and alpine areas, as displayed in Table 5.5.

In general, the climate of an alpine area can be characterised as follows: a low or moderate seasonality of daylight, higher ultraviolet rates, a low mean annual temperature, with a low seasonality, a lower atmospheric pressure and lower oxygen tension than the Arctic, a highly variable rainfall precipitation regime, according to altitude, and a highly variable snowcover, according to altitude and topography and the pattern of durability (Griggs, 1930; Russell, 1948; Bliss, 1956; Hanson, 1958; Johnson & Billings, 1962; Mani, 1962, 1978; Billings, 1964, 1983 Billings & Mooney, 1968; Yoshino, 1975; Prime, 1978; Price, 1981).

Compared to adjacent lowland areas, the alpine zone in Korea has several environmental characteristics which relate to the above and these may be summarised as follows. They have intensive solar radiation including ultraviolet, low oxygen and CO^2 pressures, a low mean annual temperature and a considerable diurnal variation of temperature, prolonged snowy spells in winter, a high frequency of strong wind, and intensive frost action. This is the case for all mountain areas, from the far north of the peninsula, to Cheju Island.

Many of these climatic features are responsible for influencing the growth and distribution of alpine plants in Korea. This is certainly the case in respect of the evergreen arctic-alpine and alpine species, with which this chapter is particularly concerned. However, this is not to deny the importance to plant growth and survival of other environmental parameters, such as soil type, parent material, topography and biotic conditions. Billings (1985) has suggested that it is particularly important to determine the tolerance ranges of plants in respect to climate, since they help to explain the operation of ecosystems in a changing biosphere, and to predict possible future geographical range patterns (see also Melzack & Watts, 1982a,b). Tolerance ranges are also vital in the reconstruction of patterns and processes of plant distribution in the past, since these often are a direct result of environmental fluctuations.

Temperature, and evergreen arboreal arctic-alpine and alpine plants

Consideration is given here to the effect of temperature on the tolerance ranges of evergreen arboreal arctic-alpine and alpine plants. This arises in part out of the fact that the data base for temperature in Korea is much better than that of other climatic parameters. Many workers believe that it is the average environmental (including climatic) conditions which are responsible for the growth and distribution of plants; however, in reality it is often the extreme climatic events which are more important for them

(Walter, 1963; Levitt, 1980). This latter view is the one accepted here, and it is further explained below in respect of the effects of high summer temperatures and low winter temperatures on evergreen plant distribution in Korea.

High summer temperature patterns

A good deal of research has found a negative correlation between the distribution of arctic and alpine plants, and the presence of high summer temperatures. Some of the major influences arising from these are as follows; first, the distribution of several alpine plants seems to be directly controlled by the influence of high summer temperatures (Wilson, 1956; Damman, 1965, 1976); second, the distribution of the arctic-alpine *Oxyria digyna* is restricted whenever carbohydrate sources are over-exploited, and these commonly happen under conditions of high summer temperatures (Bjorkman & Holmgren 1961; Mooney & Billings, 1961, 1965; Stewart & Bannister, 1973; Billings, 1981); third, a diminution of plant growth rates can be created when high temperatures intensify the rates of daily respiration in a plant (Budyko, 1974; Yim & Kira, 1975; Kira, 1977; Fitter & Hay, 1981); fourth, such higher respiration rates in alpine plants make them more susceptible to metabolic imbalances (Higgins & Spomer, 1976; Box, 1981a; Crawford, 1989); and fifth, many alpine plants, e.g. *Diapensia lapponica* and *Dryas octopetala*, directly suffer from abnormally high summer temperatures (Pennington, 1974; Day & Scott, 1964; Walton, 1984). Also, it should be noted that many alpine plants cannot tolerate a prolonged daylight regime during periods of high summer temperatures (Rousseau, 1953).

Intensive research has been carried out on the relationships between the daily maximum temperatures and the distribution of alpine plants in Scandinavia and the British Isles (Dahl, 1951, 1953, 1955, 1963, 1964, 1966, 1986; Conolly, 1961; Conolly & Dahl, 1970; Gauslaa, 1984, 1985; Crawford, 1989). According to Dahl (1951), summer temperatures (mean daily maxima) are responsible for determining the southern, and lower altitudinal distribution limits of many alpine vascular plants in the northern hemisphere. The reasons for this may be varied. First, there may be a rapid increase in the respiration rate beyond a certain point on the temperature scale, which may destroy plant cellular structures; second, alpine vascular plants tend to grow well when they are kept within only very limited ranges of night and day temperatures; and third, high summer temperatures are more advantageous to the growth of warmth-loving plants, in comparison to alpines, and this in turn tends to exclude alpine plants from such areas through competition.

However, the tolerance to high summer temperatures within the same alpine species can vary in different countries (Conolly & Dahl, 1970). For example, *Dryas octopetala* can withstand extreme temperatures of 27°C in Scandinavia, 25°C in England, Wales and Ireland, but only 23°C in Scotland; *Diapensia lapponica* can withstand extreme temperatures of 26°C in Scandinavia, but only 18.3°C in Newfoundland. *Phyllodoce caerulea* can maintain its growth up to 27°C in Scandinavia, *Rhododendron lapponicum* up to 24°C, also in Scandinavia, and *Vaccinium uliginosum* up to 18.3°C in Newfoundland (Dahl, 1951, 1953; Damman, 1965; Conolly & Dahl, 1970). It is likely that these differences in the levels of the limiting isotherms for growth between species in Scandinavia, Britain and Newfoundland may be attributable to the modifying effect of oceanity on temperatures, and species adjustment to these, within the latter two areas (Conolly & Dahl, 1970). According to Precht *et al.* (1973), active heat injury arising from high temperatures takes place at 47°C to 54°C for alpine dwarf shrubs, and at 44°C to 50°C for evergreen conifers. The upper limit of photosynthesis can be distinguished as being at 42°C to 47°C for arctic and alpine dwarf shrubs, and at 36°C to 38°C for boreal conifers (Kappan, 1981). It is only rarely that these temperatures are achieved under natural conditions.

Looking at this problem in another way, Yim (1977), in Korea, suggested that the mean value of a warmth index (i.e., in this case, the sum of monthly mean temperatures over 5°C) as applied to several alpine species might explain the location of their altitudinal limits, albeit in a very general way. The relevant means and standard deviations of each alpine species studied by Yim (1977) are as follows: e.g. 58.9°C (±19.6°C) for

Pinus pumila, 62.8°C (±6.9°C) for *Picea jezoensis*, 71.1°C (±12.9°C) for *Taxus cuspidata*, 72.3°C (±10.7°C) for *Abies nephrolepis*, and 93.2°C (±10.9°C) for *Pinus koraiensis*. For comparison, equivalent values for more warmth-loving evergreen plants are *Pinus thunbergii* at 114.5°C (±4.4°C), and 112.8°C (±5.0°C) for *Camellia japonica*. According to Noh (1983), the average sum of the mean daily temperatures in alpine-species environments in Korea for the annual growing season display a similar pattern; thus *Taxus cuspidata* (2016°C), *Pinus koraiensis* (2307°C), *Sabina sargentii* (2310°C), *Picea jezoensis* (2545°C), *Abies koreana* (2784°C) and *A. holophylla* (2803°C) in comparison with a value for the environment of the temperate evergreen conifer *Pinus densiflora*, of 2962°C.

Low winter temperature patterns

The importance of low winter temperature for the survival of plants and their regional growth patterns has been demonstrated widely (see Melzack & Watts, 1982a,b in respect to *Taxus baccata* in the U.K.). In general, alpine plants are well adapted to grow at low temperatures, and have a lower optimum temperature for shoot growth than temperate species (Chapin & Shaver, 1985). A wide range of thermal amplitudes has been reported in connection with tolerance to low winter temperatures for the survival of alpine plants. For example, Levitt (1966) has suggested that winter hardiness in the genera *Rhododendron*, *Empetrum* and *Abies* normally maximises at −20°C to −30°C; and equivalent values for the species *Pinus pumila* and *Juniperus sibirica* in East Asia may be in excess of −40°C. However, Sakai & Otsuka (1970) reported that some tissue damage may begin to occur in *Pinus pumila* at temperatures of −30°C to −25°C; and for *Emptrum nigrum* and *Vaccinium vitis-idaea*, the equivalent values would be −20°C to −10°C. General tissue death may eventuate in *Empetrum nigrum* when temperatures fall to −45°C or below, and this will be initiated for *Taxus cuspidata* at −30°C to −20°C (Agricultural Research Service, 1965). For *Taxus cuspidata*, too, some differences may be noted in respect of the point below which cold hardiness no longer preserves particular organs from damage; for leaves, this is reached at −40°C, and for roots at −12°C (Zehnder & Lanphear, 1966; Mityga & Lanphear, 1971). Tissue damage may commence in *Rhododendron dauricum* at −20°C to −15°C, and in *Diapensia lapponica* at −25°C to −70°C (Sakai & Otsuka, 1970).

There is little doubt that, since many lowland plants in temperate areas become vulnerable to tissue damage from cold weather at considerably higher temperatures than those described above, arctic-alpine plants have a distinct competitive advantage in highland regions, due to this one fact alone; conversely, they are at a competitive disadvantage generally in lowland regions adjacent to high land, although they may exist there under conditions which are particularly ‘open’, and favourable for their survival (e.g. some riverbanks, shore-lands etc.) (Dahl, 1951, 1953; Walton, 1984).

Temperature data in Korea

In this section, the relationships between high summer temperatures and the lower limits of alpine species in Korea are discussed. Reference is also made to work carried out in attempts to understand the relationship between low winter temperatures, and the upper limits of alpine species in Korea.

According to Bliss (1985), summer day-time air temperatures near the surface, where the alpine plants grow, are often considerably higher than those at standard weather bureau heights (1.2 metres), and this is clearly important for plant growth. This work, however, has employed standard meteorological data from a range of sites, from 1931 to 1960, at 1.2 metres in height (Central Meteorological Office, 1968), because of the general absence of weather data closer to ground level in the mountains of Korea. From these data, which come from lowland weather stations in the main, some of which lie within the general vicinity of mountain sites, it is then possible to calculate lapse rates, and accordingly temperature data for different altitudes upslope may be extrapolated. Most of the meteorological stations in Korea lie between sea level and 100 m a.s.l. (Table 5.6), and accordingly, for convenience, all lapse-rate calculations have been taken to begin at 100 m a.s.l. These data then

Table 5.6. Estimated temperatures in Korean mountains.

Site No.	Name of mountain	Altitude of mt. (m)	*Mean daily max. temp. of Aug. (°C)	*Mean daily min. temp. of Jan. (°C)	Related meteorological station
1	Jung	1003	27.4	−17.9	Kyungwon, Onseung, Chongseong
2	Songjin	1146	24.5	−13.0	Unggi
4	Chayoo	1500	27.6	−19.6	Moosan
5	Paikdoo	2744	26.8	−24.5	Samsoo, Moosan, Chunggangjin
7	Mantap	2205	27.3	−15.9	Myongcheon
11	Chilbo	906	27.2	−13.9	Myongcheon, Dancheon
13	Hoochi	1335	23.6	−24.0	Pungsan
14	Seungjuk	1970	29.3	−23.2	Chosan
15	Ro	2240	28.2	−26.8	Kanggae
16	Pirae	1470	29.5	−20.0	Changseong
17	Keumpae	1676	30.1	−13.1	Shinheung
19	Peenanduk	1316	29.8	−20.1	Heecheon, Changseong
20	Nangrim	2002	29.9	−16.6	Heecheon, Shinheung
21	Myohyang	1909	29.6	−20.1	Heecheon
22	Sasoo	1747	28.4	−15.0	Youngwon, Hamheung
23	Haram	1485	28.2	−18.9	Yangduk
24	Chuae	1530	28.8	−12.6	Koksan, Tongcheon
25	Keumkang	1638	27.8	−11.9	Koseung, Hoeyang
26	Koowol	954	29.3	−11.3	Eunyool
28	Myulak	816	29.9	−11.4	Chaeryong, Kaeseong
29	Changsoo	711	30.2	−12.2	Chaeryong
32	Sooyang	945	29.5	−10.4	Haeju, Changyeon
33	Sorak	1708	29.0	− 9.7	Injae, Yangyang
34	Whaak	1468	32.1	−14.5	Whacheon
40	Odae	1563	30.7	−10.5	Kangnung, Hongcheon
42	Sori	472	29.8	−12.3	Kaeseong, Kapyong
44	Taeji	1300	29.4	−12.1	Pyongchang
45	Yongmoon	1157	30.8	−11.7	Yangpyong
49	Chiak	1288	30.7	−12.7	Wonju
51	Taebaik	1561	31.0	−11.6	Youngwol
55	Ilwol	1219	29.6	− 7.2	Cheongsong, Woljin
58	Sokri	1058	30.5	− 9.8	Poun
60	Kaeryong	845	30.7	− 8.6	Kongju
66	Palgong	1192	31.5	− 6.9	Seonsan, Youngcheon
68	Kaya	1430	31.3	− 7.9	Keochang, Seonsan
70	Dukyoo	1508	31.3	− 9.8	Mooju, Keochang
74	Kaji	1240	30.6	− 5.7	Youngcheon, Ulsan
77	Naejang	763	30.8	− 6.2	Cheongup
79	Paikyang	750	30.8	− 6.0	Cheongup, Changseong
80	Chiri	1915	30.9	− 6.6	Namwon, Hamyang
83	Moodeung	1187	31.1	− 4.8	Kwangju
119	Halla	1950	29.7	+ 1.9	Cheju
145	Seongin	984	27.4	− 1.9	Ulreung

* Temperature at about 100 m a.s.l. and calculated based on data by C.M.O. (1968).
Site numbers refer to Table 2.6, p. 32.

may be compared with those for tissue damage indicated in the previous section.

In East Asia, the temperature lapse rates are greater for winter northerly-wind conditions than for summer southerly-wind situations, and the highest lapse-rate values tend to be reached during the winter in eastern Siberia and Japan (Yoshino, 1966a; Barry, 1981). For East Asia, generally, Yoshino (1966b) has calculated that a lapse rate of −0.61°C per 100 metres may be applicable for most situations (particularly in Japan), and a value of −0.55°C per 100 metres is regarded as being the equivalent in Korea (Yim & Kira, 1975). This work has used the latter figure for the Korean climatic regime in summer; and in winter, a lapse-rate value of −0.65°C per 100 metres is considered more appropriate.

Accordingly, temperature variations for different altitudes have been calculated on the basis of the above lapse rates from a range of standard meteorological stations located near each mountain. Two criteria have been considered as regards the selection of relevant meteorological stations in this analysis. First, the nearest meteorological station to the mountain always has been employed. Second, the mean values of the nearest neighbouring meteorological stations (usually one or two, but occasionally three, depending on circumstances) have been used as a basis for calculation, if no meteorological station is available within 50 km of the mountain. For convenience, all calculations are taken to have a base line at 100 m a.s.l., as indicated above.

High summer temperatures and the lower distributional limits of species

Details of the mean daily maximum temperatures for August, as they relate to the lower limits of alpine species, and of the mean daily minimum temperatures for January, as they relate to the upper limits of alpine species, on 43 Korean mountains, are given in Table 5.7.

Out of the 41 selected Korean evergreen arboreal arctic-alpine and alpine species, 24 species, with relevant altitudinal distributional data, are analysed. The relationships between the lower limit of arctic-alpine and alpine species and the mean daily maximum temperatures for August display three characteristic features.

First, the lower limit of evergreen broadleaved arctic-alpine plants is found within a range of relatively low mean daily maximum temperatures for August, viz. *Dryas octopetala* var. *asiatica* (16.7–23.3°C), *Ledum palustre* subsp. *decumbens* (17.5°C), *Rhododendron aureum* (18.0–25.2°C), *Ledum palustre* var. *minus* (18.6°C), *Phyllodoce caerulea* (18.8°C), *Vaccinium vitis-idaea* var. *minus* (19.1–24.1°C), *Ledum palustre* var. *diversipilosum* (19.4–24.1°C), *Rhododendron dauricum* (19.5–26.3°C), *R. parvifolium* (19.8°C), *Diapensia lapponica* subsp. *obovata* (20.4°C), *Empetrum nigrum* var. *japonicum* (20.9°C) and *Rhododendron fauriae* form. *rufescens* (22.7–25.7°C). The relatively large individual ranges of maximum temperatures for some species in this list simply reflects the tendency for these species to grow widely over different parts of the Korean mountain system, whereas species with more restricted distributions will have more clearly defined limits of maximum temperature. Nevertheless all these evergreen broadleaved arctic-alpine species appear to be very sensitive to high summer temperature regimes, and most species have very narrow amplitudes of tolerance to warm summer conditions. Intolerance to high summer temperatures is best expressed in reduced levels of competition as compared to other more warmth-loving species, once the limits of tolerance to August temperature levels have been crossed.

Second, the lower limit of evergreen coniferous alpine and subalpine plants is not so clearly expressed in terms of the mean daily maximum temperatures for August. Thus, the lower limits of *Juniperus sibirica* are characterised by daily maximum temperatures for August of from 15.2–31.0°C, those of *Pinus pumila* are set at 16.9–23.4°C, *Sabina sargentii* at 17.8–25.7°C, *Pinus koraiensis* at 18.7–31.0°C, *Picea jezoensis* at 19.9°C –25.8°C, *Abies nephrolepis* at 20.3C–27.0C, *A. holophylla* at 20.3–30.8°C and *Taxus cuspidata* at 21.0–27.2°C. This suggests that evergreen coniferous alpine and subalpine plants are in general more adaptable to warm summer temperatures than are the evergreen broadleaved arctic-alpine plants, and also that they have relatively wide thermal amplitudes. This adaptation to warm summer temperatures encourages these species to grow generally in alpine and subalpine zones below that of the evergreen broadleaved arctic-

Table 5.7. Temperature ranges at the vertical limits of species in Korea.

Species name	Range	Lower limit		Upper limit		No. of
	National minimum & maximum (m)	Range of mean daily max. temp. Aug.	Average of mean daily max. temp. Aug.	Range of mean daily min. temp. Jan.	Average of mean daily min. temp. Jan.	Mountains
Taxus caespitosa	1590	–	21.3	–	−18.8	1
T. cuspidata	300–1900	21.0/27.2	24.8	−31.8/−7.1	−21.7	14
Abies holophylla	100–1500	20.3/30.8	27.2	−30.7/−7.3	−19.6	33
A. koreana	500–1900	23.7/28.6	25.4	−18.3/−9.8	−14.1	5
A. nephrolepis	700–2200	20.3/27,0	24.7	−38.5/−18.2	−25.8	22
Picea jezoensis	700–2300	19.9/25.8	23.7	−33.0/−18.2	−28.1	15
P. koraiensis	450–1600	19.8/25.4	23.1	−32.1/−20.1	−25.0	3
Pinus koraiensis	100–1900	18.7/31.0	26.4	−32.1/−12.1	−21.7	33
P. pumila	900–2200	16.9/23.4	20.8	−33.0/−20.1	−27.2	9
Thuja koraiensis	400–2500	20.2/26.8	24.0	−33.0/−19.5	−25.1	13
Juniperus sibirica	800–2300	15.2/23.8	19.2	–	–	3
Sabina sargentii	700–2300	17.8/25.7	21.9	−31.8/−10.1	−20.4	10
Dryas octopetala var. *asiatica*	1300–2200	16.7/23.3	20.0	–	–	2
Diapensia lapponica subsp. *obovata*	1800–1950	–	20.4	–	−10.1	1
Emptrum nigrum var. *japonicum*	1700–1900	–	20.9	–	− 9.8	1
Ledum palustre var. *diversipilosum*	800–2000	19.4/24.1	21.5	−39.2/−22.9	−29.5	5
L. palustre var. *minus*	1600–2300	–	18.6	–	−38.8	1
L. palustre subsp. *decumbens*	1800–2540	–	17.5	–	−40.4	1
Phyllodoce caerulea	1540–1800	–	18.8	–	−35.6	1
Rhododendron aureum	800–2300	18.0/25.2	21.6	−39.2/−19.5	−27.5	9
R. dauricum	300–2000	19.5/26.3	22.7	−32.1/−10.1	−24.5	8
R. fauriae form. *rufescens*	700–1700	22.7/25.7	23.9	−26.1/−20.1	−23.4	6
R. parvifolium	800–1100	–	19.8	–	−30.5	1
Vaccinium vitis-idaea var. *minus*	1000–2200	19.1/24.1	20.5	−39.2/−22.3	−28.9	9

N.B. Unit for temperature is °C. All calculations involve the use of the lapse rate formulae, as indicated on p. 132.

alpine plants. This propensity parallels that found among similar evergreen coniferous subalpine plants elsewhere in the world, as for example *Pinus sylvestris* (Scots pine), which grows equally well in northern Scotland as in the warmer southern parts of Russia; and *Pinus ponderosa* (yellow pine), which may be found in both the northern and the much warmer southern extremities of the continental U S.A.

Third, the lower limits of endemic evergreen coniferous subalpine species may be found in a relatively narrow zone in which mean daily maximum temperature ranges for August are, for *Picea koraiensis* (19.8–25.4°C), *Thuja koraiensis* (20.0–26.8°C), *Taxus caespitosa* (21.3°C) and *Abies koreana* (23.7–28.6°C). These endemic evergreen coniferous subalpine plants grow in a slightly warmer summer climate than that which is normal for both the evergreen broadleaved arctic-alpine and alpine plants, and the evergreen coniferous alpine and subalpine plants in general. Endemic evergreen coniferous subalpine plants are, accordingly, less sensitive to high summer temperature conditions than are the other above-mentioned evergreen coniferous plants. An important biogeographical consideration which has not previously been reported is that the distinct disjunctive distribution of many of these endemic species both horizontally and vertically is undoubtedly accentuated because of their relatively narrow thermal amplitude in a place such as Korea, which has an extremely rugged and mountainous terrain.

Overall, evergreen broadleaved arctic-alpine and alpine species are confined to the alpine zone with cool summer temperatures, because of their vulnerability to high summer temperatures. Evergreen coniferous alpine and subalpine plants are found within both the alpine to subalpine zones, and rarely even in the lower montane zones, in which summer temperatures are relatively high. Endemic evergreen coniferous subalpine species grow in the transitional area between the subalpine and montane zones. A comparative analysis of the mean daily maximum summer temperatures at the lower altitudinal limit of *Dryas octopetala* in Scotland (23°C) by Conolly & Dahl (1970), and *Dryas octopetala* var. *asiatica* in Korea (16.7–23.3°C), that of *Diapensia lapponica* in Scandinavia (26°C), in Newfoundland, Canada (18.3°C) by Dahl (1951, 1953) and Damman (1965), and of *Diapensia lapponica* subsp. *obovata* in Korea (20.4°C) displays a good level of coincidence to each other. However, in the case of *Phyllodoce caerulea* there are big gaps between the equivalent temperature limits in Scandinavia (27°C) and Korea (18.8°C), which still need further explanation. A similar pattern has been noticed in *Diapensia lapponica* between Scandinavia and Newfoundland. The sensitivity of evergreen coniferous plants to high summer temperature noted by Yim (1977) and Noh (1983) also corresponds quite well with this independent data noted above.

Low winter temperatures and the upper distributional limits of species

Similarly, three altitudinal patterns can be seen from the relationships displayed between the upper limits of arctic-alpine and alpine plants and the mean daily minimum temperatures of January.

First, the upper limit of evergreen broadleaved arctic-alpine plants expands into areas with extremely low mean daily minimum temperatures for January, i.e. *Ledum palustre* subsp. *decumbens* (−40.4°C), *Vaccinium vitis-idaea* var. *minus* (−39.2°C to −22.3°C), *Ledum palustre* var. *diversipilosum* (−39.2°C to −22.9°C), *Rhododendron aureum* (−39.2°C to −19.2°C), *Ledum palustre* var. *minus* (−38.8°C) and *Phyllodoce caerulea* (−35.6°C). Clearly, evergreen broadleaved arctic-alpine species are very adaptable to cold temperature conditions, and hence these species dominate the alpine zone.

Second, the upper limits of evergreen coniferous alpine and subalpine plants are to be found mainly in areas in which the very cold mean daily minimum temperatures for January are slightly higher than the preceding examples. Thus, *Abies nephrolepis* grows in mean daily minimum temperatures for January of −38.5°C to −18.2°C, *Pinus pumila* at −33.0°C to −20.1°C, *Picea jezoensis* at −33.0°C to −18.3°C, *Pinus koraiensis* at −32.1°C to −12.1°C, *Sabina sargentii* at −31.8°C to −10.1°C, *Taxus cuspidata* at −31.8°C to −7.1°C and *Abies holophylla* at −30.7°C to −7.3°C. In other words, these evergreen coniferous alpine and subalpine plants also adapt well to cold winter temperatures, but they tend to occupy alpine and subalpine belts below the zone of evergreen broadleaved arctic-alpine plants above.

Third, two types can be noticed in connection with the upper limits of endemic evergreen coniferous subalpine plants; first, a cold winter-adapted set of species, including *Thuja koraiensis* (mean daily minimum temperature for January of −33.0°C to −19.5°C) and *Picea koraiensis* at −32.1°C to −20.1°C; and second, species adapted to a relatively mild winter temperature regime, such as *Taxus caespitosa* (−18.8°C) and *Abies koreana* (−18.3°C to −9.8°C).

Overall, the evergreen broadleaved arctic-alpine plants are well capable of expanding into alpine zones with extremely cold winter temperatures, due to their generally high levels of winter hardiness. The evergreen coniferous alpine and subalpine plants are found within the alpine to subalpine zones usually below the former group, but are also capable of withstanding quite cold winter temperatures. The endemic evergreen coniferous subalpine plants grow in zones some of which have very cold winter temperatures, though others have relatively mild winter temperatures. A comparative analysis of the mean daily minimum winter temperatures (or cold damage-occurring temperatures) at the upper altitudinal limits of *Pinus pumila* in Japan (−30°C to −25°C) by Sakai & Otsuka (1970), and that in Korea (−33.0°C to −21.0°C); and that of *Taxus cuspidata* in the U.S.A. (−30.0°C to −20.0°C) by the Agricultural Research Service (1965), and that in Korea (−31.8°C to −7.1°C) also coincides quite well.

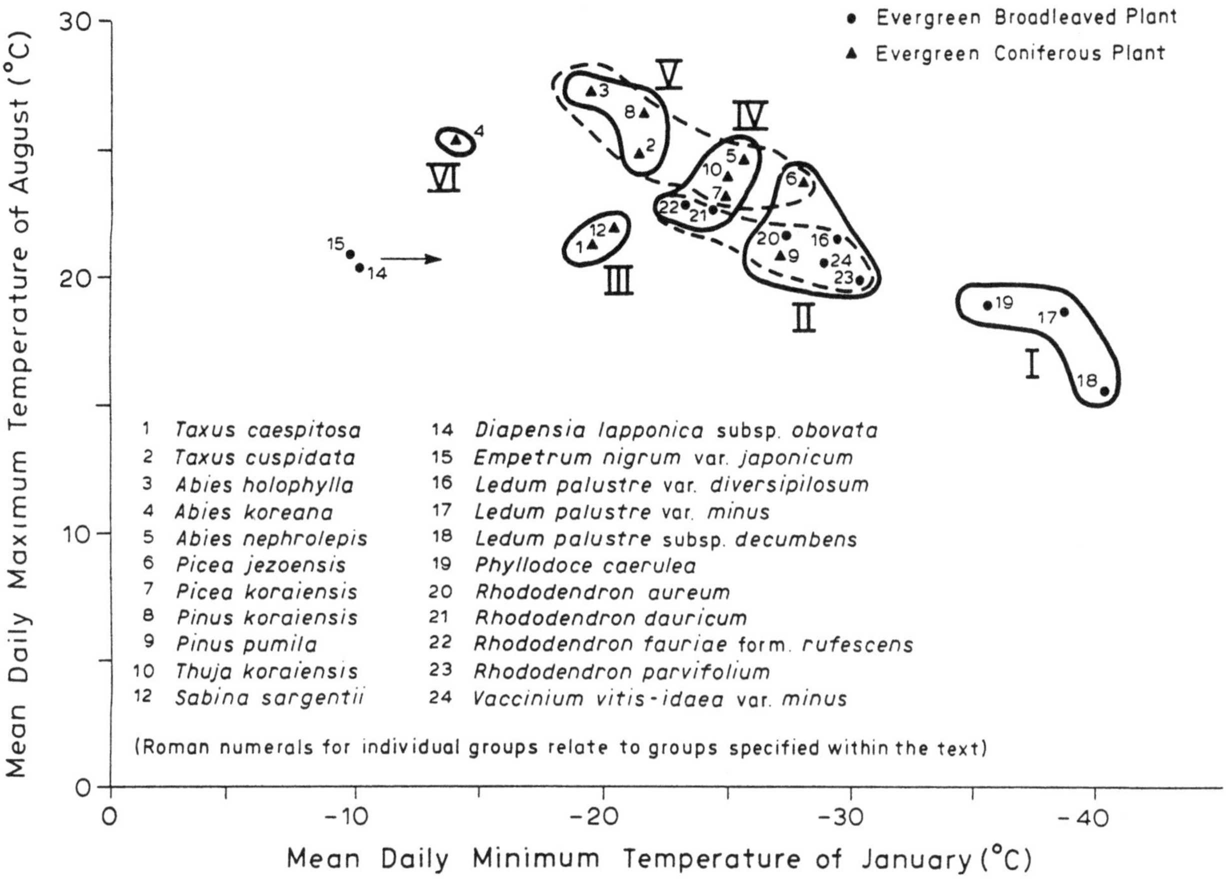

Figure 5.9. Averages of the mean daily maximum temperatures for August, and the mean daily minimum temperatures for January.

Mean temperatures values at the lower and upper limits

If one plots the mean daily maximum temperatures for August as they relate to the lower limit of selected arctic-alpine and alpine species against the mean daily minimum temperatures for January as they relate to the upper limit of arctic-alpine and alpine species, six distinct groups of species are displayed (Fig. 5.9).

First, three evergreen broadleaved arctic-alpine and alpine species belong to the group of plants most sensitive to high summer temperatures (viz. 17.5°C to 18.8°C for the mean daily maximum temperatures for August), but at the same time they are most tolerant to low winter temperatures (viz. −40.5°C to −35.6°C for the mean daily minimum temperature for January). These are *Ledum palustre* subsp. *decumbens*, *L. palustre* var. *minus* and *Phyllodoce caerulea*.

Second, four evergreen coniferous alpine and two evergreen broadleaved alpine species belong to the high-summer sensitive group (viz. 19.8°C to 23.7°C for the mean daily maximum temperatures for August), while at the same time they are relatively tolerant of low winter temperatures (viz. −30.5°C to −27.2°C for the mean daily minimum temperatures for January). These include *Ledum palustre* var. *diversipilosum*, *Rhododendron aureum*, *R. parvifolium*, *Vaccinium vitis-idaea* var. *minus*, *Picea jezoensis* and *Pinus pumila*.

Third, an evergreen coniferous alpine plant, *Sabina sargentii*, and an endemic evergreen coniferous alpine plant, *Taxus caespitosa*, also are sensitive to high summer temperatures (viz.

21.3°C to 21.9°C for the mean daily maximum temperatures for August), but are less tolerant to low winter temperatures (viz. −20.4°C to −18.8°C for the mean daily minimum temperatures for January).

Fourth, two evergreen broadleaved alpine species, one evergreen coniferous alpine species and two endemic evergreen coniferous alpine species are characterised by slightly greater tolerance overall to high summer temperatures (viz. 22.7°C to 24.7°C for the mean daily maximum temperatures for August), and are still quite tolerant to low winter temperatures (viz. −25.8°C to −23.4°C for the mean daily minimum temperatures for January). These are *Rhododendron dauricum*, *R. fauriae* form. *rufescens*, *Abies nephrolepis*, *Thuja koraiensis* and *Picea koraiensis*.

Fifth, three evergreen coniferous alpine species belong to a regime of relatively high summer temperatures (viz. 24.8°C to 27.2°C for the mean daily maximum temperature of August), and tolerate less cold winter conditions (viz. −21.7°C to −19.6°C for the mean daily minimum temperature of January). These are *Abies holophylla*, *Pinus koraiensis* and *Taxus cuspidata*.

Finally, one endemic evergreen coniferous plant, *Abies koreana*, belongs to the most adaptive group to high summer temperatures but is sensitive to low winter temperatures (viz. −14.1°C for the mean daily minimum temperatures for January).

Two evergreen broadleaved arctic-alpine species, *Empetrum nigrum* var. *japonicum*, and *Diapensia lapponica* subsp. *obovata*, which occur in the range of 20.4°C to 20.9°C for high summer temperatures, and −10°C to −9.8°C for low winter temperatures may be joined to the first group, if the upper limits of their range, which are not available in Korea, are considered. Though the present work has designated six groups, another alternative subjective designation may yield, similarly, five divisions which divide the existing three groups numbered II, IV and V into two separate groups, one of which contains evergreen coniferous plants (*Taxus cuspidata*, *Abies holophylla*, *A. nephrolepis*, *Picea jezoensis*, *P. koraiensis*, *Pinus koraiensis* and *Thuja koraiensis*), and the other evergreen broadleaved plants (*Ledum palustre* var. *diversipilosum*, *Rhododendron aureum*, *R. dauricum*, *R. fauriae* form. *rufescens*, *R. parvifolium and Vaccinium vitis-idaea* var. *minus*), as well as one evergreen coniferous species *Pinus pumila* (Fig. 5.9). This latter designation clearly fits with previous results in respect of temperature ranges. Overall, accordingly, evergreen broadleaved arctic-alpine and alpine species belong to the high summer-temperature sensitive group, and evergreen coniferous alpine species to the high summer-temperature adaptive group. On the other hand, the designated six groups reflect very accurately the results of both low-winter and high-summer temperatures for the 24 species.

The range of mean daily maximum temperatures for August at the lower limits of arctic-alpine and alpine species and that of the mean daily minimum temperatures for January at the upper limits of arctic-alpine and alpine species further yield distinct thermal ranges of arctic-alpine and alpine plants in Korea. Five groups can be delimited on this basis, and these are presented in Fig. 5.10.

First, very large thermal ranges are found among the group of evergreen coniferous alpine plants, e.g. *Abies holophylla* (No. 3) and *Pinus koraiensis* (No. 8). These species clearly have a broad ecological amplitude, which enables them to grow widely in alpine, subalpine and even lower montane zones in Korea.

Second, five species have a relatively large thermal range, and are characterised in particular by a lower tolerance to very high summer temperatures, e.g. *Taxus cuspidata* (No. 2), *Sabina sargentii* (No. 12), *Rhododendron dauricum* (No. 21), *Abies nephrolepis* (No. 5) and *Rhododendron aureum* (No. 20). This may give rise to their occurrence in both alpine and subalpine zones in Korea, and to their absence on the lower montane slopes.

Third, there is a sharp decrease of thermal range in the case of *Vaccinium vitis-idaea* var. *minus* (No. 24), *Ledum palustre* var. *diversipilosum* (No. 16), *Picea jezoensis* (No. 6), *Thuja koraiensis* (No. 10) and *Pinus pumila* (No. 9). This signifies a relatively narrow ecological amplitude for these species, and, as a result, confines them accordingly to a comparatively small vertical zone in comparison with the two previous groups.

Fourth, two endemic evergreen coniferous species display a small thermal range, namely *Picea*

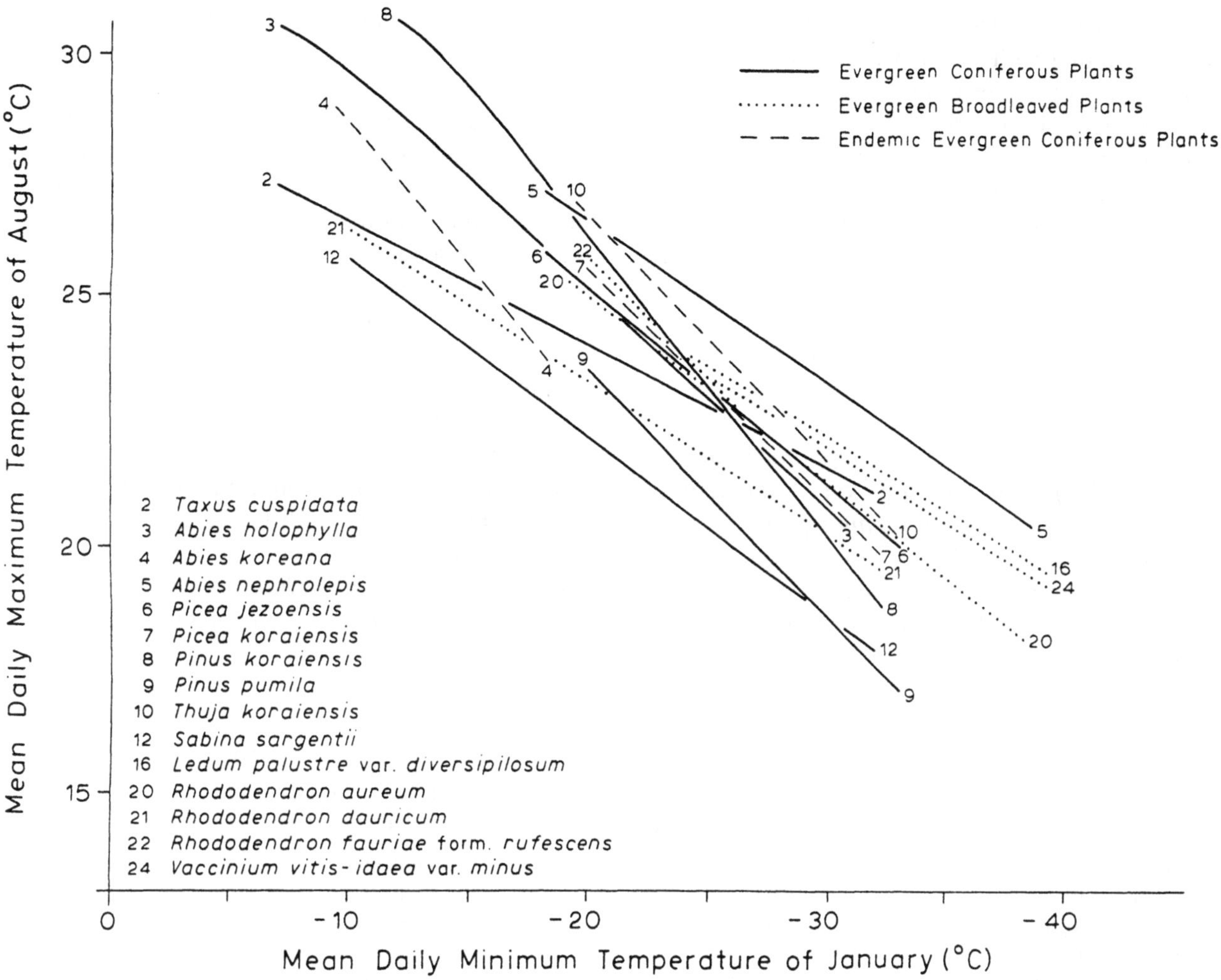

Figure 5.10. Thermal ranges (upper and lower limits) of arctic-alpine plants in Korea.

koraiensis (No. 7) and *Abies koreana* (No 4). This may at least in part explain the notably horizontally disjunctive distribution of these endemic species in Korea.

Finally, the very small thermal range of *Rhododendron fauriae* form. *rufescens* (No. 22) also probably explains in itself the very disjunctive distribution of this species, within the northern part of Korea.

Overall, the disjunctive distribution of evergreen broadleaved arctic-alpine and alpine species in the alpine zones of Korea is due in large measures to their sensitivity to high summer temperatures, and the relatively narrow ecological amplitudes. In contrast, the wide spread of evergreen coniferous alpine species in the alpine, subalpine and even lower montane zones of Korea is due to their greater adaptation to high summer temperatures and their development of extremely broad ecological amplitudes. The disjunctive distribution of endemic evergreen coniferous alpine species in the subalpine zones is due to their adaptation to mild summer temperatures, and at the same time a tolerance to anything other than relatively low winter temperatures (see Table 5.8).

Results and discussion

The selected evergreen arboreal arctic-alpine and alpine plants in Korea consist of 41 species, 18 genera and 10 families. The dominant families are the Ericaceae (18 spp.) and the Abietaceae (9 spp.), along with the Taxaceae (3 spp.), the Pinaceae (3 spp.), the Juniperaceae (3 spp.), the Cupressaceae (1 sp.), the Rosaceae (1 sp.), the Diapensiaceae (1 sp.), the Empetraceae (1 sp.) and the Caprifoliaceae (1 sp.).

On the basis of the broad-scale distributional patterns of East Asia, these 41 evergreen arboreal arctic-alpine and alpine plants in Korea can be divided into two groups. First, there are 13 evergreen arboreal *arctic-alpine* plants, including one evergreen coniferous plant *Sabina sargentii*, and 12 evergreen broadleaved plants, viz. *Dryas octopetala* var. *asiatica*, *Diapensia lapponica* subsp. *obovata*, *Empetrum nigrum* var. *japonicum*, *Chamaedaphne calyculata*, *Ledum palustre* var. *diversipilosum*, *L. palustre* subsp. *decumbens*, *Oxycoccus microcarpus*, *O. quadripetalus*, *Phyllodoce caerulea*, *Rhododendron parvifolium*, *Vaccinium vitis-idaea* var. *minus* and *Linnaea borealis*. Another 28 species, including 18 evergreen coniferous plants, and 10 evergreen broadleaved plants, belong to the evergreen arboreal *alpine* group of plants. These 28 species include 7 endemic evergreen coniferous plants, namely *Taxus caespitosa*, *Abies koreana*, *Picea intercedens*, *P. koraiensis*, *P. pungsanensis*, *P. tonaiensis* and *Thuja koraiensis*; and one endemic evergreen broadleaved plant *Ledum palustre* var. *maximum*. The appearance of the 8 Korean endemic arboreal alpine plants (19.5% of the total evergreen arboreal arctic-alpine and alpine plants considered here) no doubt reflects the long-term isolation of species in the peninsula, and the local environmental peculiarities which have both accentuated this isolation and aided the development of genetic diversity.

Within the Korean peninsula, 7 major distributional patterns can be found from the 41 selected evergreen arboreal arctic-alpine and alpine plants. First, 8 species are present throughout the peninsula, and these include *Taxus cuspidata* (920–1507 m a.s.l., means of the lower and upper limits on a national scale), *Abies holophylla* (493–1008 m), *A. nephrolepis* (918–1507 m), *Picea jezoensis* (980–1646 m), *Pinus koraiensis* (632–1308 m), *Sabina sargentii* (1367–1650 m), *Rhododendron dauricum* (1013–1450 m) and *R. fauriae* form. *rufescens* (1050–1547 m). No endemic species are found in this group. Second, 5 species are present from the north to the midlands, and consist of *Taxus cuspidata* var. *latifolia*, *Pinus pumila* (1485–1805 m), *Thuja koraiensis* (1077–1573 m), *Rhododendron aureum* (1586–1893 m) and *Vaccinium vitis-idaea* var. *minus* (1500–1835 m). *Thuja koraiensis* is an endemic species. Third, one species, *Empetrum nigrum* var. *japonicum* (1700–1900 m), is located both in the north and in Cheju Island, but not in between. Fourth, one endemic species, *Abies koreana* (1060–1900 m), is present in the south and in Cheju Island. Fifth, 23 species are restricted solely to the north, and include *Picea intercedens*, *P. koraiensis*, *P. pungsanensis*, *P. tonaiensis*, *Pinus tabulaeformis*, *Juniperus sibirica* (1567–), *Sabina davurica*, *Dryas octopetala* var. *asiatica* (1750–), *Andromeda polifolia*, *A. polifolia* var. *acerosa*, *Chamaedaphne calyculata*, *Ledum palustre* var. *diversipilosum* (1200–1575 m), *L. palustre* var. *latifolia*, *L. palustre* var. *maximum*, *L. palustre* var.

Table 5.8. Relative tolerance of arctic-alpine and alpine plants at the vertical distributional limits in Korea.

Plants/tolerance	High summer-temperatures at the lower limit	Low winter temperatures at the upper limit
Evergreen coniferous plants	Tolerant (31.0°C to 15.2°C)	Intolerant (−38.5°C to −7.1°C)
Endemic evergreen coniferous plants	Moderate (28.6°C to 19.8°C)	Moderate (−33.0°C to −9.8°C)
Evergreen Broadleaved plants	Intolerant (−26.3°C to 16.7°C)	Tolerant (−40.4°C to −19.2°C)

minus (1600–2300 m), *L. palustre* subsp. *decumbens* (1800–2540 m), *L. palustre* var. *yesoense*, *Oxycoccus microcarpus*, *O. quadripetalus*, *Phyllodoce caerulea* (1540–1800 m), *Rhododendron parvifolium* (800–1100 m), *Vaccinium vitis-idaea* var. *genuinum* and *Linnaea borealis*. Five endemic species, e.g., *Picea intercedens*, *P. koraiensis*, *P. pungsanenensis*, *P. tonaiensis* and *Ledum palustre* var. *maximum* are exclusively found in the north. Sixth, two species, *Picea schrenkiana* and *Taxus caespitosa*, are restricted to the midlands, and of these, *Taxus caespitosa* (1590–) is an endemic species. Seventh, one species, *Diapensia lapponica* subsp. *obovata* (1800–1950 m), is restricted to Cheju Island.

The occurrence of 38 species (92.7% of these Korean evergreen arboreal arctic-alpine and alpine plants) in the north of the Korean peninsula, including the 23 species which are exclusively confined to the north, may have been encouraged by: first, the north-south linkage of mountain ranges in East Asia and within the Korean peninsula, which has enabled the frequent floristic exchanges of the Pleistocene to take place there; and second, the availability of a wide range of habitat conditions in the north both past and presently, viz., alpine, subalpine, montane, lowland, coastal and swampy areas, which have aided the survival and development of these species. The isolation of *Empetrum nigrum* var. *japonicum* in the north and in Cheju Island, and that of *Diapensia lapponica* subsp. *obovata* in Cheju Island itself, seems to have been the result of climatic changes during the Pleistocene glaciation, and especially the climatic amelioration of the post-Pleistocene phase, which has broken a presumed former wider distribution into fragments. The disjunctive distribution of *Abies koreana* in the south and in Cheju Island has no doubt been caused similarly.

In respect of the vertical distributional ranges relating to the means of the lower limits for the 23 evergreen arboreal arctic-alpine and alpine species for which data are available, four distinct divisions may be delineated. First, the altitudes of ca. 500 to ca. 1000 m a.s.l. are dominated mainly by evergreen coniferous plants; second, at ca. 1000 m a.s.l., two endemic evergreen coniferous plants, e.g. *Thuja koraiensis* and *Abies koreana* are to be found; third, from ca. 1000 to ca. 1500 m a.s.l., there is a mixture of evergreen coniferous plants and evergreen broadleaved plants; and fourth, from ca. 1500 to ca. 1800 and above, communities are dominated by evergreen broadleaved plants.

The presence of evergreen arboreal broadleaved plants mainly in the northern alpine areas, at ca. 1500 to ca. 1800 m a.s.l. or above, in which low summer temperatures and very cold winter temperatures prevail, is no doubt due to their physiognomic and ecophysiological adaptations against cold environments, as specified previously (pp. 122–127). Below the alpine zones, or in the south, where higher summer temperatures prevail, these adaptations are disadvantageous to the survival of these species. In these places, there are more evergreen coniferous plants and some evergreen broadleaved plants, which have a similar tolerance to high summer temperatures but they remain less tolerant to low winter temperatures. Evergreen coniferous plants, in general, occupy a relatively wide distributional range (both horizontally and vertically) from the north to the south, and from the lower montane to alpine zones. The lower part of the 'transitional' zone, at ca. 1000 m a.s.l., is occupied by endemic evergreen coniferous plants, which have a relatively high tolerance to high summer temperatures, and are still tolerant of low winter temperatures. Evergreen coniferous plants also extend to lower montane areas at ca. 500 to 1000 m a.s.l., where they display high-summer temperature adaptive ability, and are mixed in with deciduous broadleaved plants. In summary, in respect of tolerance to high summer temperatures at the lower vertical limit of their range, and of low winter temperatures at the upper vertical limit of their range, the evergreen arboreal arctic-alpine and alpine plants in Korea may be divided into four groups; first, the high summer-temperature-intolerant but low winter-temperature-tolerant evergreen broadleaved plants; second, a mixture of mainly evergreen broadleaved plants but also a few evergreen coniferous plants, which have an intermediate ecophysiological range between the evergreen broadleaved plants and the endemic evergreen coniferous plants; third, the endemic evergreen coniferous plants, which are moderately adaptive to both high summer temperatures and low winter temperatures; and fourth, the high summer-temperature -tolerant, but low winter-temperature intolerant evergreen coniferous plants, which are often mixed with other deciduous broadleaved plants. Further ramifications of these divisions are examined in the next chapter, which deals, at least in part, with the distribution of some of these species on South Korea's third highest mountain, Mt. Sorak.

CHAPTER 6

Alpine and subalpine vegetation on Mt. Sorak

The Mt. Sorak massif connects the northern Korean mountain ranges to those in the south, and contains a rich flora, with both northern and southern floristic elements present. Here, however, we are concerned particularly with an assessment of its alpine and subalpine vegetation, and this, it is hoped, will help in the understanding of past and present-day vegetation dynamics for many parts of Korea. The subalpine vegetation zone has been taken to refer to the transitional zone between the upper limit of tall shrubs or trees, and the continuous montane forest zone (p. 83). In the case of Mt. Sorak, evergreen coniferous and broadleaved plants and deciduous broadleaved plants, in either tree or shrub form, are found together in this zone. Within it, the occurrence of a local tree-line and krummholz also is discussed, and special interest has been placed on the presence of *Pinus pumila*, a classic northern alpine and subalpine species in East Asia, since Mt. Sorak is at the southernmost limit of its range in the Asian continent.

Environment

Mt. Sorak (Soraksan in Korean) is located to the north of the Taebaik Mountain Range, which links the peninsula's northern alpine and subalpine areas to southern Korea. It occupies the district bounded by Latitude 38°05′ 25″ to 38°12′ 36″ North, and Longitude 128°18′ 06″ to 128°30′ 43″ East, and covers ca. 373 km^2. Taecheongbong (the Huge Blue Peak, in Korean translation) forms the summit of Mt. Sorak, at 1708 m a.s.l., and it is the third highest mountain after Mt. Halla (1950 m a.s.l.) and Mt. Chiri (1915 m a.s.l.) in South Korea (Fig. 6.1, Photo 6.1).

The denomination of Mt. Sorak as a Natural Monument (No. 171) in 1965, and as a National Park (No. 5) in 1970, both by the Korean Government, and further as a Conservation Area for the Man & Biosphere Project in 1982, by UNESCO, is partly due to the large number of plant species (ca. 1013 spp.) recorded within its confines, and its relatively rich fauna, with 25 species of mammal (including the Korean panda), ca. 90 species of bird, 11 species of reptile, 9 species of amphibian, 6 species of spider, ca. 360 species of insect and ca. 40 species of fish (Korean National Park Authority, 1984; Yim & Baik, 1985).

Geologically, Mt. Sorak is composed of a core of Precambrian gneiss, with later Mesozoic sediments, Mesozoic granites and Quaternary sediments superimposed. Some remnant pieces of the original uplifted peneplain which formed it are found on the summits of the Soraksan Range, and the other ranges branching out of Taecheongbong. The eastern slope of the Sorak Range is being rapidly dissected by river erosion grading down to the nearby east coast, to form a steep and rugged topography, while the western slope displays, in contrast, a relatively gentle landscape (Chung *et al.*, 1984). The distance to the east coast is about 15 km, and that to the west coast is about 200 km. There are six river systems draining the eastern slope and five on the western slopes. The steepness of slope on the eastern side has recently accelerated soil erosion processes near the summit of Mt. Sorak.

The climatic conditions on Mt. Sorak can be estimated from two adjacent Meteorological Stations, Sokcho (38°12′ N, 128°36′ E, 25.8 m a.s.l.) on its eastern side, and Inje (38°03′ N, 128°10′ E, 199.7 m a.s.l.) on its western side (see

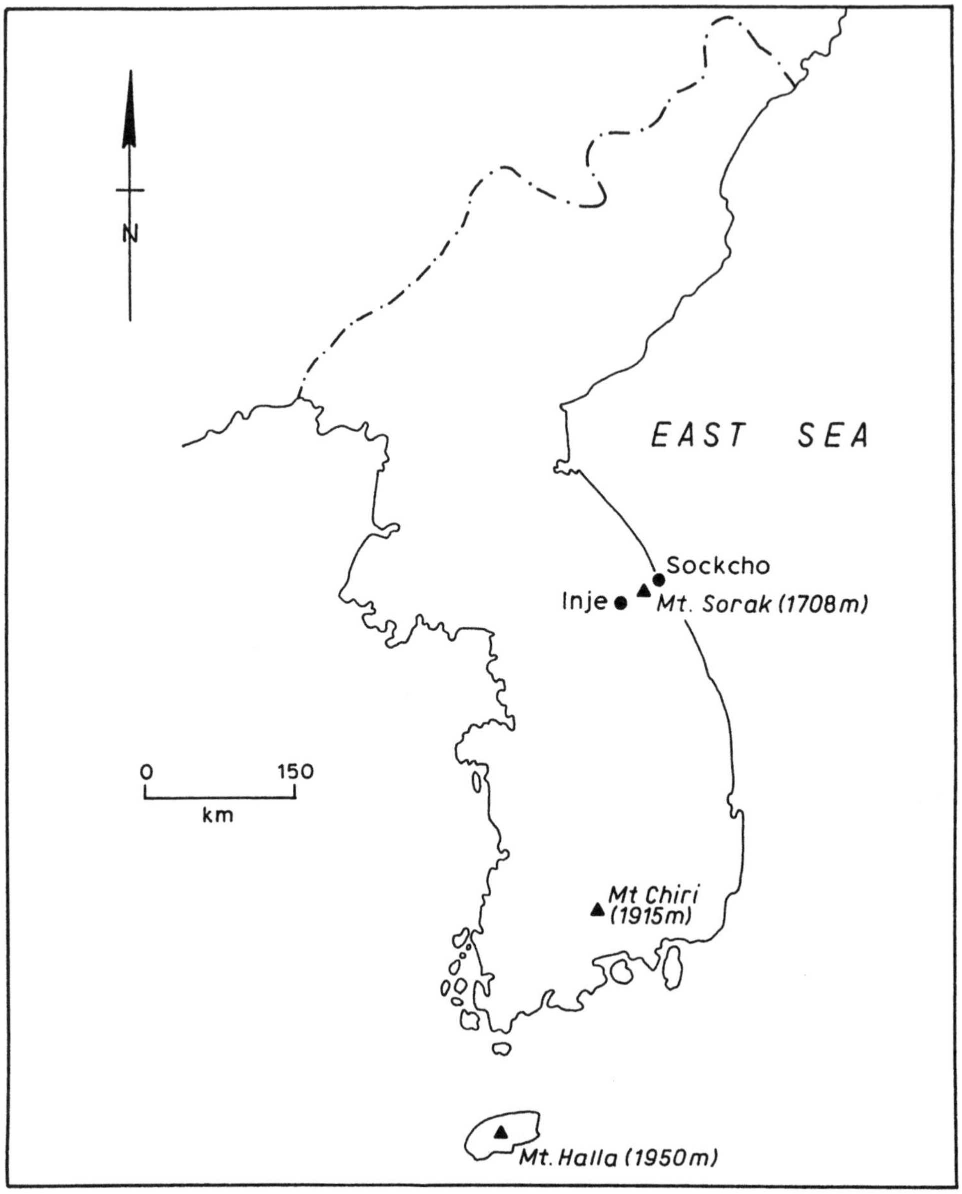

Fig. 6.1. The location of Mt. Sorak.

Fig. 6.1). Additionally, meteorological data have been collected on a short-term basis from the Taecheongbong Shelter, at the top of Mt. Sorak itself. The climatic conditions for Sokcho (1968–1986) and Inje (1971–1985) are presented in Table 6.1. Those for the Taecheongbong Shelter are more usefully reviewed later (Table 6.6, below).

It is clear from Table 6.1 that the annual and seasonal mean temperatures of Sokcho, which lies adjacent to the East Sea, are generally milder than those of Inje, which is in an inland area. The mean temperatures from May to July at Inje, however, are slightly higher than those of Sokcho. The annual and seasonal precipitation in Sokcho are also generally higher than those of Inje, but the rainfall from April to July in Inje is greater than that of Sokcho. Overall, the eastern slope of Mt. Sorak may be said to have relatively milder and more humid conditions than those present on its western slope, with the exception that, during the late spring and early summer,

Photo 6.1. Mt. Sorak (Photograph by J.M. Kim.)

temperature and rainfall on the western slope of Mt. Sorak is relatively higher, and more abundant than that on the eastern slope.

Flora and vegetation: general aspects

The first investigation of flora and vegetation on Mt. Sorak was carried out by Chung in 1923, and in 1935 Nakai stressed the significance of Mt. Sorak in East Asian biogeography by reporting the southernmost limit of *Pinus pumila* there (Chung & Lee, 1963). The total number of plant species listed for Mt. Sorak varies according to the authority; for those who have provided details, 109 families, 305 genera and 642 species were given by Chung & Lee (1959); 127 families, 376 genera, 758 species, 181 varieties and 14 forms by Chung & Lee (1963); 853 species by the Korean Ministry of Education (1967); 801 species, 4 varieties and 45 forms by Lee & Yim (1978); 852 species by the Province of Kangwon (1984); and most recently, 1013 species by Yim & Baik (1985).

As to the vegetation as a whole, Choi & Yim (1984) have listed 10 dominant species on Mt. Sorak, namely *Quercus mongolica*, *Q. serrata*, *Pinus densiflora*, *Acer pseudo-sieboldianum*, *Carpinus laxiflora*, *Styrax obassia*, *Fraxinus rhynchophyllus*, *Tilia amurensis*, *Lindera obtusiloba* and *Abies holophylla*. In the same year, however, Park (1984) reported many more common species, namely at high and middle altitudes, viz., *Pinus densiflora*, *Quercus mongolica*, *Acer formosum* var. *coreanum*, *A. mono*, *Carpinus laxiflora*, *Fraxinus rhychophylla*, *Styrax obassia*, *Lindera obtusiloba*, *Tilia amurensis* var. *barbigera*, *Rhododendron schlippenbachii*, *R. aureum*, *Thuja koraiensis*, *Pinus pumila*, *Abies nephrolepis*, *Pinus koraiensis*, *Quercus urticefolia*, *Carpinus erosa*, *Betula ermanii* var. *acutifolia*, *Populus maximowiczii*, *Cornus controversa*, *Taxus cuspidata*, *Lespideza bicolor* var. *japonica* and *Sasamorpha purpurascens* var. *borealis*.

The 65 Korean endemic species existing on Mt. Sorak (ca. 407 species are present in the peninsula *in toto*) form a high proportion of the total number found in the country as a whole. A few more (75 species) are present on Mt. Halla, but only 42 endemic species are found on Mt. Chiri

Photo 6.2. Mature *Quercus mongolica* – *Thuja koraensis* forest, Sorak National Park, ca. 1500 m. (Photograph by D. Watts.)

(Lee, 1980b, 1984; Lee & Yu, 1984; Park, 1984; Yim & Baik, 1985). Seventeen species out of the 65 species are endemic to Mt. Sorak alone. As a whole, the Korean endemic species which grow on Mt. Sorak comprise both arboreal and herbaceous plants, as indicated in Table 6.2.

The occurrence of the 65 Korean endemic species on Mt. Sorak is likely to be due, in the first instance, to the frequent floristic movements within the peninsula in the past under changing climatic conditions (see pp. 97–98). The 17 endemic species on Mt. Sorak include several arboreal plants, e.g. *Taxus caespitosa*, *Pinus densiflora* form. *erecta*, *Clematis fusca* var. *coreana*, *Pyrus ussuriensis* var. *diamantica*, *Spiraea pubescens* var. *lasiocarpa*, *Tilia megaphylla*, *Forsythia ovata*, and 10 herbaceous plants, e.g. *Anemone koraiensis*, *Vicia venosa* var. *cuspidata* (?), *Buplerum euphorbioides*, *Androsace cortusaefolia*, *Gentiana axillariflora* var. *coreana*, *Adenophora glandiflora* form. *alba*, *A. lamarkii* form. *albiflora*, *Hanabusaya asiatica* form. *albescens*, *H. asiatica* form. *intermedia* and *Leontopodium coreanum*. The appearance of these 17 indigenous endemic species on the mountain is likely to be the result of either the hybridisation of species through the interchange of plants between mountains and the adjacent plains, or the evolution of new species *in situ* within the diverse environmental and habitat conditions.

The fact that Mt. Sorak is a prime transitional zone, which contains the general southernmost limit of many northern floristic elements and the general northernmost limit of many of the southern floristic elements, emphasises its role as a confluence of the northern and southern floras in Korea, and provides invaluable biogeographical evidence as to the possible plant migrations, which were associated with the climatic oscillations of the past.

Photo 6.3. Late autumn scene of Mt. Chiri (1915 m).

Mt. Sorak forms the general southernmost limit of 16 arboreal and 24 herbaceous plants of the northern floristic element of East Asia. The 16 arboreal plants are 2 evergreen coniferous species (*Pinus pumila*, *Thuja koraiensis*; Photos 6.2. and 6.4), 1 evergreen broadleaved species (*Rhododendron aureum*; and 13 deciduous broadleaved species, e.g. *Salix maximowiczii*, *Chosenia bracteosa*, *Alnus fruticosa* var. *mandshurica* form. *normalis*, *Menispernum dahuricum* var. *pilosum*, *Crataegus maximowiczii*, *Pyrus ussuriensis* var. *diamantica*, *Rosa koreana*, *R. marretii*, *R. suavis* form. *rosea*, *Sapium japonicum*, *Arctous ruber*, *Vaccinium uliginosum*, and *Sinomalus komarovii*. The 24 herbaceous plants include *Carex lasiolepis*, *Eriophorum gracile*, *Allium senescens*, *A.*

Table 6.1. Climatic conditions of Sokcho and Inje, Korea.

Element	Station	Month					
		Jan	Feb	Mar	Apr	May	June
Mean temp.	Sokcho	-0.7	0.3	4.6	10.9	10.5	18.9
(C)	Inje	-5.6	-3.0	3.0	10.3	15.6	19.8
Precipitation	Sokcho	43.9	55.5	52.7	77.5	76.5	103.0
(mm)	Inje	15.6	16.3	37.7	81.4	89.1	116.5
Element	Station	Months					
		July	Aug	Sept	Oct	Nov	Dec	Annual
Mean temp.	Sokcho	22.7	23.8	19.7	14.7	8.0	2.3	11.8
(°C)	Inje	23.4	23.1	17.6	11.1	4.1	-2.7	9.7
Precipitation	Sokcho	218.0	296.2	191.5	82.5	57.2	33.7	1304.2
(mm)	Inje	236.4	273.2	132.6	37.4	33.8	23.0	1092.3

(sources Central Meteorological Office, 1982; Kangnung District Meteorological Service, 1986.)

Table 6.2. Korean endemic species on Mt. Sorak.

Life form	Species Composition
Arboreal Plants (25 spp.)	*Taxus caespitosa** (*ECP*), *Pinus densiflora* form. *erecta** (ECP), *Salix hallaisanensis*, *S. hallaisanensis* var. *longiflora*, *Clematis fusca* var. *coreana**, *C. trichotoma*, *Berberis koreana*, *Deutzia coreana*, *D. coreana* var. *triradiata*, *Pyrus ussuriensis* var. *diamantica**, *P. ussuriensis* var. *hakunensis*, *Spiraea pubescens* var. *lasiocarpa**, *Gleditsia japonica* var. *koraiensis*, *Lespedoza thunbergii* var. *intermedia*, *Rhamnus koraiensis*, *Tilia megaphylla**, *T. rufa*, *T. taquetii*, *Acanthopanax chiisanensis*, *Oplopanax elatus*, *Vaccinium koreanum*, *Forsythia ovata**, *Lonicera coreana*, *L. okamotoana*, *Weigelia subsessilis*.
Herbaceous Plants (40 spp.)	*Carex fusanensis*, *C. okamatoi*, *Aletris fauriei*, *Megaleranthis saniculaefolia*, *Thalictrum coreanum*, *T. rochebrunianum* var. *glandisepalum*, *Anemone koraiensis**, *Cardamine amaraeformis*, *Saxifraga oblongiflora*, *Filipendula glaberrima*, *Vicia venosa* var. *cuspidata**, *V. venosissima*, *Buplerum euphorbioides**, *Androsace cortusaefolia**, *Gentiana axillariflora* var. *coreana**, *Salvia chanroenica*, *Pseudolysimachion kiusiana* var. *diamantica*, *P. rotundum* var. *subintegrum*, *P. rotundum* var. *coreanaum*, *Scrophularia koraiensis*, *Asperula lasiantha*, *Patrinia saniculaefolia*, *Scabiosa mansenensis* form. *pinnata*, *Adenophora glandiflora*, *A. glandiflora* form. *alba**, *A. lamarkii* form. *albiflora**, *Hanabusaya asiatica*, *H. asiatica* form. *alba*, *H. asiatica* form. *albescens**, *H. asiaica* form. *intermedia**, *H. asiatica* form. *rubescens*, *H. asiatica* form. *rubropurpurea*, *Cirsium chanroenicum*, *Leontopodium coreanum**, *Saussurea diamantica*, *S. diamantica* var. *longiflora*, *S. nutans*, *Senecio koreanus*.

(*Sources*: *Lee*, 1980, 1984; *Lee & Yu*, 1984; *Park*, 1984; *Yim & Baik*,1984.)
* *Indicates endemic species of Mt. Sorak.*
ECP = Evergreen Coniferous Plant.

victorialis var. *platyphyllum*, *Lilium cernuum*, *L. distichum*, *Iris uniflora* var. *caricina*, *Silene jeniseenis*, *S. koreana*, *Anemone glabrata*, *Eranthis stellata*, *Leontice microrhyncha*, *Saxifraga punctata*, *S. reniformis*, *Bupleurum euphorbioides*, *Pedicularis manshurica*, *P. verticillata*, *Adenophora lamarkii*, *Saussurea triangulata* subsp. *manshurica*, *Androsace cortusaefolia*, *Isopyrum raddeanum*, *I. leveileanum*, *Waldsteinia ternata* and *Leontopodium leoiolepis* (Chung & Lee. 1963; Korean Ministry of Education 1967; Lee & Yu, 1984). The majority of the arboreal species are confined to high altitudes viz. *Pinus pumila* (1590–1700 m), *Thuja koraiensis* (860–1650 m), *Rhododendron aureum* (1270–1640 m), *Salix maximowiczii* (850 m), *Alnus fruticosa* var. *mandshurica* form. *normalis* (1250–650), *Crataegus maximowiczii* (1424 m), *Rosa koreana* (1520–640 m), *R. suavis* form. *rosea* (500–1550 m), *Arctous ruber* (1600–1700 m) and *Vaccinium uliginosum* (1500–1700 m) (Chung & Lee, 1965; Lee & Chang, 1984; Yim & Baik, 1985). In recent years, however, several of the 40 species described above have also been reported in iso-

Photo 6.4. The Study Area, Taecheongbong (1708 m a.s.l.), Mt. Sorak, Korea. *Betula ermanii* var. *genuina* to left (East) of ridge; *Pinus pumila*, *Rhododendron mucronulatum* var. *ciliatum* and *Quercus mongolica*, down right (West) of ridge; *Pinus pumila* and *Thuja koraiensis*, upper right (West) of ridge; tree-line of *Abies nephrolepis* (Photograph by D. Watts).

lated patches to the south of Mt. Sorak as well. For example, *Bupleurum euphorbioides* has been noted on Mt. Sokri; *Lilium cernuum* on Mts. Sokri, Wolak and Chiri, and on Ulreung Island; *Isopyrum raddeanum* on Mts. Odae and Chukryong; *Eranthis stellata* on Mts. Samak, Kunja and Mai; and *Vaccinium uliginosum* on Mt. Halla, Cheju Island (Lee & Yu, 1984; Yim & Baik, 1985).

The general northernmost limit of the 6 southern floristic elements is reached on Mt. Sorak by 3 arboreal species, viz. *Styrax japonica* (200–700 m), *Grewia parviflora* and *Triadica japonica* (180–630 m), which are present mainly in both lowland and montane zones; and 3 herbaceous species (*Megaleranthis saniculifolia*, *Phegopteris decursive-pinnata* and *Carex okamotoi*. *Styrax japonica*, however, has also recently been reported on Mt. Keumkang, which is situated to the north of Mt. Sorak (Lee & Chang, 1984; Lee & Yu, 1984; Yim & Baik, 1985).

Overall, the flora and vegetation of Mt. Sorak demonstrates three distinctive biogeographic characteristics, namely, first, a high floristic diversity for the size of the area (ca. 1013 spp. on ca. 373 km^2); second, the occurrence of many Korean endemic species (65 spp.), and especially 17 species endemic to Mt. Sorak alone; and third, the coexistence of northern and southern floristic elements. Many of the northern elements lie at, or close to, the southernmost limit to their range in East Asia, and many of the southern elements lie close to, or at the northernmost limit of their range, thus increasing the biogeographical importance of the site. Biogeographically, the disjunctive distribution of cold freshwater fishes between Northeast China and Mt. Sorak, e.g. *Brachymystax lenok*, *Oncorhynchus masou*, *Moroco lagowskii*, *M. oxycephalus*, *M.* spp. and *Barbatula toni*, also provides evidence of an exchange of biota in the past, between the northern part of East Asia and Mt. Sorak, where no connecting river channel is now available. The characteristics of both flora and fauna imply that Mt. Sorak has served as one of the major refugia for biota in the Korean peninsula, certainly during the later Pleistocene glacial phases.

Subalpine vegetation

Since Mt. Sorak occupies an extensive area, and contains a large number of species, with many differing vertical ranges, we decided to concentrate our present work on the subalpine vegetation, and especially the northern floristic elements within it, since opportunities for field work, in this remote region, were limited, and in this way, at least one plant formation would be examined in some depth.

The top of Mt. Sorak (the highest part of the mountain) has been described as a subalpine zone in Chapter 2, and a detailed examination of the flora and vegetation structures would tend to support this. However, it should be noted that there are several species present which more strictly belong to an alpine species group *per se*, including *Pinus pumila*, *Rhododendron aureum*, *Rosa koreana*, *Vaccinium uliginosum*, *Oplopanax elatum*, *Geranium maximowiczii*, *Adenophora lamarkii* and *Anemone narcissiflora*. This suggests that the vegetation on the highest part of Mt. Sorak lies close to the top of the subalpine zone, and comes very close to being an intermediate zone between true alpine vegetation and subalpine vegetation. Further, krummholz forms of both deciduous broadleaved and evergreen coniferous plants exist there. Within this context, we intend to consider the species composition of this subalpine vegetation, along with the vertical ranges of component species, their distribution, the structure and formation of the tree-line krummholz, and the relationship of all these features to past vegetation history and present environmental conditions.

Species composition

The geographical boundary of this chapter is confined to the subalpine zone of Mt. Sorak, as defined above, comprising the area above ca. 1500 m a.s.l., in which both northern and alpine plants are common.

Over the years, numerous alpine and subalpine species have been reported from Mt. Sorak (vertical ranges in parentheses). These include evergreen coniferous tree and shrub species, e.g. *Taxus caespitosa* (1650–1700 m a.s.l.), *Pinus pumila* (1500–1700), *Abies nephrolepis* (700–1700), and *Thuja koraiensis* (800–1700); deciduous broadleaved tree and shrub species, e.g. *Betula ermanii* var. *acutifolia* (700–1700), and *Rhododendron fauriae* form. *rufescens* (800–1700); and herbaceous species, e.g. *Hanabusaya asiatica*, *Geranium maximowiczii*, *Adenophora lamarkii*, and *Anemone narcissiflora*, all named by Chung & Lee (1959). Later, in 1965, these two authors added more species to their list, including many deciduous broadleaved plants, e.g. *Alnus fruticosa* var. *manshurica* (800–1700), *Quercus mongolica* var. *manshurica* (950–1700), *Clematis koreana* (1400–1600), *Rosa koreana* (–1700), *Acer tschnoskii* var. *rubripes* (750–1600), *Salix hallaisanensis* (1550–1650), *Rhododendron fauriae* form. *rufescens* (700–1700), *R. mucronulatum* var. *ciliatum* (750–1700), *Arctous ruber* (1600–1700), *Vaccinium koreana* (1300–1550), *V. uliginosum* var. *krushianum* (1500–1700), *Oplopanax elatum* (1300–1600), *Syringa elatum* (1350–1650) and *Lonicera sachalinensis* (1350–1600).

In 1967, the Korean Ministry of Education listed 5 arboreal species for Mt. Sorak, namely *Taxus cuspidata*, *Abies nephrolepis*, *Pinus pumila*, *Thuja koreansis*, and *Rhododendron fauriae* var. *rufescens*; and 23 herbaceous species, e.g. *Anemone narcissiflora*, *Aconitum napiforme*, *Bistorta vulgaris*, *Silene tenuis*, *Potentilla matsumurae*, *Sanguisorba hakusanensis*, *S. parviflora*, *Geranium dahuricum*, *G. eriostemon*, *Viola selkirkii*, *V. xanthopetola*, *Bupleurum euphorbioides*, *Cnidium tachiroei*, *Pedicularis nigrescens*, *P. verticillata*, *Scabiosa mansensis*, *Hanabusaya asiatica*, *Chrysanthemum zawadskii*, *Leontopodium leoiolepis*, *Adenophora lamarkii*, *Eriophrum polystachys*, *Allium taquetii* and *A. victorialis*.

Then, Hagman *et al.* (1978) recorded 2 evergreen coniferous plants, *Abies nephrolepis* and *Pinus pumila*; 9 deciduous broadleaved plants, e.g. *Quercus mongolica*, *Betula ermanii*, *Acer pseudosieboldianum*, *A. tschnoskii* var. *rubripes*, *Syringa wolfii*, *Vaccinium* spp., *Salix hallaisanensis*, *Rhododendron schlippenbachii*, *R. muc-

ronulatum; and 3 climbers, viz. *Tripterygium regelii*, *Clematis koreana* and *C. fusca*, all near the summit of Mt. Sorak (on Taecheongbong).

In 1984, Park reported 4 evergreen coniferous plants, e.g. *Abies holophylla*, *A. nephrolepis*, *Pinus pumila*, *Thuja koraiensis*, 1 evergreen broadleaved plant (*Rhododendron aureum*), and 5 deciduous broadleaved plants, e.g. *Quercus mongolica*, *Spiraea ulimifolia*, *Sorbaria stellipilla* var. *typica*, *Rhododendron mucronulatum* var. *ciliatum* and *R. schlippenbachii* between 1400 to 1600 m a.s.l.

More recently, Yim & Baik (1985) have listed 23 subalpine arboreal species which reach to about 1600 m a.s.l. or thereabouts, including 4 evergreen coniferous species, e.g. *Taxus caespitosa* (1590 m a.s.l.), *Abies nephrolepis* (860–1660), *Thuja koraiensis* (860–1560), and *Pinus pumila* (1590–1700); 18 deciduous broadleaved plants e.g. *Clematis koreana* (1630–1700), *Alnus fruticosa* var. *manshurica* (1250–1650), *Betula costata* (730–1660), *B. ermanii* var. *acutifolia* (260–1700), *Quercus mongolica* (170–1700), *Rosa koreana* (1520–1640), *Prunus nakaii* var. *typica* (530–1690), *Sorbus commixta* (400–1610), *Acer tschnoskii* var. *rubripes* (170–1640), *Tripterygium regelii*, (400–1610), *Salix hallaisanensis* (1560–1640), *Rhododendron mucronulatum* (280–1700), *R. mucronulatum* var. *ciliatum* (510–1700), *R. schlippenbachii* (300–1700), *Vaccinium koreanum* (530–1690), *Oplopanax elatum* (1500–1600), *Lonicera caerulea* var. *edulis* (1620–1690) and *L. sachalinensis* (1040–1610); and one evergreen broadleaved species, *Rhododendron aureum* (1270–1640).

Vertical ranges

There are several clearly-distinguishable vertical vegetation zones on Mt. Sorak. Uyeki, as early as 1933, recognised the presence of a 'warm' forest zone below 1600 m a.s.l., and a 'frigid' forest zone above 1600 m a.s.l., the latter containing arboreal species, such as *Pinus pumila*, *Abies nephrolepis*, *Thuja koraiensis*, *Alnus fruticosa* var *mandshurica* form. *normalis*; and herbaceous plants, e.g. *Hanabusaya asiatica*, *Lilium cernuum* and *Anenome narcissiflora*. Park & Hong (1959) then reported their definition of a 'warm' zone as being below 1100 m a.s.l., and of a 'frigid' zone as being above 1600 m a.s.l. In the latter, a *Thuja koraiensis – Rodgersia podophylla* var. *viritis* association, an *Abies nephrolepis – Acer tschnoskii* var. *rubripes* association, a *Quercus mongolica – Rhododendron schlippenbachii* association, and a *Betula chinensis – Rhododendron schlippenbachii* association were present. Hong (1959) also described *Rhododendron mucronulatum*, *Taxus caespitosa* and *Vaccinium koreana* as being present on Mt. Sorak, as an alpine shrub association. In contrast, Chung & Lee (1959) classified the vegetation zones of Mt. Sorak rather differently, nominating five zones, namely a 'piedmont zone' (below 400 m a.s.l.), a 'forest tree zone' (400–1200 m), a 'sub-tree zone' (1200–1500 m), a 'shrub zone' (1500–1650 m) and an 'herbaceous zone' (1650–1708 m), on the basis of physiognomy and species composition. More recently, Baik & Yim (1983) have reported a 'subarctic forest', containing a *Pinus pumila – Rhododendron mucronulatum* association, a *Thuja koraiensis – Rhododendron schlippenbachii* association, and a *Pinus pumila – Quercus mongolica* association, all present above 1400 m a.s.l. The dissimilarities in respect of these vertical zonations, by these several authorities, are due mainly to the use of the different indicator species listed above.

As determined from our own field work in the summer of 1987, and data from Chung & Lee (1965) and Yim & Baik (1985), the vertical ranges of the 24 subalpine arboreal species which reach above 1600 m a.s.l. on Mt. Sorak can be categorised into four altitudinal groups (Fig. 6.2), which may throw further light on the vertical pattern of distribution of many of the species noted above.

The first of these includes 7 subalpine species which grow *exclusively* between ca. 1500 to 1700 m a.s.l., including 2 evergreen coniferous species (*Pinus pumila*, 1590–1700 m; and *Taxus caespitosa*, 1590–1650 m on Mt. Sorak), and 5 deciduous broadleaved species, namely *Clematis koreana* (1630–1700 m), *Lonicera caerulea* var. *edulis* (1620–1690 m), *Salix hallaisanensis* (1560–1640 m), *Rosa koreana* (1520–1640 m) and *Oplopanax elatum* (1500–1600 m). The group contains 3 Korean endemic species, *Taxus caespitosa*, *Salix hallaisanensis* and *Oplopanax elatum*. The physiognomy of these species displays primarily a

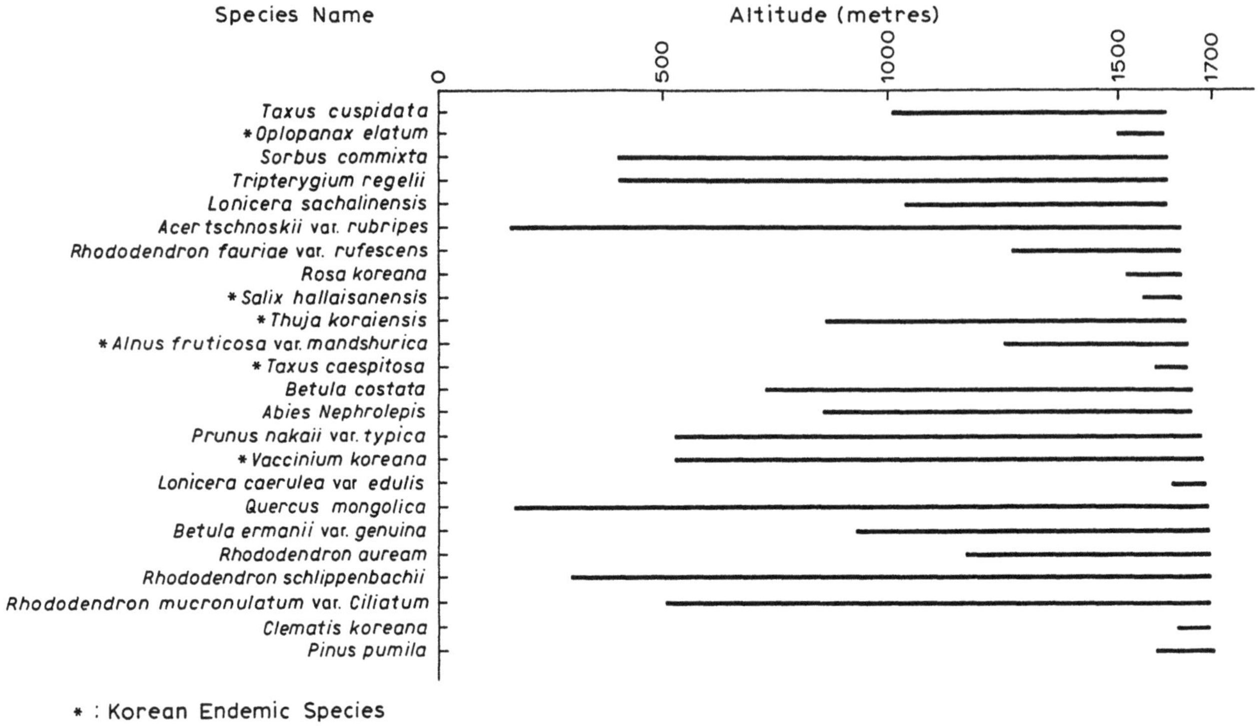

Fig. 6.2. Vertical ranges of sub-alpine species of Mt. Sorak.

shrub form; these are 4 deciduous shrubs, viz. *Lonicera caerulea* var. *edulis*, *Salix hallaisanensis*, *Rosa koreana* and *Oplopanax elatum*, 1 deciduous vine (*Clematis koreana*), 1 evergreen coniferous shrub (*Taxus caespitosa*), and 1 creeping evergreen shrub or small tree (*Pinus pumila*). Undoubtedly, this reflects a morphological adaptation to the harsh environment in the subalpine zone. The isolated occurrence of these 7 species exclusively in the subalpine zone of this mountain, i.e. at 1500 to 1700 m a.s.l., is likely to have been due to movement upslope from primary Pleistocene refuge areas at lower altitudes (see p. 97), as a result of the post-glacial climatic amelioration.

The second group contains 4 subalpine species, which are located on this mountain between 1000 to 1700 m a.s.l. It includes 1 evergreen coniferous tree, *Taxus cuspidata* (1000–1600 m); 1 evergreen broadleaved shrub, *Rhododendron aureum* (1270–1640 m); and 2 deciduous broadleaved species, *Alnus fruticosa* var. *mandshurica* (1250–1650 m) and *Lonicera sachalinensis* (1040–1610 m). This group has one Korean endemic species, *Alnus fruticosa* var. *mandshurica*, and shows an even species balance between deciduous trees, deciduous shrubs, evergreen trees and evergreen shrubs.

In the third group, 7 species have a wider vertical range of from 500 to 1700 m a.s.l. They include 1 evergreen coniferous shrub, *Thuja koraiensis*; 1 evergreen coniferous tree, *Abies nephrolepis*, and 5 deciduous broadleaved species, viz. *Betula costata* (730–1660 m), *B. ermanii* var. *genuina* (900–1700 m), *Prunus nakaii* var. *typica* (530–1690 m), *Vaccinium koreana* (530–1690 m) and *Rhododendron mucronulatum* var. *ciliatum* (510–1700 m). Two Korean endemic species, *Vaccinium koreana* and *Thuja koraiensis* grow in this group. It contains 3 deciduous broadleaved shrubs, viz. *Rhododendron mucronulatum* var. *ciliatum*, *Vaccinium koreana* and *Prunus nakaii* var. *typica*; and 2 deciduous broadleaved trees, *Betula costata* and *B. ermanii* var. *genuina*.

The fourth group shows an even wider vertical range, with species present from below 500 m to 1700 m a.s.l. Among these are 6 deciduous broadleaved species, viz. *Tripterygium regelii* (400–1610 m), *Sorbus commixta* (400–1610 m), *Rhododendron schlippenbachii* (300–1700 m), *Quercus mongolica* (170–1700 m), and *Acer tschnoskii* var. *rubripes* (170–1640 m). None of these are endemics. There are no evergreen species in this group.

At least in the case of the evergreen coniferous species in the first group, these are species with a truly narrow ecological amplitude; they may be close to their limits of survival on Mt. Sorak, and accordingly they may be regarded as being potentially endangered species, especially if current global warming trends are maintained.

From the above, it is clear that a large number of arboreal plants (ca. 24 spp.) occur in the subalpine zone of Mt. Sorak. On the ground, three major subalpine communities are recognisable; first, a *Pimus pumila* community at the summit of Mt. Sorak (Taecheongbong); second, an *Abies nephrolepis* community found to the north, northwest, south and, especially, west of Taecheongbong to a maximum extent, in the latter direction, of 15 km (Fig. 6.3); and third, a *Betula ermanii* var. *genuina* community which grows just below *Abies nephrolepis* (Fig. 6.3).

Recent mapping (Fig. 6.3) has confirmed that Taecheongbong (1708 m a.s.l.), the highest summit of Mt. Sorak, is covered by *Pinus pumila*. *Abies nephrolepis* occurs on the other mountain tops of the Mt. Sorak range, i.e. at heights somewhat below the *Pinus pumila* zone on Taecheongbong and elsewhere at 1518 m a.s.l., 1430 m a.s.l., 1421 m a.s.l. and 1319 m a.s.l. *Betula ermanii* var. genuina grows below the level of *Abies nephrolepis* on all the same peaks. The only one of these species restricted solely to the subalpine zone is *Pinus pumila*. It is perhaps worth noticing at this stage that the higher forest of these communities is dominated by a species with a very narrow ecological amplitude (*Pinus pumila*), and that the other two dominant species (*Abies nephrolepis* and *Betula ermanii* var. *genuina*) have somewhat larger amplitudes; the further significance of this is to be examined later (p. 157). Two other subalpine communities, dominated by *Thuja koraiensis* and *Ligularia fischeri*, and recognised by Yim & Baik (1985), have been included in Fig. 6.3; however, since they are of little importance overall in the general vegetation distributional pattern at the summit of Mt. Sorak, they are not considered further here.

The overall species composition, abundance and coverage of the *Pinus pumila* community, located in the subalpine area between Taecheongbong and Chungcheongbong, a subsidiary peak, then was analysed in detail so as to understand better the vegetation structure associated with it.

This work employed two field survey techniques, namely quadrat and belt (or line) transects, and the combination of the two methods has produced acceptable results.

Eight quadrats of 5 × 5 m in size were taken, so as to examine the coverage of all subalpine species within the *Pinus pumila* community. The predominant area of *Pinus pumila* between Taecheongbong and Chungcheongbong, which covers an area about 500 × 100 m in size, was divided into four sectors, and two quadrats were randomly taken from each sector (Fig. 6.4). The line (or belt) transect (viz. 30 cm width of belt) method was also employed to survey, on a species basis, the whole area of *Pinus pumila*, by taking regularly-placed sample stands. Thus the area of *Pinus pumila* was divided into 20 transect sectors, at intervals of 25 m, from Taecheongbong to the point near the second peak of Mt. Sorak (Chungcheongbong), where *Pinus pumila*'s continuous range ends (Figs. 6.4 and 6.5). The 20 line-transect surveys of this area produced a list of 25 subalpine species for the community, and the results are presented in Table 6.3.

From this, it is clear that the main components of the *Pinus pumila* community as a whole on Mt. Sorak in rank order are, other than the dominant species itself, *Thuja koraiensis*, *Rhododendron mucronulatum* var. *ciliatum*, *Quercus mongolica* and *Betula ermanii* var. *genuina*. Other species include *Abies nephrolepis*, *Taxus caespitosa*, *Acer tschnoskii* var. *rubripes*, *Rhododendron mucronulatum* and *R. schlippenbachii*. However, there are some important differences in species composition within the *Pinus pumila* community here, as between the very wind-exposed col area between Taecheongbong and Chungcheongbong, and on Taecheongbong itself on the one hand; and on Chungcheongbong on the other. In the latter, a much greater range of 'minor' species is found, including *Rhododendron schlippenbachii*, *R. aureum*, *Sorbus commixta*, *Alnus fruticosa* var. *mandshurica*, *Vaccinium koreana*, *Acer ginnala*, *Sabina sargentii*, *Vaccinium uliginosum*, *Triptery-*

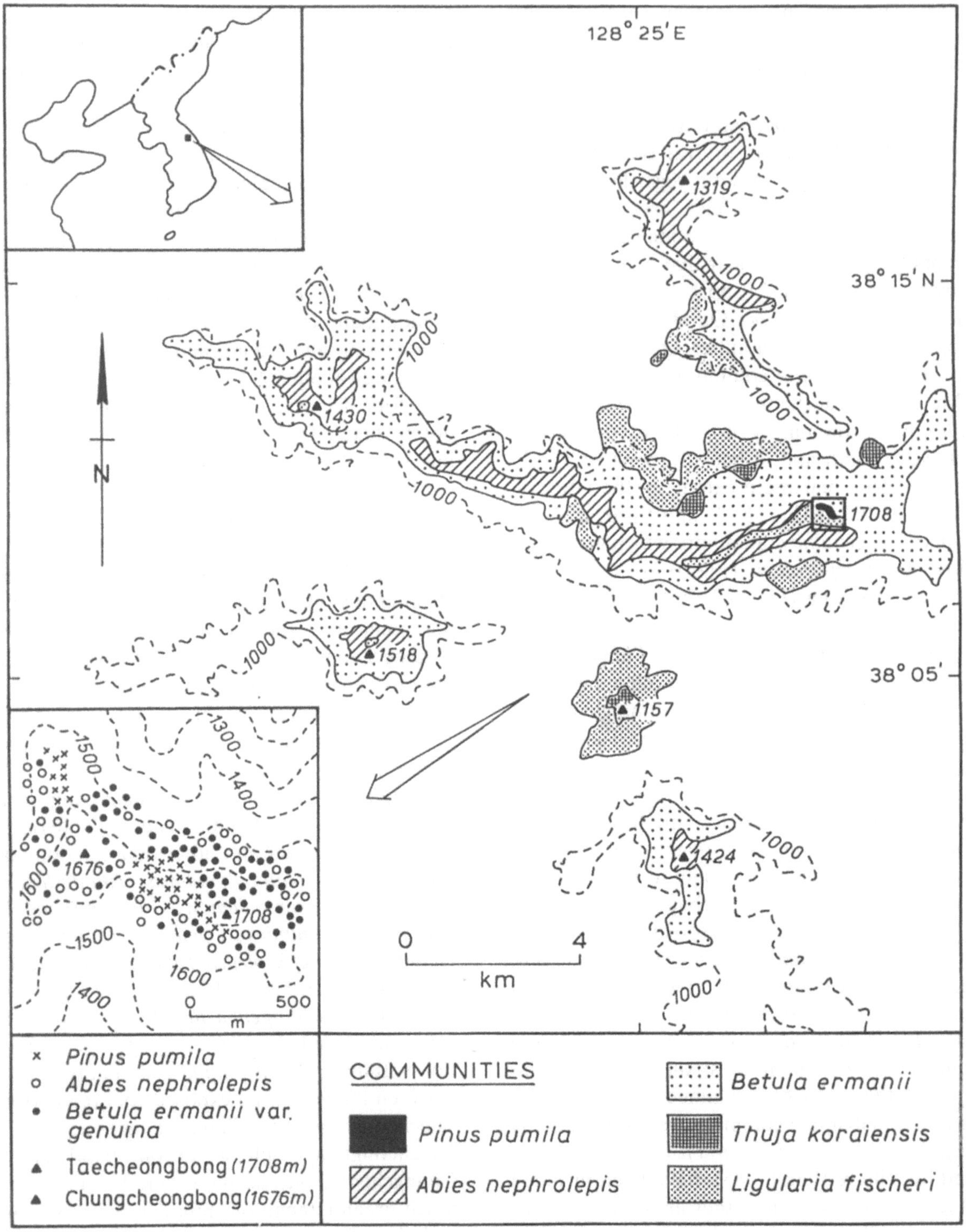

Fig. 6.3. Distribution of sub-alpine communities of Mt. Sorak.

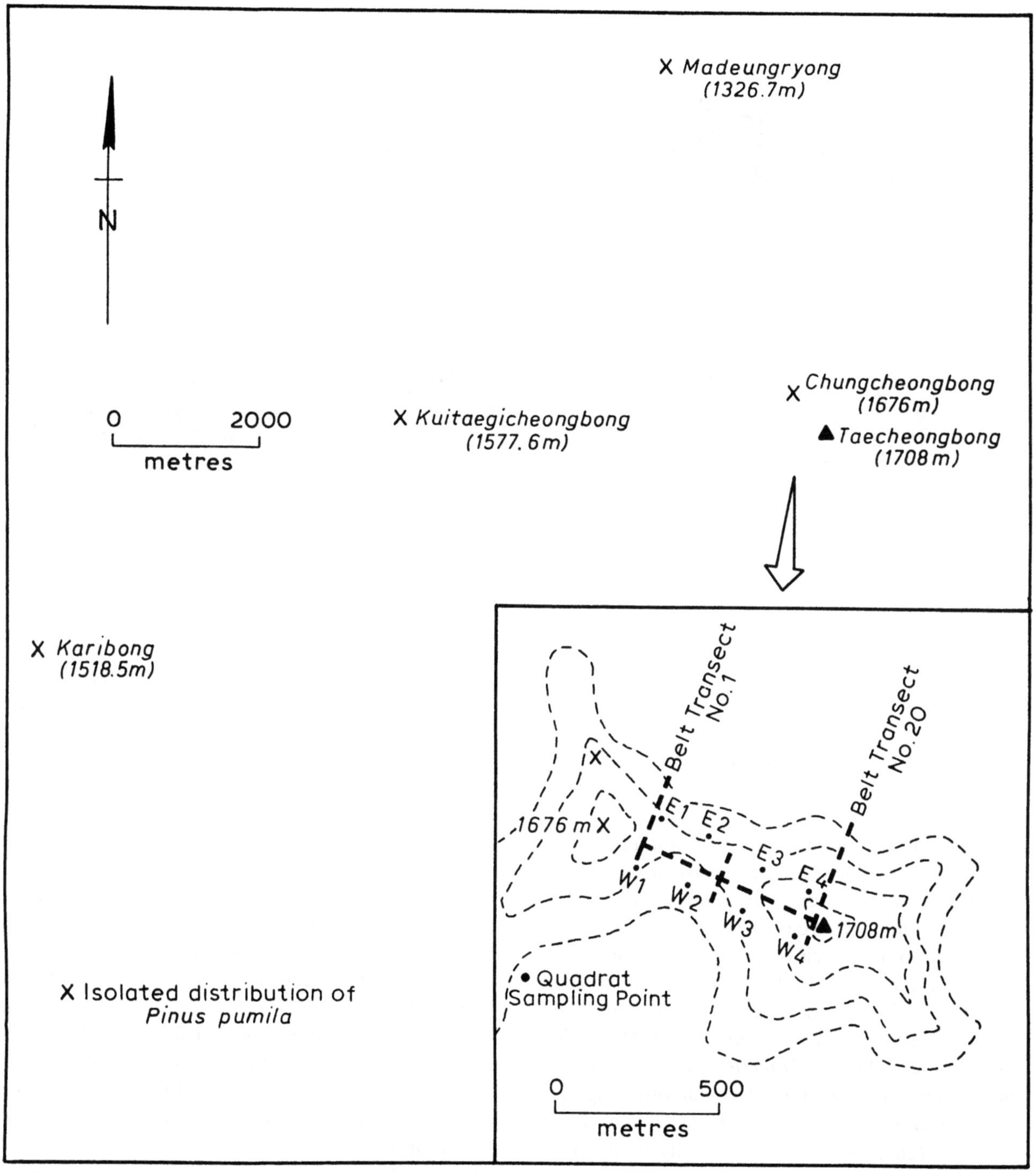

Fig. 6.4. Distribution of *Pinus pumila*, and the sampling area on Mt. Sorak.

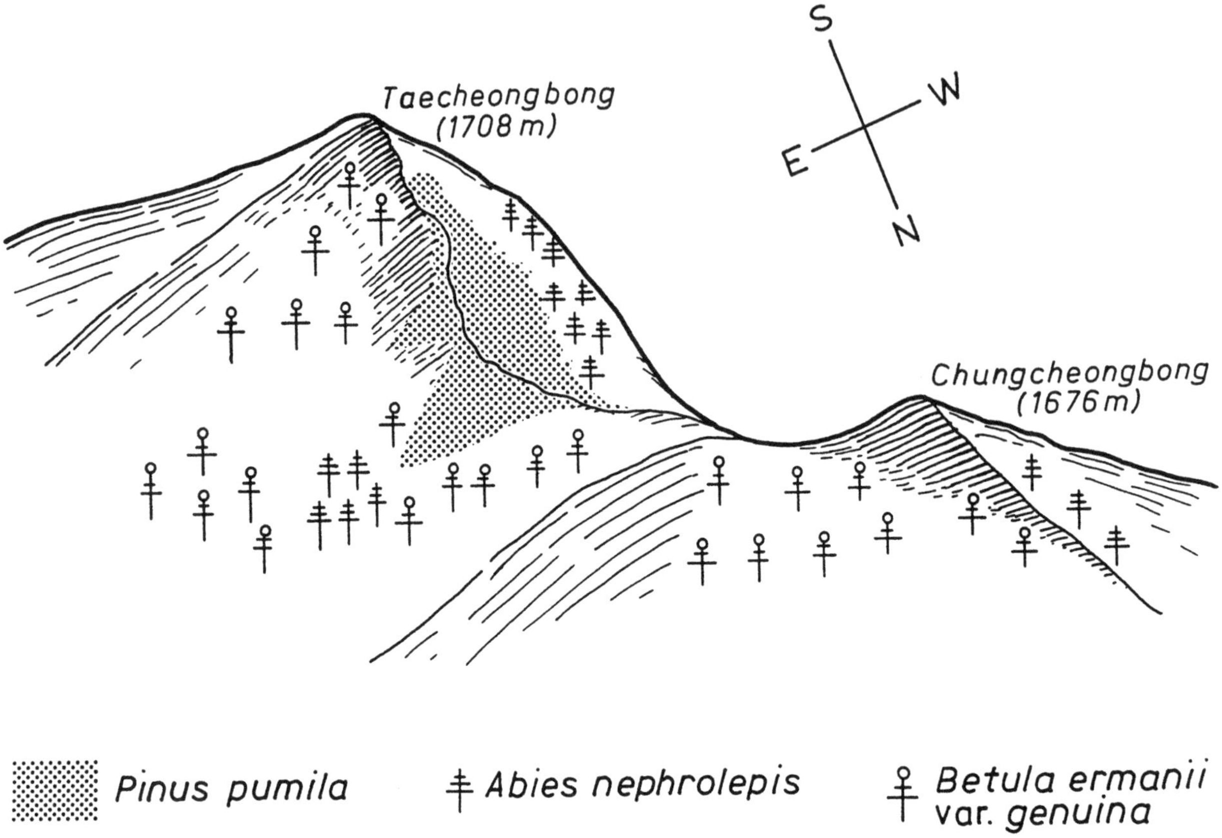

Fig. 6.5. Three-dimensional view of the distribution of *Pinus pumila* and related species on Mt. Sorak.

gium regelii, *Rosa marretii*, *Salix hallaisanensis*, *Oplopanax elatum*, *Lonicera caerulea* var. *edulis*, *Prunus nakaii* var. *typica*, *Betula costata* and *Kalopanax pictum*, whereas in the former area only 4 of those minor species are present, namely *Rhododendron aureum* (col area), *Sabina sargentii* (col area), *Vaccinium uliginosum* (col area and Taecheongbong), and *Sorbus commixta* (solely on Taecheongbong). *Quercus mongolica* is also not present in the vicinity of Taecheongbong. The Taecheongbong area in particular, then, has a very limited range of species (only 8 in all) as compared to the rest of the transect (see Table 6.3). Different plant groupings are found in *Pinus pumila* thicket in other parts of East Asia; thus *Betula middlendorffi*, *Alnus fruticosa*, *Rhododendron chrysanthum*, *R. redowskiana* and lichens in the Okhotsk, USSR (Suslov, 1961; Tseplyaev, 1965); *Empetrum nigrum* var. *japonicum*, *Vaccinium vitis-idaea*, *V. ovalifolium*, *Rhododendron aureum*, *Sorbus sambucifolia*, *Rubus pedatus* and *Ledum palustre* var. *diversipilosum* in Honshu and Hokkaido, Japan (Kobayashi, 1967, 1971; Kawano, 1971; Numata *et al.*, 1912; Yoshino, 1978); *Alnus maximowiczii*, *Sorbus commixta*, *S. sambucifolia*, *S. matsumurae*, *Acer tschnoskii* under moister conditions in Japan; and *Empetrum nigrum* var. *japonicum*, *Loiseleuria procumbens*, *Arceria nana* and *Diapensia lapponica* subsp. *obovata* in particularly windswept habitats in Japan (Ishizuka, 1974). On a physiognomic basis, the *Pinus pumila* community on Mt. Sorak is dominated by shrubby forms, as is the case for the dominant species itself. *Thuja koraiensis*, *Rhododendron* spp., *Quercus mongolica*, *Betula ermanii* var. genuina, *Taxus caespitosa* and *Acer tschnoskii* var. *rubripes*, and the krummholz form of *Abies nephrolepis* are present.

The random quadrat samples (5 × 5 m in size: Fig. 6.4) show the coverage of species in the subalpine area of *Pinus pumila*, and results are presented in Table 6.4. It is clear that this method

Table 6.3. Species composition and abundance of the *Pinus pumila* community on Mt. Sorak.

Species	Transects																					
	1	2	3	4	5	6	7	8	9	10	11	12	13	14	15	16	17	18	19	20	Total	Rank
Pinus pumila	3	9	7	11	13	15	17	13	22	48	45	36	30	22	15	12	11	11	20	13	373	1
Thuja koralensis	–	3	–	–	–	8	7	–	31	56	44	37	30	8	–	–	–	–	11	9	244	2
Rhododendron mucronulatum var. *ciliatum*	–	–	2	6	12	1	2	13	5	17	36	40	7	–	–	13	11	25	24	19	233	3
Quercus mongolica	3	10	4	5	19	11	6	8	11	2	14	–	20	38	14	–	–	–	–	–	165	4
Betula ermanii var. *genuina*	2	5	6	6	7	4	3	6	7	9	7	3	9	2	5	11	9	11	11	14	137	5
Abies nephrolepis	–	–	–	–	–	–	–	–	15	4	2	5	6	8	5	6	5	8	12	6	82	6
Taxus caespitosa	1	2	4	9	10	9	3	8	7	3	4	5	–	4	–	–	–	–	–	–	69	7
Acer tschnoskii var. *rubripes*	1	8	1	7	6	7	1	3	–	1	–	3	–	3	–	–	3	–	–	–	44	8
Rhododendron mucronulatum	2	–	1	–	3	4	2	6	–	3	–	7	–	–	–	–	–	–	–	–	32	9
R. schlippenbachii	2	2	–	–	–	2	5	2	9	–	3	–	–	–	–	–	–	–	–	–	25	10
R. aureum	1	3	2	1	1	3	5	2	1	–	–	–	1	1	–	–	–	–	–	–	21	11
Sorbus commixta	–	3	3	2	2	–	–	4	–	–	–	–	–	–	–	–	–	–	–	2	16	12
Alnus fruticosa var. *manshurica*	2	3	–	–	3	3	–	3	–	–	–	–	–	–	–	–	–	–	–	–	14	13
Vaccinium koreana	–	–	–	–	1	5	–	4	–	–	4	–	–	–	–	–	–	–	–	–	14	13
Acer ginnala	2	3	–	–	3	–	5	–	–	–	–	–	–	–	–	–	–	–	–	–	13	14
Sabina sargentii	–	–	–	–	–	1	–	3	–	2	2	1	3	–	–	–	–	–	–	–	12	15
Vaccinium uliginosum	–	1	–	4	–	–	–	1	–	–	–	–	–	2	–	–	–	4	–	–	12	15
Tripterygium regelii	1	6	2	1	1	–	–	–	–	–	–	–	–	–	–	–	–	–	–	–	11	16
Rosa marretii	–	1	2	1	–	4	–	–	–	–	–	–	–	–	–	–	–	–	–	–	8	17
Salix hallaisanensis	–	–	–	2	1	–	–	–	–	–	–	–	–	–	–	–	–	–	–	–	4	18
Oplopanax elatum	–	1	1	1	1	–	–	–	–	–	–	–	–	–	–	–	–	–	–	–	4	18
Lonicera caerulea var. *edulis*	–	–	–	–	2	1	–	–	–	–	–	–	–	–	–	–	–	–	–	–	3	19
Prunus nakaii var. *typica*	–	1	–	2	–	–	–	–	–	–	–	–	–	–	–	–	–	–	–	–	3	19
Betula costata	1	–	–	–	1	–	–	–	–	–	–	–	–	–	–	–	–	–	–	–	2	20
Kalopanax pictum	–	1	–	–	–	–	1	–	–	–	–	–	–	–	–	–	–	–	–	–	2	20

N.B. Transect No. 1 begins near Chungcheongbong, in which the continuous distribution of *Pinus pumila* ends at its northernmost point, up to No. 20 near Taecheongdong at ca. 1700 m a.s.l.

of sampling has produced a more restricted species list than that of the belt-transect method, though all the main species are represented. The coverage of *Pinus pumila* (24.9%) places this species again as dominant, that for other species being *Thuja koraiensis* (16.7%), *Rhododendron mucronulatum* var. *ciliatum* (9.3%), *Betula ermanii* var. *genuina* (7.5%), *Acer tschnoskii* var. *rubripes* (5.0%), *Quercus mongolica* (3.6%), *Abies nephrolepis* (3.3%), *Rhododendron mucronulatum* (2.0%), *R. aureum* (1.3%), and *Taxus caespitosa* (0.8%). The rest of the area (ca. 21.0%) is occupied by rocky or open spaces as well as by several minor herbaceous species*. The relationships between altitudinal range and percentage of cover in the *Pinus pumila* community show that those species restricted to the highest altitudes have the highest coverage, viz. *Pinus pumila* (1590–1700 m a.s.l.; 24.9% of coverage), *Thuja koraiensis* (700–1600 m a.s.l.; 16.7%), *Rhododendron mucronulatum* var. *ciliatum* (510–1700 m a.s.l.; 9.3%) and *Betula ermanii* var. *genuina* (900–1700 m a.s.l.; 7.5%). In general, one may say that this high mountain area with its hostile environment is occupied only by those species with a highly adaptable ecophysiology against the harsh environment. In Japan, Ito & Nishikawa (1976) have recognised six phases of cyclical change of the *Vaccinio-Pinetum pumilae*, viz. the initially bare land, the pioneer lichen-dominated phase, the building phase, the mature phase dominated by *Pinus pumila*, the degenerating phase and the final terminating phase, producing bare land again after a disintegration of the *Pinus pumila* colony. *Pinus pumila* on Mt. Sorak seems to be

* These include the spring-flowering *Veratrum glandiflorum*, *Erythronium japonicum*, *Geranium eriostemon* var. *megalanthum*, *Potentilla dickinsii* var. *breviseta* and *Adenophora lamarkii*.

Table 6.4. Coverage of the *Pinus pumila* community on Mt. Sorak.

Species	Quadrats								Total species	Mean species	Rank
	E1	W1	E2	W2	E3	W3	E4	W4	No.(coverage)	No.(coverage)	
Pinus pumila	3(18)	4(26)	5(31)	3(26)	6(41)	4(36)	–	2(21)	27(119)	3.4(24.9)	1
Thuja koraiensis	4(14)	1(40)	7(24)	–	7(26)	5(17)	3(48)	–	27(133)	3.4(16.7)	2
Rhododendron mucronulatum var. *ciliatum*	2(6)	2(4)	–	6(10)	3(11)	–	7(43)	–	20(74)	2.5(9.3)	3
Betula ermanii var. *genuina*	1(7)	–	2(13)	2(5)	–	2(7)	3(17)	2(11)	12(60)	1.5(7.5)	4
Acer tschnoskii var. *rubripes*	7(16)	4(10)	–	3(5)	–	–	2(9)	–	16(40)	2.0(5.0)	5
Quercus mongollca	–	3(11)	–	3(18)	–	–	–	–	6(29)	0.8(3.6)	6
Abies nephrolepis	–	–	–	3(9)	2(7)	2(7)	–	–	7(23)	0.9(3.3)	7
Rhododendron mucronulatum	4(11)	3(5)	–	–	–	–	–	–	7(16)	0.9(2.0)	8
Taxus caespitosa	1(3)	4(7)	–	–	–	1(4)	–	–	6(14)	0.8(1.8)	9
Rhododendron aureum	1(3)	–	3(7)	–	–	–	–	–	4(10)	0.5(1.3)	10
Sorbus commixta	–	–	1(5)	–	–	–	–	–	1(5)	0.1(0.6)	11
Rhododendron schlippenbachii	–	1(2)	–	2(3)	–	–	–	–	3(5)	0.4(0.6)	11
Alnus fruticosa var. *mandshurica*	–	–	–	–	–	–	1(4)	–	1(4)	0.1(0.5)	12
Sabina sargentii	–	–	–	–	–	–	1(3)	–	1(3)	0.1(0.4)	13
Rosa marretii	–	–	–	–	–	(3)	–	–	1(3)	0.1(0.4)	13
Vaccinium koreana	–	1(2)	–	5(9)	–	–	–	–	6(11)	0.8(0.4)	13
Tripterygium regelii	–	2(3)	–	–	–	–	–	–	2(3)	0.3(0.4)	13

N.B. Numbers in parentheses represent the coverage ot species in percentages (%); and the other values are the number of species found in each quadrat.
E represents sampling plot from eastern slope of mountain.
W represents sampling plot from western slope of mountain.

in a stage between the mature and degenerating phases, though there is as yet no clear evidence which would confirm this beyond any shadow of doubt.

Overall, *Pinus pumila*, *Thuja koraiensis* and *Rhododendron mucronulatum* var. *ciliatum* are the major components of the area in terms both of species abundance and coverage. The major species on the wind-exposed western slopes of Mt. Sorak (Taecheongbong) include *Pinus pumila*, *Rhododendron mucronulatum* var. *ciliatum*, *Quercus mongolica* and *Abies nephrolepis*, while those on the wind-protected eastern slopes consist of *Betula ermanii* var. *genuina*, *Pinus pumila*, *Thuja koraiensis* and *Acer tschnoskii* var. *rubripes*. These species differences between the eastern and western slopes demonstrate the ecological importance of wind and snow-cover on vegetation in this region, and will be further examined later.

Though Yim and Baik (1985) reported only three major subalpine communities on Mt. Sorak, i.e. *Pinus pumila*, *Abies nephrolepis* and *Betula ermanii* var. *genuina*, from our field work it is quite clear that, as far as the *Pinus pumila* community is concerned, the situation is not as simple as they suggested, there being considerable spatial variability of species therein. The major species patterns of variability, as determined from the 20 belt transects, are presented in Table 6.5. This shows that the *Pinus pumila* community has, as major co-dominants, *Quercus mongolica* near Chungcheongbong (1676 m a.s.l.). *Rhododendron mucronulatum* var. *ciliatum* near Taecheongbong (1708 m a.s.l.), and *Thuja koraiensis* in the middle of the transect. The different associate-species composition of the *Pinus pumila* community from Chungcheongbong to Taecheongbong requires further micro-scale work in the future, in order to better understand its full relevance. However, it is worth mentioning at this stage that two of the major component species in these transects have unexpectedly large vertical amplitudes on Mt. Sorak, i.e. *Rhododendron mu-*

Table 6.5. Major species composition, as compiled for each of the 20 belt transects on Mt. Sorak.

Transect No.	Major species	Secondary species
1	*Pinus pumila*	*Quercus mongolica*
2	*Quercus mongolica*	*Pinus pumila*
3	*Pinus pumila*	*Betula ermanii* var. *genuina*
4	*Pinus pumila*	*Taxus caespitosa*
5	*Quercus mongolica*	*Pinus pumila*
6	*Pinus pumila*	*Quercus mongolica*
7	*Pinus pumila*	*Thuja koraiensis*
8	*Pinus pumila*	*Rhododendron mucronulatum* var. *ciliatum*
9	*Thuja koraiensis*	*Pinus pumila*
10	*Thuja koraiensis*	*Pinus pumila*
11	*Pinus pumila*	*Thuja koraiensis*
12	*Rhododendron mucronulatum* var. *ciliatum*	*Thuja koraiensis*
13	*Pinus pumila*	*Thuja koraiensis*
14	*Quercus mongolica*	*Pinus pumila*
15	*Pinus pumila*	*Quercus mongolica*
16	*Rhododendron mucronulatum* var. *ciliatum*	*Pinus pumila*
17	*Pinus pumila*	*Rhododendron mucronulatum* var. *ciliatum*
18	*Rhododendron mucronulatum* var. *ciliatum*	*Pinus pumila*
19	*Rhododendron mucronulatum* var. *ciliatum*	*Pinus pumila*
20	*Rhododelldron mucronulatum* var. *cillatuln*	*Betula ermanii* var. *genuina*

N.B. Transect No. 1 begins near Chungcheongbong (1676 m a.s.l.), at which the continuous distribution of *Pinus pumila* ends at its northernmost point; transects continue up to No. 20 near Taecheongbong at ca. 1700 m a.s.l.

cronulatum var. *ciliatum* (510–1700 m a.s.l.) and *Quercus mongolica* (170–1700 m.s.l.).

Overall, the heavy presence of these two particular species in the *Pinus pumila* community suggests that they may have advanced upslope very quickly from lower altitudes as a result of the post-glacial warming, and that they are beginning to compete with *Pinus pumila*, which in turn is struggling to survive. The further expansion of these two species might place *Pinus pumila* itself into an even more endangered situation than it is in at present; however, more detailed field work is required to establish the precise competitive and environmental relationships which exist between these three species.

Tree-line

Classification

Over the world as a whole, various terms and categories have been suggested to describe the upper limit of tree growth, or vegetation zonation near the tree-line. Some of these, in general order of ascending altitude, are as follows; the forest line (where the continuous forest stops), the grove line (where the trees cease to form communities, of larger or smaller size), the tree-line (where the arboreal species cease to form trees), and the absolute tree-line, i.e. where tree species disappear altogether, even as krummholz communities (Rydberg, 1913); the economic limit of forest, the biological limit of forest, the tree-line, and the altitudinal limit of arboreal species (Hustrich, 1953); the upper forest limit, i.e. timberline, which is the line joining the highest occurrence of closed forest, the combat zone (Kampfzone), where the forest opens out and large upright trees occur only singly or in small groups, the tree-line or tree-limit, defined as a line joining the uppermost erect trees, and the krummholz zone, where only low-growing stunted forms exists, which either hug the ground surface or attain at most the height of the snow cover (Tranquillini, 1976a); and the timberline or the economic forest line, beyond which commercial cutting of trees endangers natural reproduction, the physiognomic forest line, which is the limit of the forest itself where it reproduces itself only in 'good' years, and the tree-line, which is the absolute poleward or upper limit of trees (Pruitt, 1978). In general, however, one may say that the term 'timber-line' means the upper limit of continuous growth of trees with economic value, and 'tree-line' represents the logical upper limit of closed tree growth. The term 'tree-line' in this work refers to the actual upper limit of trees *per se*, beyond which arboreal tree species are unable to develop an erect tree form (though shrubs may grow in erect form above the line). Above this line, too, krummholz forms of trees may exist.

Tree-line and species

In Eurasia, tree-line genera consist of *Larix*, *Pinus*, *Betula*, *Populus*, *Fagus*, *Picea* and *Juniperus* (Hare, 1954; Hedberg, 1972). More especially in East Asia, where the altitudes of the tree-line are placed at ca. 1300 to 1500 m a.s.l. in the Sikhote-Alin Range of the Soviet Far East, near the Korean border at 50° N to 45° N, *Betula ermanii* generally predominates, though in some areas it is replaced by an admixture of *Picea jezoensis*, *Abies nephrolepis* and *Larix* spp. (Gorchakovsky & Shiyatov, 1978; Arno, 1984). In Japan, the tree-line occurs at ca. 1400 to 1500 m a.s.l. in Hokkaido at 45° N to 42° N, and at ca. 1900 to 2000 m in Tohoku at 41° to 38°; here, *Pinus pumila* takes over from other trees, such as *Betula ermanii*, *Alnus maximowiczii*, *Picea* spp., *Abies* spp. and evergreen broadleaved dicotyledon shrubs, at the tree-line (Ohwi, 1965; Kawano, 1971; Numata, 1972, Jinno & Suzuki, 1973; Ishizuka, 1974; Maekawa, 1974; Shidei, 1974; Ito & Nishikawa, 1977; Yoshino, 1978; Franklin *et al.*, 1979; Hämet-Ahti, 1979; Koizumi, 1979; Kojima, 1979; Specht, 1979). In the Asian tropics, the timber-line lies at ca. 4000 m a.s.l. in the case of Malaysia (Van Steenis, 1962; Troll, 1968). According to Arno (1984), the fundamental difference between timberlines on mountains in humid tropical regions, as compared with those in the temperate latitudes, is that the tropics lack a cold and warm seasonality. Instead, their cold altitudinal zones are controlled by the year-round presence of nightly frost or near-freezing temperatures. Tropical timberlines are usually formed by a dense evergreen forest consisting of numerous species of broadleaved trees.

In northern Korea, the tree-line on Mt. Paikdoo at 42° N (2744 m a.s.l.) occurs at ca. 2000 m a.s.l. From 1750 m to the tree-line at 2000 m a.s.l. is a narrow belt of *Larix gmelini* forest, intermixed with *Betula ermanii* and *Rhododendron* spp. Above the tree-line, the ground is covered by alpine shrubs, consisting of *Rhododendron* spp., *Vaccinium* spp., *Dryas octopetala* var. *asiatica*, *Phyllodoce caerulea* and others (Takahasi, 1944; Wang, 1961; Zhao, 1986; Xu *et al.*, 1987). According to Lautensach (1945), on southern slopes the upper tree-line lies at 2200 m a.s.l. on Mt. Kwanmo, at 2150 m a.s.l. on Mt. Maengbu, and at 1950 m a.s.l. on Mt. Paikdoo. The theoretical climatic tree-line on Mt. Sorak is between 1700 to 1800 m a.s.l. (see p. 86). In southern Korea, the tree-line on Mt. Halla (1950 m as.l.) of Cheju Island at 33° N lies at ca. 1850 to 1900 m a.s.l., in which zone *Abies koreana* and *Betula ermanii* form a shrubby krummholz community (Song & Nakanishi, 1985).

Tree-line and environment

There are numerous theories relating to the causal relationships between tree-line and environmental conditions in temperature regions. Some of these are as follows. First, it is assumed that certain minimum high summer temperatures are required for the growth of trees at high altitudes; consequently, beyond a certain altitudinal level, daily summer temperatures are not sufficient to allow for tree growth. In general terms, the dividing line between tree growth and non-growth corresponds to an annual mean temperature of 10°C, though there are some variations away from this (Swan, 1967; Troll, 1973; Ives, 1978; Tranquillini, 1979; Costin, 1981; Kirkpatrick, 1982; Sage, 1986; Grace, 1987). Second, the general year-round presence of heat deficiency at high altitudes may limit tree growth (Daubenmire, 1954; Bliss, 1963; Holtmeier, 1980). Third, winter dessiccation may restrict tree growth at high altitudes (Tranquillini, 1963, 1964, 1976a,b, 1979, 1980; Wardle, 1965, 1971; Baig & Tranquillini, 1976; Krebs, 1985). Fourth, the presence or absence of tree growth may relate to a range of other phenomena, such as snow cover, wind, cryopedogenic activities, anthropogenic activities and fire (Billings, 1964; Jahn, 1970; Rapp, 1970; Hedberg, 1972; Troll, 1973; Higgins

& Spomer, 1976; Green, 1983; Agee & Smith, 1984).

The meso- and microscale distributional patterns of plants which occur in alpine environments may also be caused by a range of environmental factors, including atmospheric moisture availability, topography, soil pH, soil water potential, wind and snow cover. These two latter features require some further elaboration.

In the case of wind, one may say that there are many positive aspects of the effect of wind on plant growth, such as the encouragement of the physical exchanges of heat, carbon dioxide and water vapour between the atmosphere and vegetation (Warren Wilson, 1957; Grace, 1977; Larsen, 1980). High winds in winter and spring, however, have two strongly negative effects; physical injury (abrasion), and the inception of physiological drought. The patterns of physical injury caused by high winds may include a general deformation of the tree, the abrasion of twigs projecting above snow cover, as ice crystals and snow particles are blown on to them, other mechanical injuries, the wind-training of branches, and, at times, defoliation (Daubenmire, 1954; Spomer, 1962; Tranquillini, 1964; Troll, 1968; Watts, 1971; Savile, 1972; Zwinger & Willard, 1972; Yoshino, 1973, 1975; Grace, 1977, 1981; Slavik, 1977; Wade & Hewson, 1979; Baumgartner, 1980; Etherington, 1983; Cherov, 1985). Physiological drought induced by high winds, particularly in the early spring when species change from a cold hardiness to a growth phase (Melzack & Watts, 1982a,b), is expressed by the dessiccation of trees and particularly their transpiring surfaces, leading to cell damage and, on occasions, cell death. These dessiccating and cooling effects of wind increase proportionately with an increase in wind velocity. Increasing wind speed causes an increase in the convective transfer of heat away from the leaf, and a lowering of leaf temperature. This further creates an increase in the plant's transpiration rate, and a subsequent demand for more water from the ground, which is often difficult to meet at the end of the cold hardiness phase (Benninghoff, 1952; Shetler, 1963; Spomer, 1964; Bryant & Scheinberg, 1970; Courtin & Mayo, 1975; Barry, 1979, 1981; Tranquillini, 1982; Moggi, 1985; Krebs, 1985; Sutcliffe, 1986).

In general, snow cover, on the summits of major mountains, depends largely on topographic features, and factors relating to wind exposure. Snow cover can affect the growth of plants in two ways, favourably and unfavourably. Favourable effects, if the plant is totally covered by snow, include the protection of it from excessively low temperatures, and drying and/or abrasion by high winds; and if not, the absorption and retention of environmental moisture, and/or wind-blown minerals, which may enhance the plant's nutrient stock. At the same time, the existence of a snow cover over a plant does not prevent the continuation of photosynthesis, especially in the case of evergreen plants (Gjaerevoll & Bringer, 1965; Richardson & Salisbury, 1977; Speirs, 1979; Laegaard, 1980; Jonasson, 1981; Ford, 1982; Walton, 1982; Dahl, 1986). On the negative side, the unfavourable effects of snow cover include a tendency for the encouragement of fungal diseases, the curtailment of the plant's growing season, a restriction on the production of vegetative propagules, and some physical damage associated with snow-weight pressure (Holtmeier, 1973; Zwinger & Willard, 1972; Maekawa & Shidei, 1974; Laegaard, 1980; Raven & Walters, 1984).

Although detailed research into the effects of environment on vegetation forms in the subalpine zone of Mt. Sorak remains a matter for the future, field observations undertaken by the authors from August 1987 to December 1987 relating to the presence and structure of vegetation and the tree-line here strongly suggest that there are three major local environmental influences on plant growth: topography/aspect; a high frequency of strong winds, especially in winter; and differential snow cover. The situation is summarised in Fig. 6.6.

It has already been indicated that the theoretical climatic tree-line for Mt. Sorak is between 1700 to 1800 m a.s.l. On the northern and eastern sides of Mt. Sorak, the actual climatic tree-line, beyond which arboreal tree species are unable to develop an erect tree form, approaches this theoretical limit, in the sense that erect trees impinge on the summit of the mountain. However, this pattern is modified on the slope of Taecheongbong, where the actual local tree-line falls to ca. 1650 m a.s.l. Further, on the saddle between Taecheongbong and Chungcheongbong, it falls to

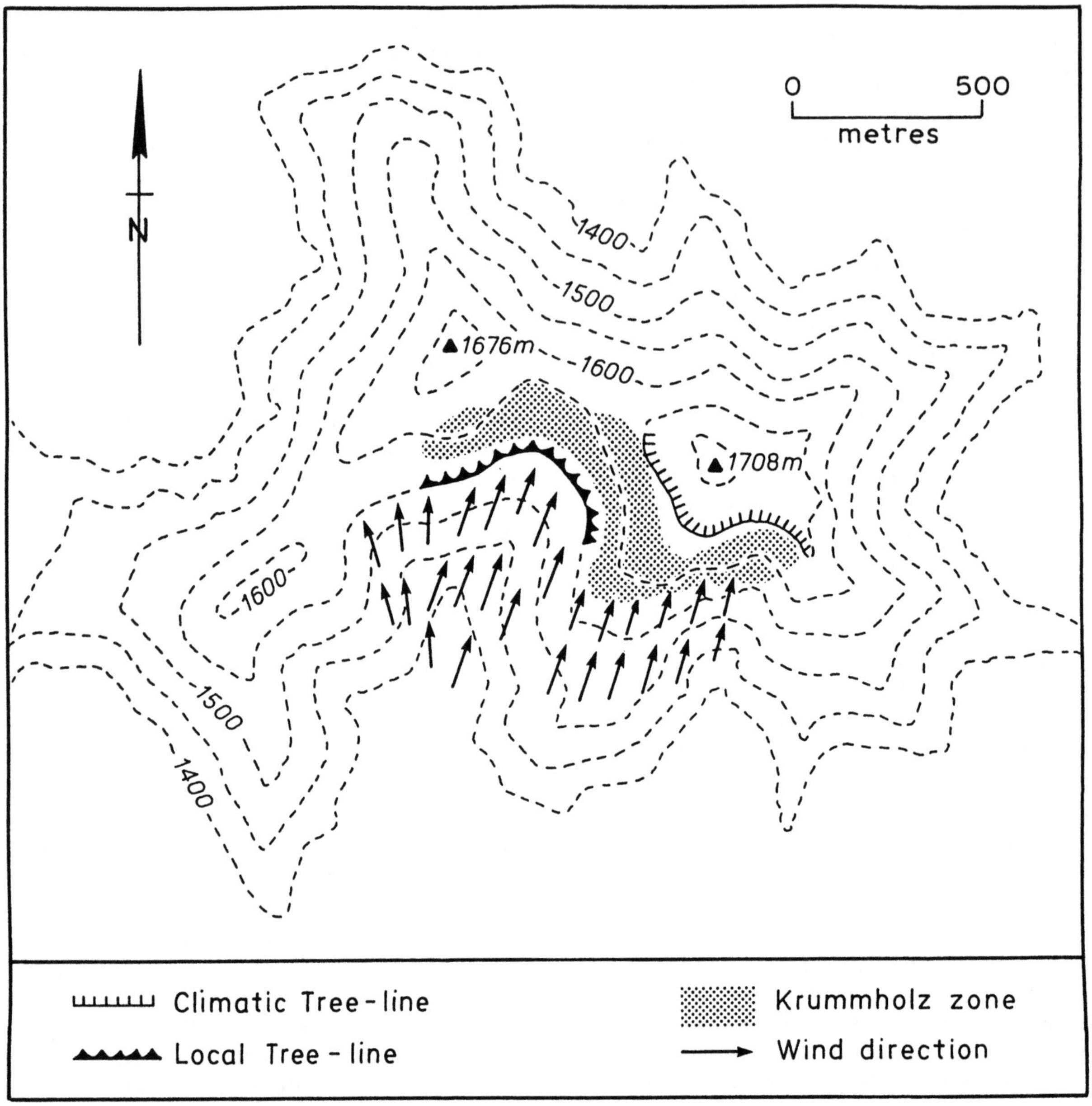

Fig. 6.6. Sub-alpine vegetation, tree-line and winter wind direction on Mt. Sorak.

ca. 1550 m a.s.l. The area above the tree-line on Mt. Sorak is dominated by a *Pinus pumila* community (pp. 149–156), and the tree-line is mainly formed by isolated *Abies nephrolepis*. The transitional community between the tree-line and the continuous forest zone present at altitudes of ca. 1550 to ca. 1650 m a.s.l. on the western and south-western slopes of Taecheongbong and below 1708 m a.s.l. on the eastern and northern slopes includes *Abies nephrolepis*, *Betula ermanii* var. *genuina*, *Rhododendron mucronulatum* var. *ciliatum* and *Quercus mongolica*; and these species together form a krummholz zone.

On the leeward northern and eastern slopes of Taecheongbong, the terrain is covered by *Betula ermanii* var. *genuina*, *Abies nephrolepis*, *Pinus pumila* and others. In general, the leeward slopes near to the summit are occupied mainly by a *Betula ermanii* var. *genuina* community (see Fig. 6.3), and show no clear tree-line. The absence of a tree-line on these northern and eastern slopes may be due to: first, the protection of communities from freezing northwesterly high winds during the winter by topographic features; second, the steep slopes here, which seem to prevent a heavy accumulation of snow blown by the winter winds from the mountain tops, and which thus might cause snow avalanches and consequently tree destruction; and third, the early melting of accumulated snow cover in the spring, particularly on the sunny easterly slopes, which would extend the growing season considerably.

In contrast, on the windward southern and western slopes of Taecheongbong, where local westerly and north-westerly winds from the deep valley-gorge below appear to prevail especially during the winter (as deduced from the growth direction of creeping *Pinus pumila* stems and the malformation of vegetation or krummholz), *Pinus pumila* is the main cover, along with *Abies nephrolepis*, *Rhododendron mucronulatum* var. *ciliatum*, *Thuja koriaensis*, and *Quercus mongolica*, and a local tree-line is displayed. The depression of the tree-line at ca. 1550 m a.s.l. on the saddle ridge between Taecheongbong (1708 m a.s.l.) and Chungcheongbong (1676 m a.s.l.), is likely to be the result of the so-called 'summit effect' or 'top phenomenon' which relates to the effect of wind on and near the summit of mountains (Yoshino, 1973), and suggests that, on the mountain top itself, stronger wind speeds are the main cause of the lowering of the upper limits of the subalpine zone. The combination of a scarcity of snow cover and the frequency of strong local westerly winter winds on the exposed saddle ridge is, however, the major cause for the reduction in the tree-line there. Further details of this are given in the following section.

Krummholz

Classification and environment

Krummholz is a common form of tree community which is present close to the tree-line, consisting of stunted and crooked individuals of normally upright-growing tree species (Holtmeier, 1973; Marr, 1977; Wardle, 1981a; Benedict, 1984; Goudie, 1985). The krummholz shape of the tree may be due to any or all of the following factors: first, high winds on the mountain (Sekiguti, 1951; Daubenmire, 1954; Suslov, 1961; Billings, 1964; Troll, 1968; Zwinger & Willard, 1972; Noguchi, 1979; Collinson, 1980; Jones, 1983; Pears, 1985); second, resulting from the high winds, above-normal winter dessiccation rates and abrasion rates from ice particles, and/or an excessive accumulation of glazed ice particles causing physical damage (Holroyd, 1970); third, the interaction of wind, snow and winter dessiccation (Wardle, 1968, 1974, 1981b; Barry, 1981); and finally, general physical damage caused by sand or snow abrasions, frost damage and snow pressure (Daly, 1984; Ohsawa, 1984).

Krummholz trees may take various forms. First, they may give rise to tree carpets (Putman, 1948; Oka, 1972; Grant & Mitton, 1977; Ives, 1980). Second, a range of wind-shaped trees may develop. There are several consequences of this. Trunks may remain vertical, but in some cases, branches are bent drastically to the leeward of the tree by the prevailing winds during their growing period. Then, in the case of winter-wind-shaped trees, trunks also remain vertical, but branches are severed by the effects of winds carrying snow and frozen rain. In severe cases, trunks and branches are both deformed by strong prevailing winds, especially during the growing season (Yoshino, 1964, 1966b, 1973, 1975). Third, variable forms may occur, with upright trees from a few hundred metres below the timberline, partly upright forms present immediately below the timberline, and a prostrate form above the timberline (Terri, 1969).

Krummholz on Mt. Sorak

To illustrate the krummholz types further, one may refer to the physiognomy of *Betula ermanii* var. *genuina* on the eastern and southeastern slopes of Taecheongbong, and that of *Abies nephrolepis* at similar altitudes, on the southwestern slope of the same peak, which show remarkable contrasts with each other (Fig. 6.5). According to Suslov (1961), *Betula ermanii* can endure a more severe climate than any *Picea* or *Abies* species, and it can grow accordingly in rocky places where other trees cannot. It is one of the most widespread species in many arctic-alpine communities in East Asia. The species also tends to replace conifers on steeper slopes, which have more restricted moisture availability and younger soils than elsewhere; and it also seems to be more tolerant of heavy snow. Therefore, *Betula ermanii* appears to be a good indicator of the subalpine forest type, and often covers both gentle and steep slopes in the subalpine to alpine zones (Kawano, 1971; Numata *et al.*, 1972).

On the leeward, eastern side of Taecheongbong on Mt. Sorak, which opens to the East Sea, *Betula ermanii* var. *genuina* grows up to 2 to 4 metres high, and the krummholz form is hardly found (Fig. 6.6). This results from it being protected from severe westerly and south-westerly winter winds by higher land to the windward, and it may also be that relatively higher rainfall prevails on the eastern slopes of Mt. Sorak as compared to the western ones.

On the windward southwestern slopes of the saddle between Taecheongbong and Chungcheongbong, and close to the summit of Taecheongbong, krummholz dominated by *Abies nephrolepis* at 1550 to 1650 m a.s.l., along with mat-formed *Quercus mongolica*, *Thuja koraiensis*, *Taxus caespitosa*, *Acer* spp., and *Rhodondendron* spp., and creeping *Pinus pumila* is formed (Fig. 6.3). Unlike the situation on the eastern and southeastern slopes of Taecheongbong, *Betula ermanii* var. *genuina* also forms part of this krummholz.

Indeed, three types of *Abies nephrolepis* krummholz can be noticed there, based on the formation processes and mechanisms of this species, on a seasonal basis. First is a prevailing-wind-shaped type, without serious malformation of both trunk and branches, which has adapted to the prevailing wind during the growing season, especially in spring (Fig. 6.7a). This is especially common on the wind-exposed mountain top and ridge. Second, an occasionally-high-wind-shaped type may be seen, which has formed through the collapse of the main trunk, by an unusual gale, such as a summer typhoon (Fig. 6.7b). This has no particular distributional preference on Mt. Sorak, and its occurrence seems almost to be on a random-chance basis. Third, a winter-wind-shaped type which has been shaped through the desiccation of, and/or abrasion of projecting branches above the winter snow cover, by driven ice crystals and snow particles during the late autumn and winter, also is present (Fig. 6.7c). The presence of this type is determined by the existence of shelter from prevailing winds. The levels of snow depth then can build up sufficiently for this type to form.

Overall, frequent mesoscale prevailing local winds, which are blown from the western lowlands toward the high-ground saddle between Taecheongbong and Chungcheongbong, may be predominantly responsible for the occurrence and distribution of krummholz trees here, as well as for their shape and direction. First, slightly crooked krummholz is formed by the prevailing wind during the growing season (mainly spring and early summer). Second, the unusual summer gust associated with typhoons and similar storms is responsible for the broken-trunk form of krummholz. Third, mat, cushion, table or flagged krummholz of *Abies nephrolepis* on Mt. Sorak is, on the other hand, more likely to be due to the existence of high winds in winter, which blow up along the deep gorge on the southwestern slopes of the Taecheongbong-Chungcheongbong ridge (i.e. Sorak Waterfall Gorge). Most of the *Abies nephrolepis* krummholz is found on the southwestern slopes of the Mt. Sorak peaks, where these winds cause the greatest destruction. It is also noticed that most branches of *Abies nephrolepis* krummholz, and of *Pinus pumila* are facing northeast, which is to be expected from a prevailing south-westerly wind. However, different shapes of krummholz forms can also be seen, in cases where separate developmental stages are in progress, or where they are the product of more than one of the seasonal stages mentioned above.

The summit of Mt. Sorak (Taecheongbong) and the saddle between Taecheongbong and Chungcheongbong, are hostile and severe habitats for plant growth for several reasons. The 'top phenomenon' (see p. 161) causes a lowering of air and soil temperatures, especially during the winter season. Then, too, fallen snow tends to be blown away and the ground suffers accordingly both from drought and severe cold in the early spring. Also, a high level of evapotranspiration is found, and is especially intense over the mountain summit and the main ridge, in which areas winds are accelerated by the topographic disposition, and can easily kill plants by dehydration. Finally, these strong winds on the exposed high land also

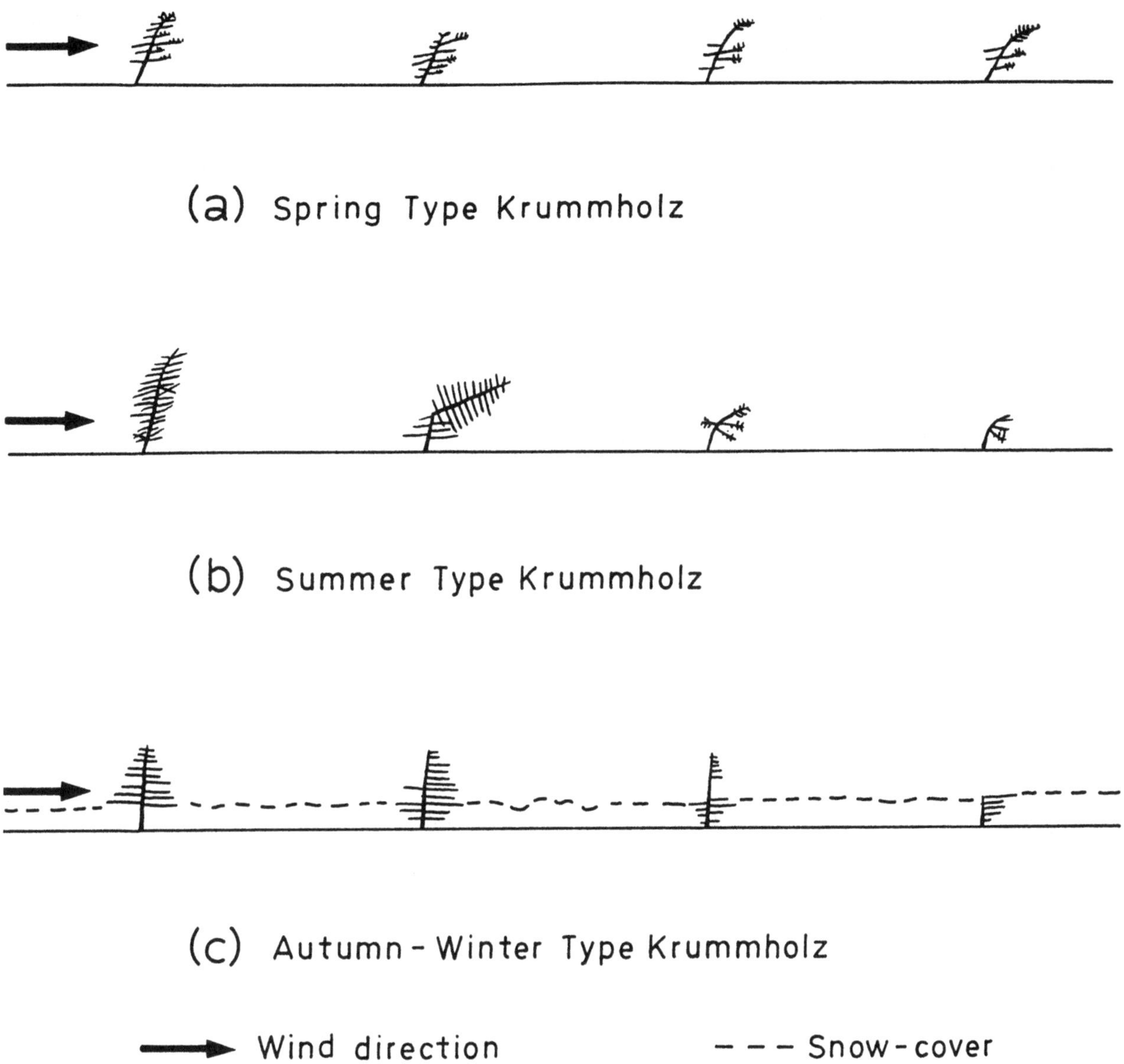

Fig. 6.7. Krummholz types of *Abies nephrolepis* on Mt. Sorak.

have removed humus and top soil, and consequently have brought sterile edaphic conditions. All these controls tend further to bring an added competitive advantage to cryophilous plants, such as *Pinus pumila*, *Abies nephrolepis*, *Betula ermanii* var. *genuina*, *Thuja koraiensis*, etc., as opposed to thermophilous plants.

***Pinus pumila* on Mt. Sorak**

Though 24 arboreal plants inhabit the subalpine zone (above ca. 1600 m a.s.l.) on Mt. Sorak, this work concentrates on the *Pinus pumila* community *per se* (see p. 148) in the main because it is, on this peak, close to the southernmost limit of its range in the Eurasian continent. Also, it is one of the very few species on Mt. Sorak whose vertical range is limited completely to high altitudes. In addition, it shows unique physiognomic adaptations to the local environmental conditions. The horizontal and vertical ranges of *Pinus pumila* in East Asia and Korea, its phytosociological structures, ecological characteristics and relationships with the local environment on Mt. Sorak, all are discussed below.

Botanical description and distributional range

Pinus pumila is a shrub, 0.5–2 m in height, with its main branches extending horizontally up to a maximum of 8/9 m, creeping close to the ground surface. Its winter buds are narrowly conic, about 0.6 cm long, with a sharp-pointed apex, and closely-pressed scales. Its needles are arranged in fives, 3.8 to 5.1 cm long, with entire or faintly-toothed margins. Its cones are ovoid, very shortly stalked, and rarely open to liberate the seeds; they are ca. 5 cm long, and orange-brown when ripe. Scales are few, about 1.5 cm broad, with dark-coloured, triangular and reflexed tips. Seeds are about 0.6 cm long, pear-shaped and wingless (Dallimore & Jackson, 1954; Ohwi, 1965).

From the end of the Tertiary, and possibly earlier, *Pinus pumila* was present in the undergrowth of highland coniferous forests in the temperate zone of East Asia, and occupied an intermediate position between taiga and mountain tundra assemblages during the Sartan (Wurm or Devensian: see Table 3.3) glaciation in Siberia (Tikhomirov, 1946; Mirov, 1967; Powers, 1973; Buks, 1984).

The present horizontal distribution of *Pinus pumila*, as first indicated by Wilson (1916, 1918), extends from Korea and Japan, to Lake Baikal, Sakhalin, the Okhotsk coast, the Kurile Islands, Kamchatka and Manchuria. *Pinus pumila* has been reported from 35°20′ N to 70°31′ N, and from 105° E to 190° E, in northern Mongolia, the vicinity of Lake Baikal, the vicinity of the East Siberian Arctic Ocean, eastern Siberia, the Soviet Far East, the Sikhote-Alin Mountains, Kamchatka, Sakhalin, the Kurile Islands, Hokkaido and Honshu in Japan, Manchuria and Korea (Hultén, 1927; Tatewaki, 1935; Tikhomirov, 1946; Dallimore & Jackson, 1954; Wu, 1956; Den Ouden & Boom, 1965; Critchfield & Little, 1966; Mirov, 1967; Kawano, 1971; Kobayashi, 1971; Numata *et al.*, 1972; Nikolskaya & Pedorova, 1973; Ishizuka, 1974; Mezhennyi, 1979; Iwata, 1986). The horizontal range of the species is presented in Fig. 6.8.

Within this region, the lower altitudinal limit of *Pinus pumila* varies from sea level in Sakhalin, 600 to 800 m a.s.l. in the Okhotsk and Magadan, 1400 m a.s.l. in the Verkhoyanskii mountains, Russia, and at 2300 to 2700 m a.s.l. on Mt. Hakusan, Japan. The upper altitudinal limit of *Pinus pumila* occurs at 3200 m a.s.l. in some parts of northern Eurasia (Wilson, 1916; Tseplyaev, 1965; Mirov, 1967; Ueno, 1979). According to Iwata (1986), the zone above the upper limit of *Pinus pumila* in Japan corresponds to the middle periglacial zone of the European and Himalayan mountains, where a partly vegetation-covered patterned ground occurs.

In Korea, *Pinus pumila* grows from ca. 41° N to 38° N, and is present on Mts. Kwanmo, Pukpodae, Nampodae, Mantap, Ro, Puksobaik, Puchonkowon, Maengbu, Sasoo, Myohyang, Pohyun, Pirae, Seungjuk, Nangrim, Haram, Keumkang and Sorak (Lautensach, 1945; Chung & Lee, 1965, Kim *et al.*, 1981). The vertical range of *Pinus pumila* varies according to the particular mountain; on Mt. Mantap it is at 2000 to 2200 m a.s.l., 1500 to 1600 m a.s.l. on Mt. Seungjuk, 1600 to 1900 m a.s.l. on Myohyang, 1400 to 1750 m a.s.l. on Mt. Sasoo, 900 to 1680 m a.s.l. on Mt. Keumkang and 1500 to 1700 m a.s.l. on Mt. Sorak. The presence of this species down to 900 m a.s.l. on Mt. Keumkang might be due to local environmental factors, such as geology, topography, etc., but this needs further clarification. It is also present at 2000 m a.s.l. on Mt. Ro, at 1350 m a.s.l. on Mt. Pirae and at 1450 m a.s.l. on Mt. Haram (Fig. 6.9).

The Pinus pumila community

The autecology of *Pinus pumila* has been widely investigated in Russia, Japan and Korea, and the species is known to be one of the most typical indicators of subalpine vegetation (Tikhomirov, 1946; Shidei, 1963; Vaskouskiy, 1964; Kobayashi, 1967, 1971; Kotlyarov, 1978; Okitsu & Ito, 1984; Park, 1985). Its community relationships on Mt. Sorak already have been presented in the previous section (see pp. 148–157).

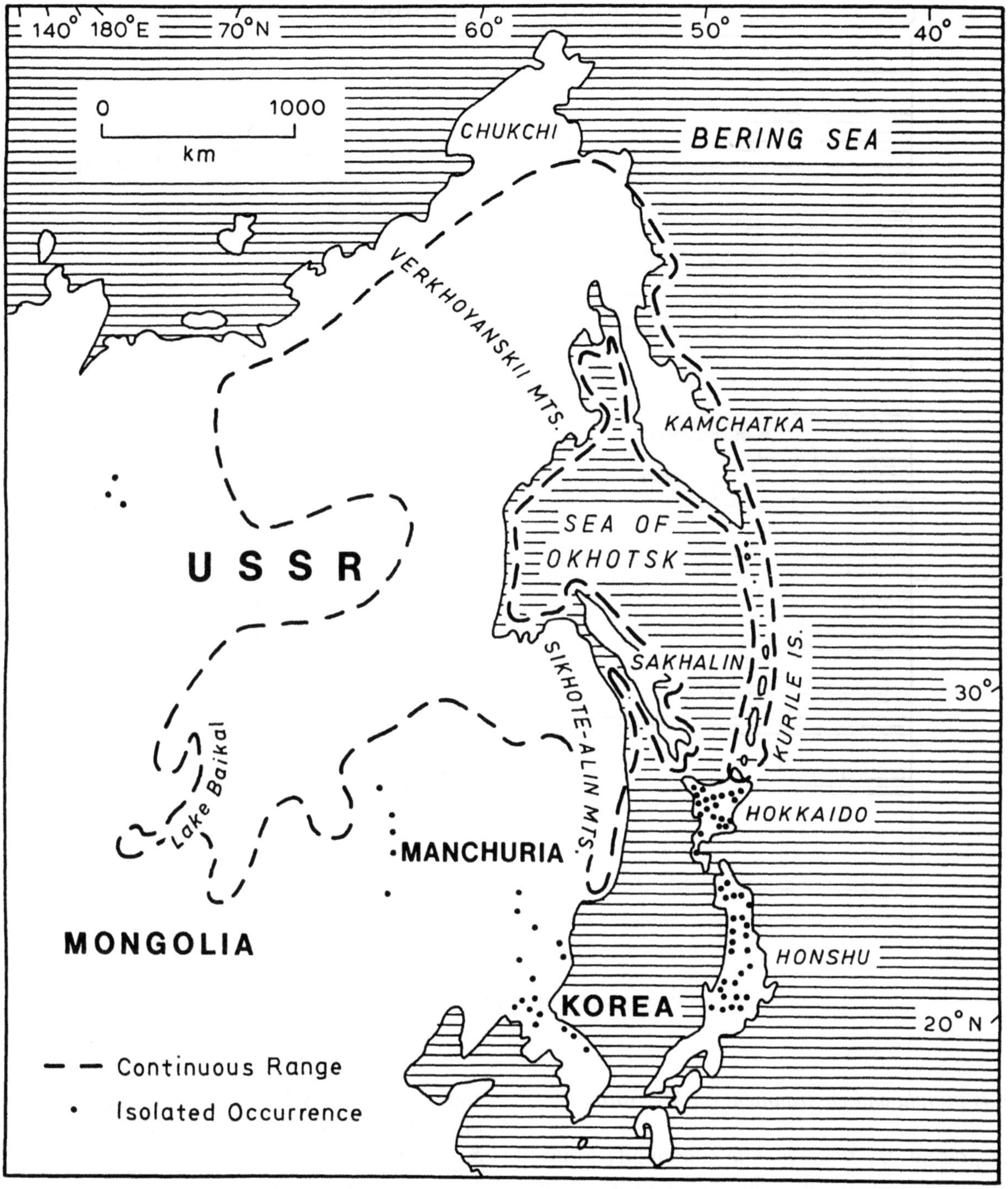

Fig. 6.8. Distribution of *Pinus pumila* in East Asia.

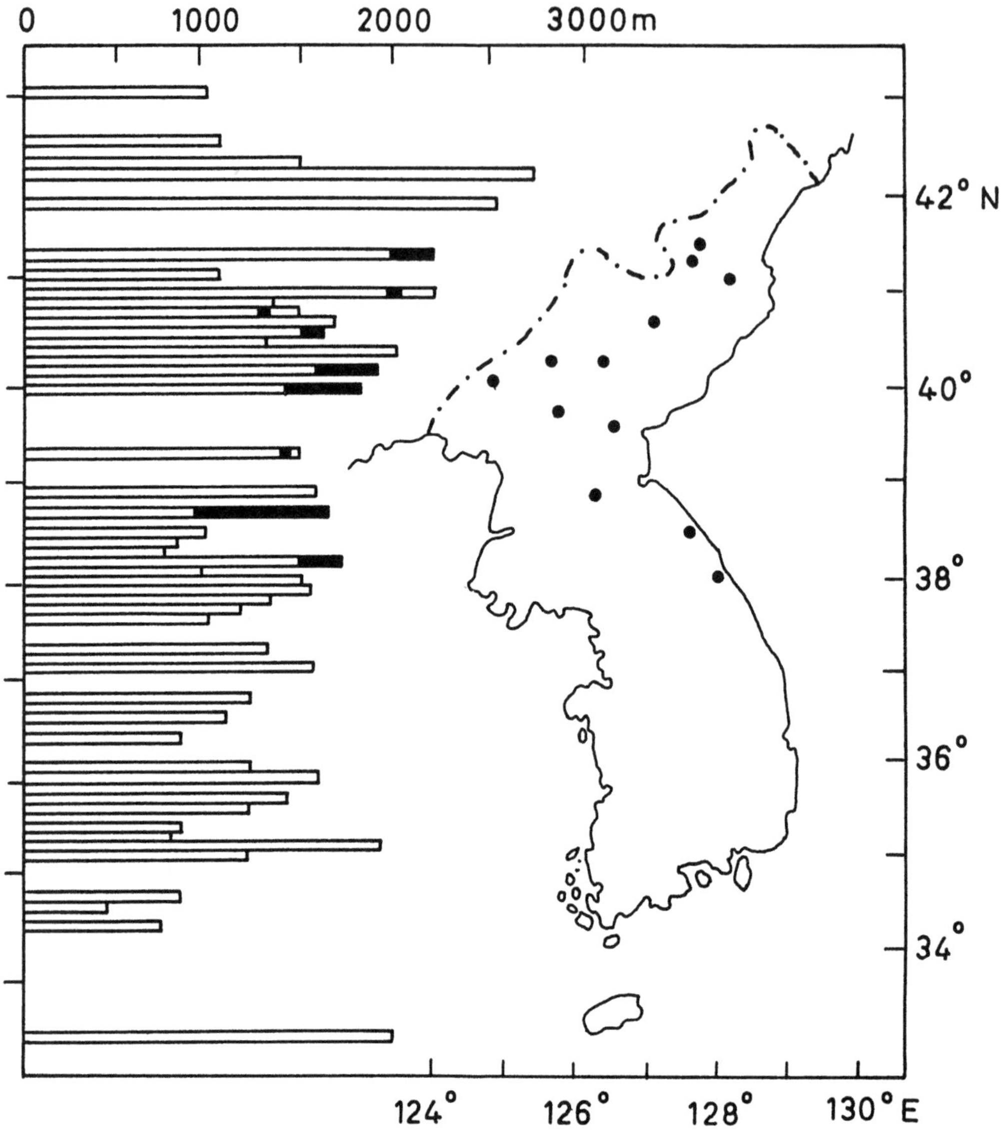

Fig. 6.9. Horizontal and vertical ranges of *Pinus pumila* in Korea.

Pinus pumila and environment on Mt. Sorak

The present-day *Pinus pumila* community in the Korean peninsula shows a disjunctive distribution from the north to the midland regions (see Fig. 6.9). The means of the vertical ranges for *Pinus pumila* in Korea, averaged from the ranges found on all peaks on which it is present, are 1485 m a.s.l. for the lower limit and 1805 m a.s.l. for the upper limit.

The species is very cold-resistant, and it can tolerate annual *mean* temperatures of −12°C, and some seasonal variations of up to −70°C (Tseplyaev, 1965; Sakai & Otsuka, 1970; Sakai & Kurahashi, 1975; Yoshida, 1986). Even after its needles freeze, the species can remain unharmed (Photo 6.5). This enables it to be classified as a chinophobous (snow-disliking) species (Tseplyaev, 1965; Hämet-Ahti *et al.*, 1974; Ishizuka, 1974). This unusual feature, which is not found among other subalpine species on Mt. Sorak, both gives it a competitive advantage in areas clear of snow during the winter, and at the same time enables it to resume growth more

Photo 6.5. Iced leaves of *Pinus pumila* at Taecheongbong, Mt. Sorak.

quickly in the spring than other species which remain covered by snow until later in the season (Canaday & Fonda, 1974; Emerick & Webber, 1982). It is indeed the case that it is found on Mt. Sorak in those areas in which snow does not accumulate (see Fig. 6.6).

The present-day southernmost distributional limits of *Pinus pumila* in Korea, on the top of Mt. Sorak, are undoubtedly due to a combination of the unique ecology of the species as indicated above, and its response to the local environment. *Pinus pumila* occupies open, windy, cold and dry exposed areas of the subalpine zone of Mt. Sorak (see Fig. 6.6), and regenerates mainly by vegetative propagation through adventitious roots, as well as (more rarely) by seedling establishment.

The local climatic conditions may be exemplified further by reference to the one high-altitude meteorological station which has been established in Korea, at Taecheongbong Shelter (ca. 1700 m a.s.l.; Table 6.6).

From this table it is clear that four months of the year (December–March inclusive) have mean maximum temperatures of below 0°C, and the same four months have mean minimum temperatures of less than −10°C. At the same time, extremes of temperature minima reach below −20°C between November and March inclusive. On the other hand, the number of snowy days throughout the winter is relatively few; accordingly, bearing in mind the high frequency of high winds in this area in winter, snow accumulation is unlikely to be great in the immediate vicinity of Taecheongbong. The same strong winds also mean that wind-chill temperatures are considerably reduced as compared to the official record given above. On the other hand, summers on the top of Taecheongbong can be warm and wet, and maximum temperatures may reach 30°C in August.

Looked at in another way, and using the lapse rates indicated previously (see p. 132), the mean minimum temperatures for January for the lower limit of *Pinus pumila* on Mt. Sorak at ca. 1550 m a.s.l. are −16.4°C, and the mean maximum temperatures for August are 21.5°C. The comparison of the mean daily maximum temperatures of August for the whole of Korea (16.9°C to 23.4°C) with those on Mt. Sorak (18.2°C to 19.1°C) suggests that the temperature range for the lower

Table 6.6. Climatic data from Taecheongbong Shelter at ca..1700 m a.s.l.

Climate	Month					
	Jan	Feb	Mar	Apr	May	June
Mean minimum temp.	−17.3	−16.5	−10.2	−4.5	3.3	7.8
Mean maximum temp.	−7.1	−5.6	−1.3	5.1	12.2	15.7
Temperature range						
(Min.)	−30.2	−28.0	−24.0	−12.0	−12.0	0
(Max.)	−4.0	−8.0	8.0	15.0	28.0	24.0
No. of snowy days	3.5	6.0	4.7	3.5	1.5	–

Climate	Month						
	Jul	Aug	Sept	Oct	Nov	Dec	Annual
Mean minimum temp.	11.6	12.6	6.9	0.1	−7.0	−13.1	−2.2
Mean maximum temp.	17.5	19.1	13.0	8.2	0.6	−3.9	6.1
Temperature Range							
(Min.)	3.0	7.0	−7.0	−13.0	−22.0	−25.0	−30.2
(Max.)	28.0	30.0	22.0	17.0	12.0	18.0	30.0
No. of snowy days	–	–	0.3	1.3	4.7	4.0	2.5

(Data from January 1985 to March 1988.)
N.B. Unit for temperature is °C.

limit of *Pinus pumila* on Mt. Sorak is found within the national mean value.

For a postulated continuous distribution of *Pinus pumila* in Korea a fall of around 9°C for the mean annual temperature is required. This would bring the lower limit of *Pinus pumila* down close to sea level. In other words, when the temperature has deteriorated by around 8 or 9°C from that of today, the cryophilous *Pinus pumila* will have expanded both downward and southward to form a continuous coverage. Also, the present continuous range of *Pinus pumila* might have been moved ca. 1200 km southwards to Mt. Sorak, on which the species currently grows disjunctively, in the case of a 9°C deterioration of temperature during the Pleistocene glacial phases (see Fig. 6.8). Similar movements and migrations of *Pinus pumila* from the arctic zones of Siberia into Sakhalin and the Kurile Islands during Pleistocene cold phases have been postulated by Nikolskaya and Fedorova (1973). The further, small-scale disjunctive distribution of *Pinus pumila* within the boundaries of Mt. Sorak itself, i.e. on Kuitaegicheongbong at 1577 m a.s.l., at 6 km west of Taecheongbong, on Karibong at 1518 m a.s.l., at 11 km southwest of Taecheongbong, and Madeungryong at 1326 m a.s.l., at 5 km north of Taecheongbong (see Fig. 6.4), also suggest the former, relatively recent presence of a continuous range of this species on this mountain, which presumably dates to a time shortly before the post-Pleistocene climatic amelioration, at say ca. 10,000 years B.P. All the evidence now suggests that *Pinus pumila* is close to the climatic limit of its further continued existence on Mt. Sorak, and only a small additional rise in temperature over present-day levels may be sufficient to render the species no longer viable in this locality.

Pinus pumila thicket on the ridge between Taecheongbong and Chungcheongbong is particularly well adapted to the severe environmental conditions. Both the height of this species, and the area covered by it, decline with an increase in wind exposure. On the windward southwestern ridge between Taecheongbong and Chungcheongbong, *Pinus pumila* is present in a prostrate form near the ground, at 5–10 cm, while the species on the leeward northeastern slope grows up to 2–3 m in height from the ground (Fig. 6.10; Photo 6.6). Under favourable experimental conditions, it has been known to reach 4 m above ground surface (Dmitrieva, 1978). The thickets on the leeward side, however, still display a predominantly creeping form, mainly because of the pressure of wind-blown snow accumulation, and the whirling wind over the mountain top and ridge (Fig. 6.10).

In general, on the windward southwestern side,

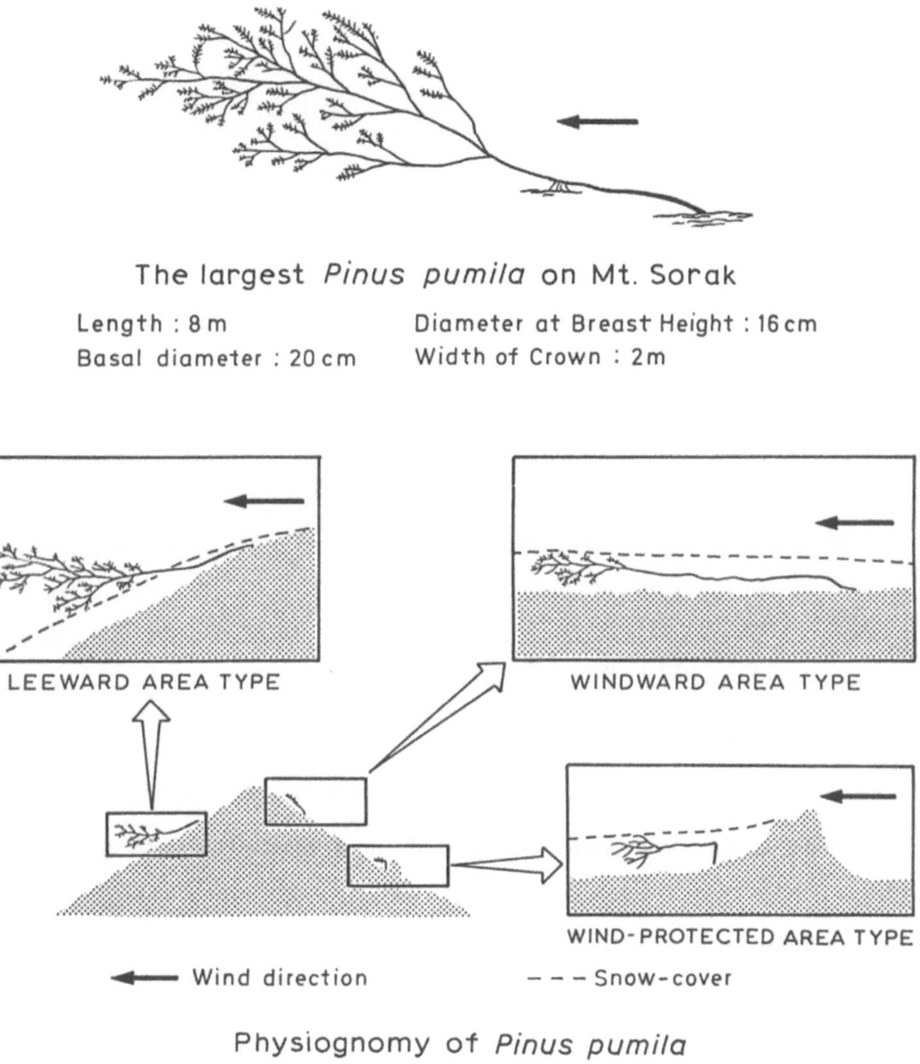

Fig. 6.10. Krummholz types of *Pinus pumila* on Mt. Sorak, Korea.

coverage by *Pinus pumila* is greater than on the leeward eastern side (see Table 6.4). This fits in with previous indications that *Pinus pumila* has a competitive advantage over other species on the windward side close to summit col. Individuals of *Pinus pumila* in this locality grow straight out of shattered rock, and there is very little soil to speak of.

According to Ito and Nishikawa (1976), frequent human trampling on its habitat, caused by tourism, threatens the cyclical development of *Pinus pumila* in Japan. A similar pattern can be noticed on Mt. Sorak. There has been a steady increase of tourists on this mountain from 1977 to 1986 (Table 6.7). Visitors to Mt. Sorak, mainly during the summer (37%) and spring (33%), of which about 3 to 5% climb up to Taecheongbong, appear to have restricted both the development and regeneration of this species.

One can make out a good case for protecting *Pinus pumila* on Mt. Sorak for several reasons. First, it is a good indicator species of past vegetation change in East Asia and Korea. Second, Mt. Sorak lies at the southernmost limit of its

Photo 6.6. Creeping *Pinus pumila* community, Mt. Sorak, Korea. (Photograph by D. Watts)

Table 6.7. The number of visitors on Mt. Sorak.

No. of tourists	1977	1978	1979	1980	1981	1982	1983	1984	1985	1986
(unit: 1000 persons)	796	777	1229	1410	1778	2438	2319	2244	2284	2304

(Source: Mt. Sorak National Park Authority, 1987.)

range, and it is biogeographically quite important there. Third, its morphology and physiognomy are unique in the present-day harsh environment near the top of Mt. Sorak. And finally, it is struggling to survive at the top end of its vertical range on this mountain; survival is likely to become an increasing problem as further climatic amelioration, associated with the greenhouse effect, begins to take place.

Results and discussion

The floristic diversity on Mt. Sorak (ca. 1013 spp.) is due in part to the size of the mountain, and its range of habitats, which accommodate an equally wide range of species, from warmth-loving plants to those of cold environments. A relatively large number of Korean endemic plants exist on Mt. Sorak (ca. 65 spp.). The appearance of 17 local endemic plants found only on Mt. Sorak is of interest as well. Moreover, the coexistence of 35 northern and 5 southern floristic elements on the mountain at their southernmost and northernmost limits respectively, adds to the floristic diversity.

In terms of altitudinal range, four groups may be recognised from the 24 subalpine arboreal species which reach above 1600 m a.s.l. on Mt. Sorak. These are, first, those species which grow exclusively between 1500 and 1700 m a.s.l.; second, those species which grow between 1000 m and 1700 m a.s.l.; third, species occurring between 500 to 1700 m a.s.l.; and finally, species growing between 100 and 1700 m a.s.l. In the first group, 7 species are restricted exclusively to the subalpine zone between 1500 and 1700 m a.s.l., and of these, three species are endemic to Korea; all of these plants have a small altitudinal range, and comparatively narrow ecological amplitudes.

From the 24 subalpine species on Mt. Sorak which grow to altitudes of above 1600 m a.s.l., four major subalpine communities may be delimited. These are dominated respectively by *Pinus pumila*, *Abies nephrolepis*, *Thuja koraiensis* and *Betula ermanii* var *genuina*. The isolated or disjunctive occurrence of these four species both within the Korean peninsula as a whole (Figs. 5.2, 5.3 and 6.9), and the boundaries of Mt. Sorak itself (Fig. 6.4), suggests the likelihood of a former continuous distribution of these species during cooler Pleistocene phases, and their intermountain migration at times during periods of climatic deterioration and subsequent climatic amelioration. In other words, as the climate deteriorated during the cold phases of the Pleistocene, these cryophilous plants seem to have shifted downwards from the mountains and southwards from the north, and the former isolated cold-loving mountain vegetation then joined together, and expanded its distribution further towards the south. A fall of ca. 9°C would have been required to allow the lower distributional limit of all these four plants to reach sea-level. On the other hand, the climatic amelioration during the warm phases of the Pleistocene, especially during the post-glacial warm period, has resulted in the retreat of the previously wide-ranging cryophilous vegetation upslope, along with a concomitant expansion of thermophilous vegetation also upslope. At present, the cryophilous plants are again isolated from each other on mountain tops, both on a meso- and a macroscale.

On Mt. Sorak, the leeward eastern slopes near to Taecheongbong (i.e. the summit of Mt. Sorak) are occupied mainly by *Betula ermanii* var *genuina*, and show no clear tree-line. However, the windward western slopes of Taecheongbong, and the areas along the saddle between Taecheongbong (1708 m a.s.l.) and Chungcheongbong (1676 m a.s.l,), both display a local tree-line at ca. 1550 m a.s.l., comprised mainly of *Abies nephrolepis* and a few other species, such as *Rhododendron* spp., and *Betula ermanii* var. *genuina*; areas above the tree-line are dominated by *Pinus pumila*. The climatic tree-line, beyond which arboreal tree species are generally unable to develop an erect tree form mainly as a result of prevailing climatic conditions, occurs at ca. 1700 m, though a further local tree-line occurs at ca. 1650 m a.s.l. on the windward western slope of Taecheongbong. The additional 100 m lowering of the local tree-line on the exposed col between Taecheongbong and Chungcheongbong down to ca. 1550 m a.s.l. is caused by the effects of strong winds on physical and physiological structures of plants at the col, and the damage which results from this.

On the leeward eastern slopes of Taecheongbong, *Betula ermanii* var. *genuina* grows up to 2 or 4 m in height, and the krummholz form of this species is hardly found. On the windward southwestern slopes of the saddle between Taecheongbong and Chungcheongbong, and close to the summit of Taecheongbong, however, a krummholz comprising *Abies nephrolepis* is found at 1550 m to 1650 m a.s.l., along with mat-formed *Quercus mongolica*, *Betula ermanii* var. *genuina*, *Thuja koraiensis*, *Taxus caespitosa*, *Acer* spp., *Rhododendron* spp., and *Pinus pumila*. Three types of *Abies nephrolepis* krummholz have been recorded, based on their modes of formation, and their season of formation. First, a prevailing-wind-shaped type without serious malformation of both trunk and branches (slightly crooked krummholz) has adapted to prevailing strong winds during the growing season, mainly in spring. Second, an occasionally-high-wind-shaped type (the broken-trunk form of krummholz) has formed through the collapse of the main trunk, by an unusual gale, such as a summer typhoon; and third, a winter-wind-shaped type (mat, cushion, table or flagged krummholz) has been shaped through the dessiccation and/or abrasion of projecting branches, above a winter snow-cover, by driven ice crystals and snow particles during the late autumn and winter. The prevailing mesoscale local winds, which blow from the western lowlands and adjacent gorge towards the high ground, are undoubtedly responsible for the general occurrence and distribution of krummholz, as well as its overall shape, and directional formation.

The high ground near to the summit of Taecheongbong, and the saddle between Taecheongbong and Chungcheongbong, is unsuitable for warmth-loving plants in general. It is an area from which snow is largely blown clear during the winter months, and this doubly emphasises its coldness and aridity during both winter and spring; edaphic conditions also are close to sterile. Under these conditions, *Abies nephrolepis* krummholz there gives way to krummholz dominated by *Pinus pumila*.

In East Asia, *Pinus pumila* grows widely, being found from 32°21′ N to 70°31′ N and from 105° E to 190° E; in Korea, however, on Mt. Sorak, it lies at the extreme southern limit of its range. The component species of the Mt. Sorak *Pinus pumila* community include *Thuja koraiensis*, *Rhododendron mucronulatum* var. *ciliatum*, *Quercus mongolica*, *Betula ermanii* var. *genuina* and others. The number of species present on the wind-exposed southwestern and western slopes of Taecheongbong, and on the col between Taecheongbong and Chungcheongbong, is fewer than that found in the community on the wind-protected eastern slopes of Taecheongbong. The major species present on the wind-exposed western slopes include *Pinus pumila*, *Rhododendron mucronulatum* var. *ciliatum*, *Quercus mongolica* and *Abies nephrolepis*, while those on the wind-protected eastern slope are dominated by *Betula ermanii* var. *genuina*, along with *Pinus pumila*, *Thuja koraiensis* and *Acer tschnoskii* var. *rubripes* as subordinates.

On the windward slopes, *Pinus pumila* shows distinctive morphological adaptations against freezing and dessiccation. It tends to form a creeping trunk near to the ground surface, and only rarely exceeds a height of more than 5 cm above ground level; it has a compact crown, and a dense thicket of twigs and leaves, which together seem to contain the available heat, and to avoid excessive evapotranspiration; and, also, the presence of adventitious roots increases the chances of successful vegetative regeneration of the species in this harsh, short growing-season environment.

Temperature ranges for the vertical limits of *Pinus pumila*, as taken from 9 Korean mountains, show that the lower limit of the species is present within the range of 16.9°C to 23.4°C, for the mean daily maximum temperature in August. On Mt. Sorak, the August mean daily maximum temperature ranges for its lower limit (18.2°C to 19.1°C) fall towards the lower end of the national range. Nevertheless, all the evidence suggests that *Pinus pumila* on Mt. Sorak is close to the limit of its further, continued existence, and that only a small additional rise in temperature may be sufficient to render the species no longer viable in this locality. *Pinus pumila*, and its habitat on Mt. Sorak, are also in danger from excessive human trampling damage. Since the species is a good indicator species for past vegetation change in Korea, and it is currently dominant in the harshest environments on this mountain, there is a good case to be made for making it a protected species, and for ensuring that the small community which it dominates is preserved, especially bearing in mind the current projected increase in global temperatures ('the greenhouse effect') and the continued increase in visitors anticipated on this mountain.

CHAPTER 7

Conclusions

Within this study, eight major biogeographic regions in the Korean peninsula have been delineated, mainly by the use of a TWINSPAN programme, and based on the known distribution of 204 evergreen coniferous, broadleaved and bamboo species on 146 sample sites (88 inland and 58 island sites). These eight biogeographic regions are: a Northern Alpine Region (I), a North-South Subalpine Region (II), a Midland Mountain Region (III), a Southern Mountain Region (IV), a Mid-Western Insular Region (V), a Southern Insular Region (VI), a West-South-Eastern Insular and Associated Inland Region (VII) and a North-South Disjunctive Region (VIII). A comparative analysis of each biogeographic division, and its component plant groups, its species composition, its physiognomy and the three-dimensional distributional patterns of species found therein, has provided invaluable information as to the nature and history of the biogeography of Korea.

The dominance of arctic-alpine and alpine evergreen broadleaved plants at the altitude of 1500 m a.s.l. to 1800 m a.s.l. in the Northern Alpine Region (I) (e.g. *Ledum palustre* var. *maximum*, *L. palustre* var. *minus*, *L. palustre* var. *procumbens*, *L. palustre* subsp. *decumbens*, *Phyllodoce caerulea* and *Rhododendron aureum*), with their diverse life forms, has been demonstrated to indicate that first, the northern parts of Korea have maintained frequent floristic exchanges with neighbouring northern circumpolar areas, notably during the Pleistocene period; second, that both lowland and montane zones of this region have served as glacial refugia for arctic-alpine and alpine plants during the Pleistocene glacial phases; third, that the upslope retreat of these plants to alpine zones since the last glaciation, due mainly to climatic amelioration during the Holocene but also to the associated biological competition from more warmth-loving plants, has now confined many of these species to the alpine zones of northern Korea; and fourth, that the morphological adaptations to these plants to severe environmental conditions are the result of long-term gradual developments which have safeguarded the survival of these species in such a harsh cryo-climatic area.

The admixture of arctic-alpine and alpine evergreen broadleaved plants from the alpine and subalpine zones, northern evergreen coniferous plants from the subalpine and montane zones, and southern bamboos and southern evergreen broadleaved plants from the montane and lowland zones ensures that the North-South Subalpine Region of Korea (II) contains a wide range of life forms. Within this region, it is suggested that diversity is the result of former active and frequent floristic movements between the Korean peninsula and neighbouring East Asian countries, and also within the Korean peninsula, especially along the north-south Hamkyung Mountain Range in the north, and the Taebaik Mountain Range in the south; and second, today, the segregation to some extent of these floras, in different altitudinal zones. From this, a complex vegetation history is postulated for this region, which also reflects its diverse environmental conditions.

The sharp decrease in the total number of species (16 spp.), and in the range of their life forms, within the Midland Mountain Region (III), as compared to those found in the North-South Subalpine Region (II) (28 spp.) and in the adjacent Southern Mountain Region (IV) (33 spp.), and also the clear vertical divisions of range within it (viz. northern evergreen coniferous plants are confined to the montane zone, and southern evergreen coniferous and broadleaved plants to the lowland zone) suggest that this region is probably a major inland transitional zone of vegetation, intermediate between predominantly northern and southern floristic elements.

The coexistence of northern evergreen coniferous plants, southern evergreen coniferous plants, southern evergreen bamboos and southern evergreen broadleaved plants, all of which have a wide range of life forms, in the Southern Mountain Region (IV) indicates, first, that the high mountains present in the south have enabled some formerly much more widespread northern subalpine species to survive the present period of climatic amelioration by moving upslope; second, that the congregation of southern floristic elements both in the lowlands and montane zones of this region represents the influence of the presently-moderate climates; and third, that the dominance of southern evergreen coniferous plants, southern evergreen broadleaved plants and southern evergreen bamboo species and the appearance of southern vines, allocates this region as being 'southern' in character, and moreover probably implies previous strong floristic connections with tropical and subtropical areas of East Asia as well. Many of these southern species still have a large distributional range, which extends to the southern parts of Japan and China.

The occurrence of northern evergreen coniferous plants and southern evergreen broadleaved plants, and the even proportions of different life forms within the Mid-Western Insular Region (V), infer that this region is an insular transitional zone of vegetation, intermediate between northern and southern floristic elements.

The presence of the greatest number of evergreen plants in the Korean peninsula (77 spp.) in the Southern Insular Region (VI), including still some arctic-alpine evergreen broadleaved plants (e.g *Diapensia lapponica* subsp. *obovata*, *Empetrum nigrum* var. *japonicum* and *Rhododendron dauricum*), northern evergreen coniferous plants, southern evergreen coniferous plants, southern evergreen broadleaved plants, and the southern bamboo species on Mt. Halla (1950 m a.s.l.) in Cheju Island, with a highly complicated range of life forms, reflects more than any other single factor the diverse environmental conditions available. These include a vertical range pattern extending from cold alpine to warm coastal zones on Mt. Halla alone. Further to this, the isolated distributions of arctic-alpine and alpine evergreen broadleaved plants and evergreen coniferous plants on the higher mountains of this region again suggest that these plants occurred on a much broader scale within the whole Korean peninsula during the Pleistocene glacial phases, and have retreated upslope to the present-day alpine zone because of the general Holocene climatic amelioration. On another level, the appearance of the evergreen endemic coniferous plant *Abies koreana* on the southernmost fringes of the peninsula mainland as well as on Cheju, may provide an indicator of a previous land connection having been established between the mainland and Cheju at a time of former low sea-level, possibly during the penultimate glacial phase.

The existence of a large number of species (41 spp.), including northern evergreen coniferous plants at 1200 to 1900 m a.s.l. on Mt. Chiri, and southern coniferous and broadleaved plants, and southern evergreen bamboo species in both its montane and lowland areas, characterise the West-South-Eastern Insular and Associated Inland Region (VII). These features reflect, first the existence of relatively mild climatic conditions and, secondly the unexpected abundance (in terms of the limitations of the environment) of southern evergreen broadleaved plants and evergreen bamboos in the inland parts of this region.

A high proportion of endemism (60.0%), along with a high degree of physiognomic diversity for the small number of species present (5 spp.), is particularly noteworthy in the North-South Disjunctive Region (VIII).

Within the peninsula, the general absence of any primary data, such as fossil pollen evidence, which relates to vegetation history, has been emphasised. This is particularly so in respect of much of the Pleistocene period. However, some scattered direct and much indirect data enable one to reconstruct in some detail the origins, migrations and formation processes of the present-day Korean biogeographic divisions, and indeed their vegetation history, especially from the Oligocene to the present-day. The first appearance of many of the present-day floristic genera indeed dates back to the Oligocene (ca. 40 to 25 million years B.P.), and includes thermophilous evergreen broadleaved genera, such as *Myrica*, *Ficus*,

Hedera and other temperate deciduous broadleaved plants. Such genera grew at up to 4 degrees north of their present distributional limits, implying that the climate of the Oligocene was warmer than that of today.

The temperate climate of the Oligocene and the Lower Miocene of Korea (ca. 25 million years B.P.) subsequently warmed still further up to the Middle Miocene. Then, the appearance of additional thermophilous broadleaved plants, e.g. *Cinnamomum*, *Cyclobalanopsis* and *Ilex*, in the north of the peninsula, suggests a maximum northward expansion of evergreen broadleaved vegetation for 'recent' Korean vegetation history, these genera then existing six degrees of latitude further to the north than they do today. A relatively rich flora, and the continued presence of thermophilous evergreen broadleaved plants in the south of the peninsula, suggests the maintenance of a warm climate there in the Upper Miocene (ca. 12 million years B.P.); however, by this time, some cryophilous plants (e.g. *Picea* and *Rhododendron*) had also begun to penetrate into the north, indicating that regional differences, associated with the general Miocene cooling, had already been initiated. Overall, the continued occurrence of numerous groups from the present-day flora since the Oligocene period indicates a long-term stability for Korean vegetation. Unfortunately, from the Pliocene to the Middle Pleistocene, no floristic fossil data are available from Korea, and so a hiatus in the vegetation history of the peninsula of some considerable length then begins.

It is known, however, that the presence of many cryophilous evergreen coniferous plants (e.g. *Taxus*, *Abies*, *Pinus* and *Thuja*) dates back at least to the Middle Pleistocene period (ca. 700000 to 110000 years B.P.). Certainly, the admixture of these coniferous evergreen plants with deciduous broadleaved plants indicates a probable temperate climate for much of the Middle Pleistocene. The further appearance of *Pinus* charcoal and *Artemisia* pollen in human-occupied caves at this time also possibly suggests an early-stage anthropogenic disturbance of natural vegetation during the Middle Pleistocene of Korea.

Subsequently, the presence of cold episodes during the Upper Pleistocene (ca. 110000 to 10000 years B.P.), in which the mean annual temperature of Korea is thought to have been lowered by ca. 8°C to 12°C during the penultimate glacial period, and by ca. 6°C during the last glacial phase, as compared with present-day values, caused a general expansion of deciduous broadleaved plants, and cold-loving evergreen coniferous plants (e.g. *Taxus*, *Abies*, *Picea*, *Tsuga*, *Pinus* and *Thuja*) within the peninsula. This is exemplified more particularly by the sudden appearance of cold-loving evergreen coniferous plants, such as *Pinus koraiensis*, *P. pumila*, *P. sibirica*, and species of *Pseudotsuga* and *Juniperus*, at these relatively late stages of the Upper Pleistocene. It is likely that the maximum southward expansion of cold-loving arctic-alpine evergreen broadleaved and coniferous plants in Korea occurred at such times in the Upper Pleistocene, and this is emphasised by their current disjunctive distributions following the most recent climatic amelioration.

From the available information, a tentative vegetation zonation for the last glaciation (Devensian) in the Korean peninsula may be worked out. Six zones may be delimited. These are: (i) small glaciers, confined to the northernmost high mountains; (ii) a limited alpine tundra or montane subarctic vegetation on mountains in northern Korea; (iii) an arctic-alpine vegetation on mountains from the north to the midlands generally as far south as Mt. Sorak, and including the higher part of Mt. Halla in Cheju Island as well; (iv) a steppe vegetation in the northwest, close to the present Chinese border; (v) coniferous vegetation, found most conspicuously in the midlands and the south, mainly in mountain areas; and (vi) a mixed coniferous and deciduous broadleaved vegetation in the western midlands, in the south, and the southern coastal districts. These tentative vegetation zones for the last glacial period undoubtedly show some dissimilarities from the present-day biogeographic divisions. These general discrepancies are due to the subsequent climatic amelioration, and may be summarised as follows: first, all vegetation zones associated with the Devensian glaciation have retreated northward and upward during the subsequent post-glacial phase; second the cryogenetic vegetation zones of the glacial period have been replaced by alpine and subalpine vegetation in the present-day biogeographic divisions; and third, the mixed coniferous and deciduous broadleaved vegetation of southern Korea has been replaced by much more strongly thermophilous evergreen broadleaved and coniferous vegetation, along with a range of deciduous broadleaved species. However, nearly all the Middle and Upper Pleistocene floras are still found in present-day Korean vegetation, further emphasising its recent overall stability. However, it should also be emphasised, at least in the case of arctic-alpine and alpine species, that these changes took place within the context of the evolution of a limited number of Korean endemics, which add to the distinctiveness of the vegetation today. At the end of the Devensian, the available fossil pollen data for Korea, though still limited, became slightly more substantive than formerly.

From 17000 to 15000 years B.P., these data suggest that cryophilous evergreen coniferous plants, e.g. *Picea*, *Abies*, and *Pinus* (Haploxylon), were increasingly widespread. Then, a decline of evergreen coniferous plants from 15000 to 10000 years B.P., and the disappearance from the limited pollen data of some cryophilous genera from 10000 to 6700 years B.P. (notably since ca. 8000 years B.P.), associated with a general northward retreat of cryophilous vegetation, and accompanied by the expansion of thermophilous vegetation from the south, marks the continued climatic amelioration. The dominance of warmth-loving *Pinus* (Diploxylon), and deciduous broadleaved plants, and the wide appearance of thermophilous evergreen broadleaved plants, indicates that the well-known postglacial hypsithermal period was well established from ca. 8000 to 4500 years B.P., and at this time temperatures were probably 1° to 2°C warmer than today. It was this particular climatic phase that more than any other event accelerated the northward and upward retreats of cryophilous arctic-alpine and alpine evergreen broadleaved and coniferous plants, and in turn the northward and upward expansion of temperate deciduous broadleaved plants and evergreen coniferous plants, and also thermophilous evergreen coniferous and broadleaved plants. There has also been an increasing disturbance of vegetation by man since ca. 5000 years B.P., mainly for agricultural and domestic activities (e.g. land clearance, building, firewood, charcoal etc.).

Most of the arctic-alpine and alpine plants in East Asia are thought to have evolved from a montane flora originally established in the high mountains of central Asia (and, in some cases, further afield) since the Miocene, but mainly during the Pleistocene. These plants repeatedly advanced downslope and southward as the climate deteriorated, notably during the Pleistocene glacial phases, but retreated back upslope and northward during the alternate phases of climatic amelioration. The absence of major topographic barriers, and of extensive glaciation, coupled with the presence of a north-south orientation in the mountain ranges of both East Asia and Korea, facilitated the dispersal of arctic-alpine and alpine plants southward from the circumpolar area, possibly through the Maritime Territory of Russia, and especially along the Sikhote-Alin Mountain Range, consequently giving rise to the present rich arctic-alpine and alpine flora in Korea. This migration was also aided by the lowering of sea-level during the glacial phases of the Pleistocene, especially in respect of the areas now covered by the West Sea (Yellow Sea), and the Korean Straits between Korea and Japan.

Further,in respect of the processes of arctic-alpine and alpine species migration in the past, the Korean peninsula seems to have served as one of the major refugia for these species during both the glacial and interglacial periods, because of their easy access to and from the circumpolar area of East Asia, and the diversity of its environments and habitats ranging from coastal lowlands to alpine zones, which encouraged their survival. The cold-phase Pleistocene climates in the alpine zones of Korea seem generally to have been too cold and harsh for arctic-alpine and alpine plants then to have survived at high altitudes. Those plants, therefore, had to confine themselves to lower montane and/or lowland areas during these phases, possibly including coastal areas. It is likely that the lowlands of the Korean peninsula, possibly including the coastal lowlands, accordingly effectively became *primary* refugia for many of these species during the cold phases of the Pleistocene. Then, as the climate subsequently ameliorated, arctic-alpine and alpine plants moved upslope on to the mountain tops and northwards, and the range of thermophilous temperate and warm temperate plants was then

Phogo 7.1. Some montane forests are exploited heavily. Soya fields are established frequently on steep slopes. Surroundings of the Naegog Spa, Kaema Upland, northern part of North Korea. (Photograph by M. Šrutek.)

expanded in the wake of this movement, to cover large areas of lowland and the lower mountains. The mountain tops of the Korean peninsula accordingly then proved to be important *secondary* refugia, certainly in the post-Devensian period. The present occurrence of six arctic-alpine species, in the alpine and subalpine zones of Korea, at the world's southernmost limit of their distribution (e.g. *Ledum palustre* var. *diversipilosum*, *L. palustre* subsp. *decumbens*, *Phyllodoce caerulea*, *Diapensia lapponica* subsp. *obovata* and *Empetrum nigrum* var. *japonicum*), and of another six species at the southernmost limit of their range in the East Asian continent (e.g. *Chamaedaphne calyculata*, *Dryas octopetala* var. *asiatica*, *Oxycoccus microcarpus*, *O. quadripetalus*, *Vaccinium vitis-idaea* var. *minus* and *Linnaea borealis*) further promotes the idea of the existence of refugia for these species in particular in the Korean peninsula, both in the Pleistocene glacial phases and during the postglacial period.

At present, the estimated number of Korean arctic-alpine and alpine species is 419, and this includes 75 arctic-alpine species (i.e. 17 arboreal and 58 herbaceous spp.), 239 alpine species (i.e. 108 arboreal and 131 herbaceous spp.) and 105 Korean endemic alpine species (i.e. 58 arboreal and 47 herbaceous spp.). The occurrence of such a large number of arctic-alpine, alpine and endemic alpine species, which *in toto* forms up to 10% of the total Korean flora, can be explained by the following reasons. First, there have been no catastrophic environmental events in the Korean peninsula during the Pleistocene such as the presence of a major glaciation, and consequently a long history of the flora is assured. Secondly, the unique geographic setting (a north-south linkage of mountain ranges both in East Asia and Korea) has enabled smooth migrations of arctic-alpine and alpine plants from north to south to take place, with the onset of the Pleistocene cold phases. Thirdly, the existence of ca. 79 high mountains (over 2000 m a.s.l. in height) in northern Korea, and of ca. 264 mountains of 1500 to 2000 m a.s l. in height, mainly in the north, but also in central and southern Korea, has ensured that isolated areas have been available for these species to colonise during interglacials, and the post-Devensian warming. Finally, the different climatic regimes present in the Korean peninsula today, particularly the predominantly continental one in the north, an intermediate one in the midlands, and a slightly oceanic one in the south and Cheju Island, have created a range of diverse climatic habitats in general and local environments in particular, which have encouraged the long-term survival of many of these species at all altitudes.

All the above situations (the frequent floristic movements between circumpolar East Asia and the Korean peninsula during the Pleistocene glacial phases, the existence of peninsular glacial refugia, the presence of easy access between mountains in the north and the availability of diverse habitats and environments for alpine flora, particularly in the north and the midlands of Korea) have resulted in the dominance of arctic-alpine and alpine plants in the north and midlands of Korea, albeit on a disjunctive distributional basis. Thus, ca. 71.7% of the total number of arboreal arctic-alpine and alpine plants, ca. 74.9% of the total number of herbaceous arctic-alpine and alpine plants and ca. 61.7% of the total number of endemic alpine plants are present in these two regions. The particular larger-scale disjunctive distribution of three arctic-alpine plants, i.e. *Diapensia lapponica* subsp. *obovata* between East Asia and Cheju Island, and *Empetrum nigrum* var. *japonicum* and *Luzula rufescens* between northern Korea and Cheju Island, further strengthens the idea of a previous land connection between the mainland of Korea and Cheju Island during at least some of the cold phases of the Pleistocene glaciation.

The relatively high degree of endemism found among arboreal alpine plants (32.2%) in Korea, in comparison with that in the herbaceous alpine plant group (25.1%), may be due to the poorer rate of migration of arboreal alpine plants compared with the herbaceous plants, and consequently the longer-term isolation of these flora. On the other hand, the continuous distributional pattern of many alpine perennial herbs may be the result of their ability to withstand better the severe climatic conditions of the Pleistocene glaciations, and then to respond and adapt quickly to the subsequent amelioration by spreading widely on open ground. Both processes may of course be in operation at the same time.

Further detailed studies on the evergreen arboreal arctic-alpine and alpine plants in Korea were confined to 41 species, 18 genera and 10 families present in this group. Out of these, 22 species belong to the dicotyledons and 19 species to the monocotyledons; in terms of their life form, they consist of 13 trees, 2 small trees and 26 shrubs. Their general distributional patterns can be divided into 8, viz: (i) circumpolar-arctic plants (*Vaccinium vitis-idaea* var. *minus*); (ii) circumpolar and arctic-montane plants (*Phyllodoce caerulea*); (iii) circumpolar-boreal plants (*Andromeda polifolia*, *Vaccinium vitis-idaea*, *Oxycoccus microcarpus*, *O. quadripetalus* and *Linnaea borealis*); (iv) northern-Eurasiatic plants (*Ledum palustre* var. *latifolia*); (v) circumpolar-northern boreal plants (*Chamaedaphne calyculata*, *Rhododendron aureum* and *R. parvifolium*) (vi) North Pacific plants (*Diapensia lapponica* subsp. *obovata*, *Ledum palustre* var. *diversipilosum*, *L. palustre* var. *minus*, *L. palustre* subsp. *decumbens*, *L.. palustre* var. *yezoense*, *Rhododendron dauricum* and *R. fauriae* form. *rufescens*); (vii) East Asian plants (*Taxus cuspidata*, *T. cuspidata* var. *latifolia*, *Abies holophylla*, *A. nephrolepis*, *Picea jezoensis*, *P. schrenkiana*, *Pinus koraiensis*, *P. pumila*, *P. tabulaeformis*, *Sabina davurica*, *S. sar-*

gentii, *Dryas octopetala* var. *asiatica* and *Empetrum nigrum* var. *japonica*; and (viii) Korean endemic plants (*Taxus caespitosa*, *Abies koreana*, *Picea intercedens*, *P. koraiensis*, *P. pungsanensis*, *P. tonnaiensis*, *Thuja koraiensis*, *Juniperus sibirica* and *Ledum palustre* var. *maximum*. These distributional patterns emphasise once again the overall complexity of plant groupings within the Korean peninsula, especially in terms of possible origins, even for this one limited group of arctic-alpine and alpine plant species.

Further to their broad-scale distributional patterns, 13 species of the 41 (1 evergreen coniferous plant and 12 evergreen broadleaved plants) belong to the *arctic-alpine* group of plants. The other 28 species (18 evergreen coniferous plants and 10 evergreen broadleaved plants) belong to the evergreen arboreal group of *alpine* plants. The appearance here of 8 Korean endemic arboreal alpine plants (19.5% of the total evergreen arboreal arctic-alpine and alpine plants) no doubt reflects the long-term isolation of these species, and also possibly favourable micro-environmental conditions in the areas which they inhabit.

Within the Korean peninsula, 7 distributional patterns can be noticed among these 41 evergreen arboreal arctic-alpine and alpine plants. These are: (i) species present throughout the peninsula (8 spp.); (ii) species present from the north to the midlands (5 spp.) (iii) species present in the north and in Cheju Island, but not in between (1 sp.); (iv) species present in the mainland south and on Cheju Island alone (1 sp.); (v) species restricted to the north of Korea (2 spp.); (vi) species restricted to the midlands of Korea (2 spp.) and (vii) species restricted to Cheju Island (1 sp.). Thirty-eight of these species (92.7%) occur within the north of Korea, and the pertinent reasons for this have already been specified (p. 118).

At present, four altitudinal divisions can be delineated in respect of the vertical lower limits of the 23 evergreen arboreal arctic-alpine and alpine species for which data are available. First, lower limits which lie at ca. 1500 to 1800 m a.s.l. and above are dominated mainly by evergreen broadleaved plants; second, those occurring at ca. 1000 to 1500 m a.s.l. comprise a mixture of evergreen coniferous plants and evergreen broadleaved plants; third, those at ca. 1000 m a.s.l. involve endemic evergreen coniferous plants; and those at ca. 500 to 1000 m a.s.l. are dominated mainly by more thermophilous evergreen coniferous plants. The presence of evergreen arboreal broadleaved plants, mainly in the northern alpine areas at ca. 1500 to 1800 m a.s.l or above, in which low summer temperatures and very cold winter temperatures prevail, is likely to be due to their ability to develop physiognomic and ecological adaptations against cold environments very effectively (pp. 123–127).

The thermal amplitudes of the Korean evergreen broadleaved, coniferous and endemic species, using high summer temperatures (temperatures within this section are calculated lapse-rate values: see p. 132) at the lower distributional limits, and low winter temperatures at the upper distributional limits, again display four vertical distributional divisions. First, evergreen arboreal broadleaved arctic-alpine and alpine plants, which dominate the northern and southern alpine zones (mainly ca. 1500 to 1800 m a.s.l. or above) are intolerant (=sensitive) to high summer temperatures (mean daily maximum temperature for August = 16.7°C to 22.7°C), but tolerant to low winter temperatures (mean daily minimum temperatures of January lie at −40.4°C to −35.6°C). Intolerance to high summer temperatures is best expressed in reduced levels of competition as compared to other more warmth-loving plants. However, evergreen broadleaved arctic-alpine species are very adaptable to cold winter temperatures, and hence these species dominate the Korean alpine zone. Second, at the altitudes of ca. 1000 to 1500 m a.s.l., within the north to midland zones, species display a similar intolerance to high summer temperatures (mean daily maximum temperatures for August are between 15.2°C to 21.0°C), but they are less tolerant to low winter temperatures (mean daily minimum temperatures for January lying between −38.5°C to −30.7°C). But evergreen coniferous alpine and subalpine species are in general more adaptable to warm summer temperature conditions than are the evergreen broadleaved arctic-alpine and alpine plants, and also they undoubtedly have relatively wide thermal amplitudes. This adaptation to warm summer temperatures encourages these species to grow generally, in alpine and subalpine zones, below evergreen broadleaved arctic-alpine and alpine plants. Evergreen coniferous alpine and subalpine plants also remain well adapted to cold winter temperatures. Third, a transitional zone, at ca. 1000 m a.s.l., is covered by a wide range of endemic evergreen coniferous plants, present from the north to Cheju Island; these include both a high-summer-temperature-intolerant group, and a high-summer-temperature-adaptive group (mean daily maximum temperatures for August at between 19.8°C to 23.7°C), and adaptations to low winter temperatures (mean daily minimum temperatures for January of between −33.0°C to −18.3°C). These endemic evergreen coniferous subalpine plants grow in a slightly warmer climate than that which is normal for the evergreen broadleaved arctic-alpine and alpine plants; but it can also be slightly cooler at the extremes than that of the non-endemic coniferous alpine and subalpine plants. Endemic evergreen coniferous subalpine plants are, accordingly, less sensitive to high summer temperature conditions than are other evergreen coniferous plants mentioned above. Fourth, the lower part of the subalpine, and most montane zones, lying from ca. 500 to 1000 m a.s.l., are occupied by high-summer-temperature-adaptive evergreen coniferous plants, as well as by many other deciduous broadleaved plants. The adaptive ability of these plants to withstand high summer temperatures, and also to tolerate relatively low winter temperatures as well, enables them to grow within a very wide vertical range in Korea, and they occasionally reach even the montane zone. Overall, the disjunctive distribution of evergreen broadleaved arctic-alpine and alpine species in the alpine zones of Korea is due in large measure to their sensitivity to the current high summer temperatures, and a relatively narrow general ecological amplitude. In contrast, the wide spread of evergreen coniferous alpine species in the alpine, subalpine and lower montane zones of Korea, is due to their adaptation both to high summer temperatures and low winter temperatures, and the development by them of an extremely broad ecological amplitude. The disjunctive distribution of the endemic evergreen coniferous subalpine species in the transitional area between the subalpine and lower montane zones of Korea is due to their adaptation to mild summer temperatures, and at the same time a sensitivity to anything other than relatively low winter temperatures. The full implications of this division to an understanding of the biogeography and ecology of Korea needs further study.

In order to examine in greater detail some of these relationships, field work was undertaken in the subalpine zone of Mt. Sorak, Korea, and this zone compared with previous work on Mt. Chiri and Mt. Halla. The flora and vegetation of Mt. Sorak (1708 m a.s.l.) comprises three distinct biogeographical characteristics, namely, a high degree of floristic diversity; the presence of a large number of endemic species; and the coexistence of the southernmost distributional limits of some northern floristic elements, and at the same time the northernmost limits of some southern floristic elements. Mt. Sorak accordingly lies at the fulcrum of many of the biogeographical relationships of Korea. The high degree of floristic diversity on Mt. Sorak as a whole (ca. 1013 spp.) is due, in part, to the size of the mountain, and its various habitats, which accommodate a wide range of species from warmth-loving plants to cold-loving plants. The occurrence of a relatively large number of Korean endemic species on Mt. Sorak (ca. 65 spp.) implies, as has been indicated previously, a response to the floristic movements and exchanges with other parts of the Korean peninsula in the past under changing climatic conditions. On Mt. Sorak, there are also 17 indigenous endemic plants. The coexistence of the 40 northern and 6 southern floristic elements on Mt. Sorak, at their southernmost and northernmost limits respectively, also seems to be the result of a focal confluence of the northern and southern floras in this region as climate has fluctuated in the past, probably as a result of its location on one of the main plant migration pathways in the Korean peninsula.

The occurrence of the 7 species exclusively found in the subalpine zone of Mt. Sorak, between 1500 to 1700 m a.s.l., is likely to have been one result of the post-glacial climatic amelioration. Out of the 24 subalpine species on Mt. Sorak, four major subalpine communities lie at above ca. 1600 m a.s.l. These are dominated by *Pinus pumila*, *Abies nephrolepis*, *Thuja koraiensis* and *Betula ermanii* var. *genuina*. The present isolated or disjunctive occurrence of these four species both within the Korean peninsula and also within the boundaries of Mt. Sorak reflect a former continuous distribution both locally and on a broader scale, their inter-mountain migration at times during phases of climatic deterioration, and the subsequent breakdown of a former continuous range into fragments as the climate ameliorated more recently, and intervening forest developed on adjacent montane and lowland areas. In other words, as the climate deteriorated during the cold phases of the Pleistocene, these cryophilous plants seem to have shifted downward from the mountains and southward from the north, and former isolated cold-loving mountain vegetation was joined together, and expanded farther towards the south. On the other hand, the climatic amelioration during warm phases, and especially during the post-glacial warm period, has resulted in the retreat of the previously wide-ranging cryophilous vegetation upslope and northward, but an expansion of the thermophilous vegetation at the same time. At present, the cryophilous plants are again isolated from each other, both on a meso- or a macroscale in the Korean mountains.

The leeward eastern slopes near to Taecheongbong, the summit of Mt. Sorak, are occupied mainly by *Betula ermanii* var. *genuina*, and show no clear tree-line. However, the windward western slopes of Taecheongbong, and the saddle between Taecheongbong (1708 m a.s.l.) and Chungcheongbong (1676 m a.s.l.), do display a tree-line, formed mainly by *Abies nephrolepis* and, to a lesser extent, by other species such as *Rhododendron* spp., and *Betula* spp. Areas above the tree-line are dominated by *Pinus pumila*. This local tree-line, beyond which arboreal tree species are unable to develop an erect tree form, mainly due to the extreme climatic conditions, occurs at ca. 1650 m a.s.l. on the windward western slopes of Taecheongbong. At the saddle between Taecheongbong and Chungcheongbong, in which strong local winds prevail, the local tree-line is found at a lower altitude (1550 m a.s.l.) on the windward southwestern slopes. The 100–metre lowering of the tree-line on the high-wind-exposed saddle between the two peaks is undoubtedly the result of the effect of local westerly high winds and their associated effects in winter and spring, which cause both physical and physiological damage to upright plants.

Krummholz forms of vegetation are present on the windward, western side of Taecheongbong. The absence of krummholz trees on the leeward eastern slopes is predominantly the consequence of their being protected from the main force of the hostile westerly and south-westerly local winter and spring winds, and also partly because of the relatively higher rainfall present, from summer monsoons, on these eastern slopes. On the windward southwestern slopes of the saddle between Taecheongbong and Chungcheongbong, and close to the summit of Taecheongbong, *Abies nephrolepis* krummholz is found at ca. 1550 to 1650 m a.s.l., along with mat-formed shrubs. In the case of *Abies nephrolepis*, three types of krummholz have been observed, based on their formation processes, and physiological mechanisms on a seasonal basis. The first of these is a prevailing-wind-shaped type, without serious malformation of both trunk and branches ('slightly-crooked krummholz'), which has adapted to the prevailing wind during the growing season, notably in spring. The second is an occasionally-high-wind-shaped type ('broken-trunk' form of krummholz), which has formed through the collapse of the main trunk by an unusual gale, such as a summer typhoon. The third is a winter-wind-shaped type (mat-, cushion-, table- or flagged-krummholz), which has been shaped through the dessiccation and/or abrasion of projecting branches above a snow cover, by driven ice crystals and snow particles during the late autumn and winter. Prevailing mesoscale local winds, which blow from the western lowlands up western gorges toward the high ground of Mt. Sorak, may be responsible for the general occurrence and distribution of krummholz in this area, as well as for its architecture.

The absence of all but the most cold-tolerant plants on the high ground near the summit of Mt. Sorak, and on the col between Taecheongbong and Chungcheongbong, may be due to several factors. First, the 'top phenomenon' or 'summit effect' causes a general lowering of air and soil temperatures, above those which may normally be expected, especially during the winter season. Second, fallen snow tends to blow away in this area, so that it eventually begins to suffer from serious drought in the days of early spring. Third, the generally high level of evapotranspiration on mountain summits or ridges, accentuated by the way in which winds are accelerated by topographic obstructions, may easily kill tall, warmth-loving plants by dehydration. Finally, strong winds on the exposed high land tend to remove humus and top soil, and consequently bring sterile edaphic conditions.

In the subalpine zone of Mt. Sorak, above ca. 1600 m a.s.l., *Pinus pumila*, *Thuja koraiensis* and *Rhododendron mucronulatum* var. *ciliatum* are the major components in both species abundance and coverage. However, a major difference in species representation can be seen as between the vegetation of the windward slopes, and that of the wind-protected area. The major species on the wind-exposed western slopes of Taecheongbong include *Pinus pumila*, *Rhododendron mucronulatum* var. *ciliatum*, *Quercus mongolica* and *Abies nephrolepis*, while those on the wind-protected eastern slopes consist of *Betula ermanii* var. *genuina*, *Pinus pumila*, *Thuja koraiensis*, and *Acer tschnoskii* var. *rubripes*. This species difference according to aspect is due mainly to the effects of differential exposure and snow cover in this subalpine zone. On the windward side, *Pinus pumila* shows a distinct morphological adaptation against damage by freezing and dessiccation, which includes: first, a creeping trunk near the ground, which rarely exceeds 5 cm above ground level; second, a compact crown and dense thicket of twigs and leaves, which all seem to help, at the same time, to keep whatever warmth is available, and to avoid excessive evapotranspiration; and third, the presence of adventitious roots, which increase the chances of successful vegetative regeneration of the species in this harsh area, with its short growing season.

The northern alpine and subalpine species *Pinus pumila* is found in East Asia generally from 32°20′ N to 70°31′ N and 105° E to 190° E, and the altitudinal range of the species within this broad region varies from sea level up to 3200 m a.s.l. In Korea, *Pinus pumila* grows from 40° N to 38° N on 16 mountains, mainly from 900 to 2200 m a.s.l. At present, the means of the vertical range of *Pinus pumila* in Korea, averaged from available data from the 16 Korean peaks on which it is present, are 1485 m a.s.l. for the lower limit and 1805 m a.s.l. for the upper limit. Though today *Pinus pumila* distributions are isolated from each other on mountain tops, if the mean annual temperatures were to deteriorate by ca. 8° to 9°C, as may have occurred during some of the Pleistocene glacial phases, the species would then have extended its range both downward and southward to form a continuous coverage in many areas. Later, it split into fragments, during the most recent climatic amelioration. The further microscale disjunctive distribution of the species, within the boundaries of Mt. Sorak itself, on several peaks, e.g. Taecheongbong-Chungcheongbong, Socheongbong, Kuitaegicheongbong, Karibong and Madeungryong (1708, 1676, 1600, 1577, 1518 and 1326 m a.s.l., respectively), all within 11 km of Taecheongbong, also suggests the former, relatively recent presence of a continuous distribution of *Pinus pumila* at high altitudes in this area, probably before the hypsithermal. The comparative analysis of the mean daily maximum temperatures of August in the whole of Korea (16.9°C to 23.4°C) and that on Mt. Sorak, based on local climatic data (18.2°C to 19.1°C), suggests that the temperature range for the lower limit of *Pinus pumila* on Mt. Sorak is attuned to the national mean daily maximum temperatures for August, though it is quite sensitive to high summer temperatures in general terms. Therefore only a small additional rise in temperature at the summit of Mt. Sorak may be sufficient to render this species no longer viable on this mountain, at its southernmost distributional limit worldwide.

Arising from all this, a plea must be made for more urgent research on these arctic-alpine, and alpine plant species in particular and their environments, not only for the further better understanding of biogeography in the Korean peninsula, but also as an example of what may happen to arctic-alpine, alpine plants and environments generally should the threatened continuing increase in global temperatures begin to materialise. The latitudinal location of the Korean peninsula, and its altitudinal and floristic diversity, together make it a critical site for such an evaluation, and we trust that a rapid response to this plea will be forthcoming through both national and international initiatives and funding.

APPENDIX 1

List of Korean evergreen plants

Species No.	Family Name	Species Name
1	Piperaceae	*Piper kadzura*
2	Chloranthaceae	*Sarcandra glabra*
3	Myricaceae	*Myrica rubra*
4	Fagaceae	*Castanopsis cuspidata* var. *sieboldii*
5		*Castanopsis cuspidata* var. *latifolia*
6		*Castanopsis cuspidata* var. *thunbergii*
7		*Cyclobalanopsis acuta*
8		*Cyclobalanopsis acuta* form. *subserra*
9		*Cyclobalanopsis gilva*
10		*Cyclobalanopsis glauca*
11		*Cyclobalanopsis myrsinaefolia*
12		*Cyclobalanopsis stenophylla*
13		*Cyclobalanopsis stenophylla* var. *latifolia*
14	Moraceae	*Ficus erecta*
15		*Ficus erecta* var. *longepedunculata*
16		*Ficus erecta* var. *sieboldii*
17		*Ficus nipponca*
18		*Ficus pumila* (= *stipulata*)
19	Loranthaceae	*Hypear tanakae*
20		*Scurrula yadoriki*
21		*Viscum coloratum* var. *lutescens*
22		*Viscum coloratum* form. *rubroauranticum*
23	Bifaria	*Bifaria japonica*
24	Lardizabalaceae	*Stauntonia hexaphylla*
25	Menispermaceae	*Stephania japonica*
26	Illiaceae	*Illicium anisatum*
27	Lauraceae	*Kadsura japonica*
28		*Cinnamomum camphora*
29		*Cinnamomum japonicum*
30		*Cinnamomum loureirii*
31		*Fiwa japonica*
32		*Izosta lancifolia*
33		*Machilus japonica*
34		*Machilus thunbergii*
35		*Machilus thunbergii* var. *obovata*
36		*Neolitsea aciculata*
37		*Neolitsea sericea*
38	Pittosporaceae	*Pittosporum tobira*
39	Hamamelidaceae	*Distylium racemosum* var. *latifolium*
40		*Distylium racemosum* var. *typicum*
41	Rosaceae	*Raphiolepsis liukiuensis*
42		*Raphiolepsis obovata*
43		*Raphiolepsis ubellata*
44		*Rubus buergeri*

45	Rutaceae	*Citrus aurantium*
46		*Citrus deliciosa*
47		*Citrus grandis*
48		*Citrus junos*
49		*Citrus kinokuni*
50		*Citrus medica* var. *sarcodactylus*
51		*Citrus natsudaidai*
52		*Citrus noblis*
53		*Citrus sinensis*
54		*Citrus unshiu*
55		*Zanthoxylum planispinum*
56	Daphniphyllaceae	*Daphniphyllum glaucescens*
57		*Daphniphyllum macropodum*
58	Buxaceae	*Buxus koreana*
59		*Buxus koreana* var. *elongata*
60		*Buxus koreana* var. *insularis*
61		*Buxus microphylla*
62	Aquifoliaceae	*Ilex cornuta* form. *typica*
63		*Ilex crenata* var. *microphylla*
64		*Ilex integra* var. *typica*
65		*Ilex rotunda*
66	Celastraceae	*Euonymus* (= *Masakia*) *chibai*
67		*Euonymus* (= *Masakia*) *japonica*
68		*Euonymus* (= *Masakia*) *japonica* var. *latifolia*
69		*Euonymus* (= *Masakia*) *radicans*
70	Rhamnaceae	*Sageretia theezans*
71		*Sageretia theezans* var. *tomentosa*
72	Elaeocarpaceae	*Elaeocarpus sylvestrus* var. *ellipticus*
73	Ternstroemiaceae	*Camellia japonica*
74		*Eurya emarginata*
75		*Eurya japonica* var. *aurescens*
76		*Eurya japonica* var. *integra*
77		*Eurya japonica* var. *montana*
78		*Sasakia ochnacea*
79		*Ternstroemia mokof*
80		*Thea sinensis* var. *bohea*
81	Flacourticeae	*Xylosma apatis*
82	Daphnaceae	*Daphne kiusiana*
83		*Daphne odora*
84	Elaeagnaceae	*Elaeagnus glabra* var. *euglabra*
85		*Elaeagnus glabra* var. *oxyphylla*
86		*Elaeagnus macrophylla*
87		*Elaeagnus martima*
88		*Elaeagnus submacropylla*
89	Araliaceae	*Fatsia japonica*
90		*Hedera tobleri*
91		*Textoria* (= *Dendropanax*) *morbifera*
92	Cornaceae	*Aucuba japonica* var. *typica*
93	Diapensiaceae	*Diapensia lapponica* subsp. *obovata*
94	Empetraceae	*Empetrum nigrum* var. *japonicum*
95	Ericaceae	*Andromeda polifolia*
96		*Andromeda polifolia*
97		*Chamaedaphne calyculata*
98		*Ledum palustre* var. *angustrum*
99		*Ledum palustre* var. *diversipilosum* (= *dilatum*, *nipponicum*)
100		*Ledum palustre* var. *latifolium* (= *groenlandicum*)
101	Rosaceae	*Dryas octopetala* var. *asiatica*
102		*Ledum palustre* var. *maximum*
103		*Ledum palustre* var. *minus* (= *longifolium*)
104		*Ledum palustre* subsp. *decumbens* (= var. *procumbens*)

105		*Ledum palustre* var. *yesoense*
106		*Oxycoccus microcarpus* (= *pusillus*)
107		*Oxycoccus quadripetalus*
108		*Phyllodoce caerulea*
109		*Rhododendron aureum* (= *chrysanthum*)
110		*Rhododendron dauricum*
111		*Rhododendron fauriae* var. *roseum*
112		*Rhododendron fauriae* form. *rufescens*
113		*Rhododendron micranthum*
114		*Rhododendron parvifolium*
115		*Vaccinium bracteatum*
116		*Vaccinium vitis-idaea* var. *minus*
117		*Vaccinium vitis-idaea* var. *genuinium*
118	Myrsinaceae	*Bladhia crenata* (= *lentiginosa*)
119		*Bladhia crenata* var. *taquetii*
120		*Bladhia crispa* var. *typica*
121		*Bladhia japonica* var. *typica*
122		*Bladhia villosa* var. *typica*
123	Symplocaceae	*Dicalix lucida*
124		*Dicalix prunifolia*
125	Oleaceae	*Ligustrum japonicum*
126		*Ligustrum japonicum* var. *rotundifolium*
127		*Ligustrum lucidum*
128		*Ligustrum ovalifolium*
129		*Ligustrum quihoui* var. *latifolium*
130		*Osmanthus zantaranus*
131	Loganiaceae	*Cardneria insularis*
132	Apocynaceae	*Trachelosphermum asiaticum* var. *glabrum*
133		*Trachelospermum asiaticum* var. *intermedium*
134		*Trachelospermum oblanceolatum*
135		*Trachelospermum jasminoides*
136		*Trachelospermum jasminoides* var. *pubescens*
137		*Trachelospermum majus*
138	Ascelpiadaceae	*Marsdenia tomentosa*
139	Verbenaceae	*Vitex rotundifolia*
140	Rubiaceae	*Damnacanthus indicus*
141		*Damnacanthus majus*
142		*Gardenia jasminoides* form. *graniflora*
143		*Serissa japonica*
144	Caprifoliaceae	*Linnaea borealis*
145		*Viburnum awabuki*
146	Berberidaceae	*Nandina domestica*
147	Bambusaceae	*Pleioblastus simoni*
148		*Pseudosasa japonica*
149		*Pseudosasa japonica* var. *purpurascens*
150		*Sasa coreana*
151		*Sasa kurilensis*
152		*Sasa quelpaertensis*
153		*Sasamorpha chiisanensis*
154		*Sasamorpha gracilis*
155		*Sasamorpha purprrascens* var. *borealis*
156		*Sinoarundinaria* (= *Phyllostachys*) *compressa*
157		*Sinoarundinaria* (= *Phyllostachys*) *nigra*
158		*Sinoarundinaria* (= *Phyllostachys*) *nigra* form. *punctata*
159		*Sinoarundinaria* (= *Phyllostachys*) *nigra* var. *henonis*
160		*Sinoarundinaria* (= *Phyllostachys*) *pubescens*
161		*Sinoarundinaria* (= *Phyllostachys*) *reticulata*
162	Taxaceae	*Taxus caespitosa*
163		*Taxus cuspidata*
164		*Taxus cuspidata* var. *latifolia*

165		*Torreya nucifera*
166		*Cephalotaxus koreana*
167		*Cephalotaxus nana*
168	Abietaceae	*Abies holophylla*
169		*Abies koreana*
170		*Abies nephrolepis*
171		*Picea intercedens*
172		*Picea jezoensis* (= *ajanensis*)
173		*Picea koraiensis*
174		*Picea pungsanensis*
175		*Picea schrenkiana*
176		*Picea tonaiensis*
177		*Tsuga sieboldii*
178	Pinaceae	*Pinus bungeana*
179		*Pinus densiflora*
180		*Pinus densiflora* var. *globosa*
181		*Pinus koraiensis*
182		*Pinus parviflora*
183		*Pinus pumila*
184		*Pinus rigida*
185		*Pinus tabulaeformis*
186		*Pinus thunbergii*
187	Cupressaceae	*Biota orientalis*
188		*Biota orientalis* var. *stricta*
189		*Thuja koraiensis*
190		*Chamaecyparis obtusa*
191		*Chamaecyparis pisifera*
192		*Juniperus coreana*
193		*Juniperus coreana* var. *rigida*
194		*Juniperus davurica*
195		*Juniperus sibirica*
196		*Juniperus utilis*
197		*Juniperus utilis* var. *longicarpa*
198		*Juniperus utilis* var. *modesta*
199		*Juniperus virginia*
200		*Sabina chinensis*
201		*Sabina chinensis* var. *horizontalis*
202		*Sabina pacifera*
203		*Sabina sargentii*
204		*Cryptomeria japonica*

APPENDIX 2

Arboreal alpine plants of Korea

Family	Species	Physiognomy*	Endemism
Taxaceae	*Taxus caespitosa*	ES	Endemic
	T. cuspidata	ET	
	T. cuspidata var. *latifolia*	ET	
Abietaceae	*Abies holophylla*	ET	
	A. koreana	ET	Endemic
	A. nephrolepis	ET	
	Larix olgensis	DT	
	L. olgensis var. *koreana*	DT	Endemic
	L. olgensis form. *virdis*	DT	Endemic
	Picea intercedens	ET	Endemic
	P. jezoensis	ET	
	P. koraiensis	ET	Endemic
	P. pungsanensis	ET	Endemic
	P. schrenkiana	ET	
	P. tonaiensis	ET	Endemic
Pinaceae	*Pinus koraiensis*	ET	
	P. pumila	ES	
	P. tabulaeformis	ET	
Cupressaceae	*Thuja koraiensis*	ET	Endemic
	Juniperus sibirica	ES	?Endemic
	Sabina davurica	ES	
	S. sargentii	ES	
Salicaceae	*Populus maximowiczii* var. *barbinervis*	DT	
	Salix berberifolia var. *brayi*	DS	
	S. berberifolia var. *genuina*	DS	
	S. bicarpa	DS	Endemic
	S. blinii	DS	Endemic
	S. floderusii	DT	
	S. floderusii var. *glabra*	DS	
	S. hallaisanensis	DS	Endemic
	S. hallaisanensis var. *longifolia*	DS	
	S. hultenii	DS	
	S. hultenii var. *elongata*	DT	Endemic
	S. integra	DS	
	S. kangensis	DT	Endemic
	S. metaformosa	DS	Endemic
	S. myrtilloides var. *manshurica*	DS	
	S. orthostemma	DS	Endemic
	S. pentandra var. *intermedia*	DV	
	S. pseudolineris	DT	
	S. roridaeformis	DT	
	S. rotundifolia	DS	Endemic
	S. rotundifolia var. *lanata*	DS	Endemic
	S. sericeo – cinerea	DS	Endemic

	S. sericeo – cinerea var. *lanata*	DS	Endemic
	S. sibirica var. *brachypoda*	DS	
	S. siuzevii	DT	
Betulaceae	*Alnus fruticosa* var. *mandshurica* form. *normalis*	DT	Endemic
	A. hirsuta var. *japonica*	DT	
	A. maximowiczii	DT	
	A. tinctiria var. *typica*	DT	
	A. vermicularis	DT	Endemic
	Betula chinensis	DT	
	B. chinensis var. *angusticarpa*	DT	
	B. costata	DT	
	B. cyclophylla	DS	
	B. ermanii var. *acutifolia*	DT	
	B. ermanii var. *ganjuensis*	DT	
	B. ermanii var. *genuina*	DT	
	B. ermanii var. *incisa*	DT	
	B. ermanii var. *parvifolia*	DT	Endemic
	B. fusenensis	DS	Endemic
	B. gmelini	DT	
	B. microphylla var. *coreana*	DS	Endemic
	B. paisanensis	DS	Endemic
	B. platyphylla var. *latifolia*	DT	
	B. platyphylla var. *mandshurica*	DT	
	B. schmidtii	DT	
Fagaceae	*Quercus mongolica* var. *tomentosa*	DT	
Ulmaceae	*Ulmus mandshurica*	DT	
Ranunculaceae	*Clematis chiisanensis*	DV	Endemic
	C. koreana	DV	
	C. mandshurica var. *lancifolia*	DV	Endemic
	C. nobilis	DV	Endemic
	C. nobilis var. *plena*	DV	Endemic
	C. ochotensis	DV	
	C. subtriternata	DV	Endemic
	C. subtriternata var. *tenuifolia*	DV	Endemic
Berberidaceae	*Berberis amurensis* var. *latifolia*	DS	
	B. amurensis var. *quelpaertensis*	DS	Endemic
Philadelphaceae	*Philadelphus pekinensis*	DS	
Spiraeceae	*Pentactina rupicola*	DS	Endemic
	Sorbaria stellipilla var. *glandulosa*	DS	Endemic
	S. stellipilla var. *typica*	DS	
	Spiraea media var. *sericea*	DS	
	S. pseudocrenata	DS	
	S. sylvestris	DS	Endemic
Ribesiaceae	*Ribes burejense*	DS	
	R. diacantha	DS	
	R. distans var. *typicum*	DS	
	R. glabellum	DS	
	R. horridum	DS	
	R. triste	DS	
	R. ussuriensis	DS	
Malaceae	*Crataegus maximowiczii*	DT	
	C. pinnatifida form. *betulifolia*	DT	Endemic
	C. pinnatifida var. *major*	DT	
	C. pinnatifida var. *partita*	DT	Endemic
	Sinomalus komarovi	DS	Endemic
	S. komarovi var. *major*	DS	Endemic
	S. komarovi var. *pilosa*	DS	Endemic
	Sorbus amurensis var. *rufea*	DT	Endemic

	S. chinophylla	DT	Endemic
	S. commixta	DT	
	S. sambucifolia	DT	
Rosaceae	*Dasiphora fruticosa*	DS	
	D. fruticosa var. *coreana*	DS	
	D. fruticosa var. *glabra*	DS	
	Dryas octopetala var. *asiatica*	ES	
	Rosa koreana	DS	
	R. marretii	DS	
	R. pimpinellifolia	DS	
	R. suavis var. *polyphylla*	DS	
	Rubus arcticus	DS	
	R. arcticus form. *tentipetala*	DS	
	R. idaeus var. *concolor*	DS	Endemic
Amygdalacae	*Prunus buergeriana*	DT	
	P. densiflora	DT	Endemic
Fabaceae	*Maackia fauriei*	DT	Endemic
Rhamnaceae	*Rhamnus taquetii*	DS	Endemic
Araliaceae	*Eleutherococcus senticosus*	DS	
	Oplopanax elatum	DS	Endemic
Cornaceae	*Chamaepericlimenum canadense*	ES	
	Cornus alba	DS	
Diapensiaceae	*Diapensia lapponica* subsp. *obovata*	ES	
Empetraceae	*Empetrum nigrum* var. *japonicum*	ES	
Ericaceae	*Andromeda polifolia*	ES	
	A. polifolia form. *acerosa*	ES	
	Arctous ruber	DT	
	Chamaedaphne calyculata	ES	
	Hugeria japonica	DS	
	Ledum palustre var. *latifolium*	ES	
	L. palustre var. *diversipilosum*	ES	
	L. plaustre var. *maximum*	ES	Endemic
	L. palustre var. *minus*	ES	
	L. palustre var. *nipponicum*	ES	
	L. palustre subsp. *decumbens*	ES	
	L. palustre var. *yesoense*	ES	
	Oxycoccus microcarpus	ES	
	O. quadripetalus	ES	
	Phyllodoce caerulea	ES	
	Rhododendron aureum	ES	
	R. dauricum	ES	
	R. fauriae form. *rufescens*	ES	
	R. parvifolium	ES	
	R. parvifolium var. *alpinum*	DS	
	R. parvifolium form. *albiflorum*	ES	
	R. trinerve	DS	
	R. tschnoskii	DS	
	Therorhodion redowskianum	DS	
	Vaccinium uliginosum	DS	
	V. uliginosum var. *alpinum*	DS	
	V. vitis-idaea var. *minus*	ES	
	V. vitis-idaea var. *genuinum*	ES	
Labiatae	*Thymus quinquecostatus* var. *ibukiensis*	DS	
Caprifoliaceae	*Linnaea borealis*	ES	
	Lonicera caerulea subsp. *edulis*	DS	
	L. caerulea var. *emphyllocalyx*	DS	
	L. caerulea var. *longibracteata* form. *normalis*	DS	
	L. caerulea form. *ovata*	DS	
	L. caerulea var. *venulosa*	DS	

	L. cerasoides	DS	Endemic
	L. chrysantha	DS	
	L. monantha	DS	Endemic
	L. nigra var. *barbinervis*	DS	Endemic
	L. ruprechtiana	DS	
	L. sachalinensis	DS	
	L. sachalinensis var. *barbinervis*	DS	
	L. sachalinensis var. *stenophylla*	DS	Endemic
	L. subhispida	DS	
	L. tatarinowii var. *leptantha*	DS	
	L. versicaria	DS	Endemic
	Sambacus latipinna var. *rhodoclotus*	DS	
	S. velutina	DS	Endemic
	Viburnum koreanum	DS	Endemic
	V. wrightii var. *stipellatum*	DS	Endemic
Asteraceae	*Artemisia triroba*	DS	

*Abbreviations: ES: Evergreen shrub, ET: Evergreen tree, DS: Deciduous shrub, DT: Deciduous tree.

APPENDIX 3

Herbaceous alpine plants of Korea

Family	Species	Physiognomy*	Endemism
Polygonaceae	*Bilderdykia ochotensis*	PN	
	B. pacifera	PN	
	B. vulgaris var. *yezoensis*	PN	
	Oxyria dygina	PN	
	Persicaria lyrata	AN	Endemic
	Pleuropteropyrum ochreatum	PN	
	Rhum coreanum	PN	Endemic
Alsinaceae	*Dianthus morii*	PN	Endemic
	D. repens	PN	
	D. superbus var. *speciosus*	PN	
	Minuartia arctica var. *major*	PN	Endemic
	M. laricina	PN	
	Silene fasiculata	PN	Endemic
Ranunculaceae	*Anemone narcissiflora*	PN	
	Pulsatilla nivalis	PN	Endemic
	Ranunuculus acris var. *elatus*	PN	Endemic
	R. borealis	PN	
	R. crucilobus	PN	Endemic
	R. crucilobus var. *chrysotrichus*	PN	Endemic
Helleboraceae	*Trollius hondoensis*	PN	
	T. japonicus	PN	
	T. macropetalus	PN	
Papaveraceae	*Papaver coreanum*	PN	Endemic
Brassicaceae	*Arabis lyrata* var. *kamtschatica*	BN	
	A. serrata var. *hallaisanensis*	PN	Endemic
	Cardamine bellidifolia	PN	
	C. resedifolia var. *mori*	PN	Endemic
	C. valida	BN	Endemic
	Draba incana var. *multicaulis*	BN	
Crassuluceae	*Rhodiola tachiroei*	PN	
Saxifragaceae	*Bergenia pacifica*	PN	
	Saxifraga cernua	PN	
	S. furumii	PN	Endemic
	S. laciniata	PN	
	S. octopetala var. *albiflora*	PN	
Spiraceae	*Aruncus aethusifolius*	PN	Endemic
Rosaceae	*Parageum calithifolium* var. *nipponicum*	PN	
	Potentilla bifurea var. *glabrata*	PN	
	P. matsumurae	PN	
	P. nivea var. *vulgaris*	PN	
	P. nivea form. *alpina*	PN	
	Sanguisorba alpina	PN	
	S. argutipens	PN	Endemic
	S. hakusanensis	PN	

	S. obutusa	PN	
	S. parvifolia	PN	
	Sibbaldia coreana	PN	Endemic
Fabaceae	*Astragalus adsurgens* var *alpinus*	PN	Endemic
	A. davuricus	PN	
	A. membranaceus	PN	
	A. membranaceus var. *alpinus*	PN	Endemic
	A. membranaceus var. *manshuricus*	PN	Endemic
	A. setsureianus	PN	Endemic
	A. ulginosus	PN	
	Oxytropis anertii	PN	
	O. koreana	PN	
	O. strobilacea	PN	
Geraniaceae	*Geranium dahuricum*	PN	
	C. eriostemon var. *glabrescens*	PN	Endemic
	G. eriostemon var. *hypoleucus*	PN	Endemic
	G. eriostemon var. *megalanthum*	PN	
	G. eriostemon var. *orientale*	PN	
	G. eriostemon var. *typicum*	PN	
	C. maximowiczii	PN	
	C. skikokianum var. *quelpaertense*	PN	Endemic
Euphorbiaceae	*Galarhoeus fauriei*	PN	Endemic
Hypericaceae	*Hypericum vaniotii*	PN	
Violaceae	*Viola biflora*	PN	
	V. cressa	PN	
	V. koraiensis	PN	Endemic
	V. silvestriformis	PN	
Apiaceae	*Cnidium tachiroei*	PN	
	Homopterya nakaiana	PN	Endemic
	Peucedanum coreanum	PN	Endemic
Pyrolaceae	*Moneses uniflora*	PN	
	Orthila secunda	PN	
Primulaceae	*Androsace lehmannia*	PN	
	Primula modesta var. *genuina*	PN	
	P. modesta var. *fauriae*	PN	
	P. sachalinensis	PN	
	Trientalis europaea var. *arctica*	PN	
	T. europaea var. *eurasiatica*	PN	
Gentianaceae	*Gentiana algida*	PN	
	G. jamesii	PN	
	G. makinoi	PN	
Scrophulariaceae	*Euphrasia coreanalpina*	AN	Endemic
	E. mucronulata	AN	Endemic
	E. multiflora	AN	
	Melampyrum arcuatum	AN	
	M. roseum var. *alpinum*	AN	
	Pedicularis apodochila	PN	
	P. glandiflora	PN	
	P. lunaris	PN	Endemic
	P. mandshurica	PN	
	P. nigrescens	PN	Endemic
	P. spicata	PN	
	P. sudetica	PN	
	P. verticillata	PN	
	Veronica stelleri var. *longistyla*	PN	
	V. subincanovelutina	PN	Endemic
Lentibulariaceae	*Pinguicula vulgaris* var. *macroceras*	PN	
Rubiaceae	*Asperula platyglium* var. *alpina*	PN	
	Galium pusillum	PN	Endemic
Dipsacaceae	*Scabiosa japonica* form. *alpina*	BN	Endemic

Campunulaceae	*Adenophora divaricata*	PN	
	A. lamarkii	PN	
	A. liliifolia	PN	
	A. triphylla form. *canescens*	PN	
Asteraceae	*Achillea paektussanense*	PN	
	Anaphalis sinica subsp. *morii*	PN	Endemic
	Antennaria dioica	PN	
	Artemisia angustissima	PN	Endemic
	A. laciniata	PN	
	A. sinanensis	PN	
	A. stelleriana	PN	
	Chrysanthemum zawadskii	PN	Endemic
	Eriogeron thunbergii subsp. *glabratus*	PN	
	Hieracium japonicum	PN	
	H. umbelletm subsp. *umbellatum*	PN	
	Leontopodium coreanum	PN	Endemic
	L. japonicum	PN	
	L. leoiolepis	PN	Endemic
	Ligularia fischeri	PN	
	L. jamesii	PN	Endemic
	L. stenocephala	PN	
	Saussurea manshurica	PN	
	S. rectinervis	PN	Endemic
	Senecio flammeus	PN	
	S. nemorensis subsp. *fuchsii*	PN	
Scheuchzeriaceae	*Scheuchzeria palustris*	PN	
Gramineae	*Agrostis flaccida*	PN	
	A. hiemalis	PN	
	Calamagrostis langsdorffi	PN	
	C. langsdorffi var. *ramosa*	PN	Endemic
	Deschampsia caespitosa	PN	
	Hierochloe alpina	PN	
	Lasiagrostis mongolica	PN	
	Phleum alpinum	PN	
	Poa deschampsioides	PN	Endemic
	P. glauca	PN	
	P. kanboensis	PN	
	Trisetum spicatum	PN	
Cyperaceae	*Carex capillaris*	PN	
	C. rotundata	PN	
	C. subumbellata var. *koreana*	PN	
	Eriophorum alpinum	PN	
	E. angustifolium	PN	
	E. japonicum	PN	
	E. manshuricum	PN	
	E. polystachyon	PN	
	E. vaginatum	PN	
	Kobresia myosuroides	PN	
Aceraceae	*Symplocarpus buchenensis*	PN	
Juncaceae	*Juncus brachyspathus*	PN	
	J. triglumis	PN	
	Luzula oligantha	PN	
	L. sudetica var. *nipponica*	PN	
Melanthaceae	*Tofieldia fauriei*	PN	Endemic
	T. nutans var. *rubescens*	PN	
	Veratrum maximowiczii var. *coreanum*	PN	
	V. versicolor form. *albidum*	PN	
	Zygadenus sibirica	PN	
Alliaceae	*Allium senescens*	PN	
	A. taquetii	PN	

Liliaceae	*Lloydia serotina*	PN	
Orchidaceae	*Epipactis puzensis*	PN	
	Epipogon aphyllum	PN	
	Herminium monorchis	PN	
	Orchis bracteata	PN	
	Platanthera maximowicziana	PN	
	P. sachalinensis	PN	
	Poneorchis pauciflora var. *joo-iwokiana*	PN	
Lycopodiaceae	*Lycopodium alpinum*	PN	Endemic
	L. complanatum	PN	
	L. selago	PN	
	L. sitchense	PN	
Selaginellaceae	*Selaginella rupestris* var. *shakotanensis*	PN	
	S. selaginoides	PN	
Pterdaceae	*Cryptogramma crispa*	PN	
Aspidiaceae	*Athyrium crenatum* var. *glabrum*	PN	
	A. melanolepis	PN	
	A. rupestre	PN	
	Cystopteris fragilis	PN	
	Lastrea quelpaertensis	PN	
	Woodsia glabella	PN	
	W. ilvensis	PN	
	W. saitoana	PN	

*Abbreviations: PN: Perennial, BN: Biennial, AN: Annual.

APPENDIX 4

Arboreal arctic-alpine plants of Korea

Family	Species
Cupressaceae	*Juniperus communis* var. *montana* (= *Sabina sargentii*?)
Rosaceae	*Dryas octopetala* var. *asiatica*
	Rubus articus
	Sorbus amurensis
	S. commixta
Diapensiaceae	*Diapensia lapponica* subsp. *obovata*
Empetraceae	*Empetrum nigrum* var. *japonicum*
Ericaceae	*Chamaedaphne calyculata*
	Ledum palustre var. *diversipilosum*
	L. palustre subsp. *decumbens*
	Oxycoccus microcarpus
	O. quadripetalus
	Phyllodoce caerulea
	Rhododendron lapponicum (= *parvifolium*?)
	Vaccinium uliginosum var. *alpinum*
	V. vitis-idaea var. *minus*
Caprifoliaceae	*Linnaea borealis*

APPENDIX 5

Herbaceous arctic-alpine plants of Korea

Family	Species
Polygonaceae	*Oxyria digyna*
	Rumex maritimus
Chenopodiaceae	*Suaeda maritima*
Alsinaceae	*Stellaria longiflora*
Nymphaceae	*Brasenia schreberi*
Ranunculaceae	*Anemone dichotoma*
	Ranunculus hyperboreus
Helleboraceae	*Caltha palustre* var. *sibirica*
Brassicaceae	*Turritis glabra*
Rosaceae	*Rubus chamaemorus*
Droseraceae	*Drosera anglica*
Fabaceae	*Lathyrus maritimus*
Elatinaceae	*Elatine triandra*
Violaceae	*Viola selkirki*
Pyrolaceae	*Moneses uniflora*
	Monotropa hypopitys subsp. *japonica*
	Pyrola minor
	P. secunda
Primulaceae	*Glaux maritima* (= *obtusilofolia*?)
Boraginaceae	*Hackelia deflexa*
Labiatae	*Galeopsis bifida*
	Lamium amplexicaule
Lentibulariaceae	*Pinguicula villosa* (= *vulgaris*?)
	Utricularia intermedia
Rubiaceae	*Galium asparine*
	G. spurium
Typhaceae	*Typha latifolia*
Zosteraceae	*Zostera marina*
Ruppiaceae	*Ruppia maritima*
Scheuchzeriaceae	*Scheuchzeria palustris*
Juncaginaceae	*Triglochin maritiumum* var. *asiaticum*
Gramineae	*Agrostis canina* subsp. *trinii*
	Bromus japonicus
	Calamagrostis canadensis subsp. *langsdorffii*
	Cinna latifolia
	Cynodon dactylon
	Digitaria ischaemum
	Poa glauca
	P. palustris
	P. trivalis
Cyperaceae	*Carex loliacae*
	C. pseudo-curaica
	C. rostrata
	C. rotundata

	Eriophorum alpina
Juncaceae	*Luzula rufescens*
	L. wahlenbergii
Alliaceae	*Allium schoenoprasum*
Orchidaceae	*Calypso bulbsa*
Lycopodiaceae	*Lycopodium complanatum*
Equisetaceae	*Equisetum hiemale*
Botrychiaceae	*Botrychium virginianum*
Ophioglossaceae	*Ophioglossum vulgatum*
Osmundaceae	*Osmunda cinnamomum*
Hymenophyllaceae	*Hymenophyllum wrightii*
Aspidiaceae	*Dryopteris filix-max*
	Polystichum braunii
Polypodiaceae	*Polypodium vulgare* subsp. *virginianum*

References

Academy of Sciences of the USSR, 1964. Physical geographical atlas of the world. Moscow (in Russian).

Agee, J.K. & Smith, L., 1984. Subalpine tree re-establishment after fire in the Olympic Mountains, Washington. Ecol. 65: 810–819.

Agricultural Research Service, 1965. Plant hardiness zone map. USDA Misc. Publ. No. 804, Washington D.C.

Ahti, T., Hämet-Ahti, L. & Jalas, 1986. Vegetation zones and their selections in northwestern Europe. Ann. Bot. Fenn. 5: 169–211.

Akazawa, T. 1983. An outline of Japanese prehistory. Recent Progress of Natural Science in Japan 8: 1–11.

Aleksandrova, V.D. 1980. The arctic and antarctic: their division into geobotanical areas. Cambridge Univ. Press, Cambridge.

Ann, H.S. & Lee, C.R. 1965. Nomina Plantarum Korearum. Hyanghaksa, Seoul (in Latin & Korean).

Ann, H.S., Lee, C.R. & Park, S.H. 1982. An agronomical enumeration of Korean plant resources. Ilchogak, Seoul (in Latin & Korean).

Arber, E.A.U., 1910. Plant life in alpine Switzerland. John Murray, London.

Arno, S.F. 1984. Timberline: mountain and arctic forest frontiers. The Mountaineers, Seattle.

Axelrod, D.I. 1952. Variables affecting the probabilities of dispersal in geologic time. Bull. Amer. Mus. Nat. Hist. 99: 177–188.

Baig, M.N. & Tranquillini, W. 1976. Studies on upper timberline: morphology and anatomy of Norway spruce (*Picea abies*) and stone pine (*Pinus cembra*) needles from various habitat conditions. Can. J. Bot. 54: 1622–1632.

Baik, S.D. & Yim, Y.D. 1983. Vegetation of Taecheongbong, Mt. Sorak. Kor. J. Ecol. 6: 1–13 (in Korean).

Baldwin, J.T. Jr. 1939. Chromosomes of the Diapensiaceae. J. Heredity 30: 167–171.

Balgooy, M.M.J. van 1971. Plant-geography of the Pacific. Blumea (Supplement) 6: 1–222.

Barkman, J.J. 1979. The investigation of vegetation texture and structure. In: Werger, M.J.A. (ed.), The study of vegetation, pp. 125–160. Dr. W. Junk, The Hague.

Barry, R.G. 1979. High altitude climates. In: Webber, P.J. (ed.), High altitude geoecology, pp. 55–74. AAAS Selected Symposium 12, Westview Press, Boulder, USA.

Barry, R.G. 1981. Mountain weather and climate. Methuen, London.

Baumgartner, A. 1980. Mountain climates from a perspective of forest growth. In: Benecke, U. & Morris, M.R. (eds.), Mountain environments and subalpine tree growth, pp. 27–39. For. Res. Inst. N.Z. For. Ser. Tech. Paper, No. 70, Wellington.

Beard, J.S. 1980. The physiognomic approach. In: Whittaker, R.H. (ed.), Classification of plant communities, pp. 33–64. Dr. W. Junk, The Hague.

Bell, J.N.B. & Tallis, J.H. 1973. Biological flora of the British Isles, *Empetrum nigrum*. J. Ecol. 61: 289–305.

Bell, P. & Woodcock, C. 1983. The diversity of green-plants. Edward Arnold, London.

Benedict, J.B. 1984. Rates of tree-island migration, Colorado Rocky Mountains, USA. Ecol. 65: 820–823.

Benninghoff, W.S. 1952. Interaction of vegetation and soil frost phenomena. Arctic 5: 34–44.

Berg, R.Y. 1963. Disjunctions in the Norwegian alpine flora and theories proposed for their explanation. Blyttia 21 (4): 133–177 (in Norwegian).

Berkeley, C.P. & Morris, F.K. 1927. Geology of Mongolia. Natural History of Central Asia. Vol. II. Amer. Mus. Nat. Hist., New York.

Billings, W.D. 1964. Plants and the ecosystem. Macmillan, London.

Billings, W.D. 1974. Adaptations and origins of alpine plants. Arctic & Alpine Res. 6: 129–142.

Billings, W.D. 1978a. High mountain ecosystems, evolution, structure, operation and maintenance. In: Webber, P.J. (ed.), High altitude geoecology, pp. 97–125. Westview Press, Boulder, Colorado.

Billings, W.D. 1978b. Alpine phytogeography across the Great Basin. In: Harper, K.T. & Reveal, J.L. (eds.), Intermountain biogeography, pp. 105–117. Brigham Young Univ. Press, Provo, Utah.

Billings, W.D. 1981. Plants in high places. Natural History 90: 82–88.

Billings, W.D. 1983. Plants, man and the ecosystem. MacMillan, London.

Billings, W.D. 1985. The historical development of physiological plant ecology. In: Chabot, B.F. & Mooney, H.A. (eds.), Physiological ecology of North American plant communities, pp. 1–15. Chapman & Hall, London.

Billings, W.D. & Mooney, H.A. 1968. The ecology of arctic and alpine plants. Biol. Rev. 481–529.

Billings, W.D. & Morris, R.J. 1951. Reflection of visible and infrared radiation from leaves of different ecological groups. Amer. J. Bot. 38: 327–331.

Birks, H.J.B. 1973. Modern pollen rain studies in some arctic and alpine environments. In: Birks, H.J.B. and West, R.G. (eds.), Quaternary plant ecology, pp. 143–168. Blackwell Sci. Publ., London.

Birks, H.J.B. 1976. The distribution of European pteridophytes: A numerical analysis. New Phytol. 77: 257–287.

Birks, H.J.B. & Deacon, J. 1973. A numerical analysis of the past and present flora of the British Isles. New Phytol. 72: 877–902.

Björkman, D. & Holmgren, P. 1961. Studies of climatic ecotypes of higher plants. Leaf respiration in different populations of *Solidago virgaurea*. Kung. Landbrukshog. Ann. 27: 379–304.

Bliss, L.C. 1956. A comparison of plant development in microenvironments of arctic and alpine tundras. Ecol. Mongr. 26: 303–337.

Bliss, L.C. 1962a. Caloric and lipid content in alpine tundra plants. Ecol. 43: 753–757.

Bliss, L.C. 1962b. Adaptations of arctic and alpine plants to environmental conditions. Arctic 15: 117–144.

Bliss, L.C. 1963. Alpine plant communities of the Presidential Range, New Hampshire. Ecol. 44: 678–697.

Bliss, L.C. 1966. Plant productivity in alpine environments on Mt. Washington, New Hampshire. Ecol. Monogr. 36: 125–155.

Bliss, L.C. 1969. Alpine community pattern in relation to environmental parameters. In: Greenridge, K.N.H. (ed.), Essays in plant geography and ecology, pp. 167–184. Nova Scotia Museum, Halifax.

Bliss, L.C. 1971. Arctic and alpine plant life cycles. Ann. Rev. Ecol. Syst. 2: 405–438.

Bliss, L.C. 1979. Arctic heaths. In: Specht, R.L. (ed.), Heathlands and related shrublands, pp. 415–424. Elsevier, Amsterdam.

Bliss, L.C. 1981. Evolution of tundra ecosystems. In: Bliss, L.C., Heal, O.W. & More, J.J. (eds.), Tundra ecosystems: a comparative analysis, pp. 5–7. IBP, No. 25, CUP, Cambridge.

Bliss, L.C. 1985. Alpine. In: Chabot, B.F. & Mooney, H.A. (eds.), Physiological ecology of North American plant communities, pp. 41–65. Chapman & Hall, London.

Böcher, T.W. 1972. Evolutionary problems in the arctic flora. In: Valentine, D.H. (ed.), Taxonomy, phytogeography and evolution, pp. 101–113. Academic Press, London.

Böcher, T.W. & Larsen, K. 1950. Chromosome numbers of some arctic or boreal flowering plants. Medd. om Grønland 147 (6): 1–32.

Bourdeau, P.F. 1959. Seasonal variations of the photosynthetic efficiency of evergreen conifers. Ecol. 40: 63–67.

Bowen, D.Q. 1978. Geographical perspective on the Quaternary. Progr. Physical Geogr. 33: 167–186.

Box, E.O. 1978. Ecoclimatic determination of terrestrial vegetation pysiognomy. Ph.D. Thesis, Univ. North Carolina, Chapel Hill, North Carolina.

Box, E.O. 1981a. Macroclimate and plant forms: An introduction to predictive modeling in phytogeography. Dr. W. Junk, The Hague.

Box, E.O. 1981b. Predicting physiognomic vegetation types with climatic variables. Vegetatio 45: 127–139.

Bradley, R.S. 1985. Quaternary palaeoclimatology. Allen & Unwin, Boston.

Bremer, K. 1983. Vicariance biogeography. Svensk Bot. Tidskr. 77: 33–40 (in Swedish).

Bright, D.N.G. 1928. The effects of exposure upon the structure of certain heath plants. J. Ecol. 16: 323–365.

Bryant, J.P. & Scheinberg, E. 1970. Vegetation and frost activity in an alpine fellfield on the summit of Plateau Mountains Alberta. Can. J. Bot. 48: 751–771.

Budyko, M.I. 1974. Climate and life. Academic Press, New York.

Buks, I.I. 1984. The problem of the Beringia phratry of formations of boreal-type vegetation. In: Kontrimavichus, V.L. (ed.), Beringia in the Cenozoic Era, pp. 235–243. Oxonian Press PVT Ltd., New Delhi.

Butzer, K.W. 1964. Environment and archaeology: An introduction to Pleistocene geography. Methuen & Co., London.

Butzer, K.W. 1977. Environment, culture and human evolution. Amer. Sci. 65: 572–584.

Cain, S.A. & Castro, G.M.D. 1959. Manual of vegetational analysis. Harper & Brothers Publ., New York.

Camp, W.H. 1944. A preliminary consideration of the biosystematy of *Oxycoccus*. Bull. Torrey Bot. Club 71: 426–437.

Canaday, B.B. & Fonda, R.W. 1974. The influence of subalpine snowbanks on vegetation pattern, production and phenology. Bull. Torrey Bot. Club 101 (6): 340–350.

Central Meteorological Office, 1968. Climatic table of Korea (1931–1960). Seoul (in Korean).

Central Meteorological Office, 1982. Climatic table of Korea, Vol. 1. Seoul (in Korean).

Central Meteorological Office, 1983. Climatic table of Korea, Vol. 2. Seoul (in Korean).

Central Meteorological Office, 1984. Climatic Table of Korea, Vol. 3. Seoul (in Korean).

Cha, J.W., Lee, W.G. & Lee, S.A. 1975. The climate and vegetation of Korea. Seomundang, Seoul (in Korean).

Chabot, B.F. & Billings, W.D. 1972. Origins and ecology of the Sierran alpine flora and vegetation. Evol. Monogr. 42: 163–199.

Chaloner, W.G. & Meyen, S.V. 1973. Carboniferous and Permian flora of the northern continents. In: Halle, A. (ed.), Atlas of palaeobiogeography, pp. 169–186. Elsevier Scientific Publ. Co., Amsterdam.

Chaney, R.W. 1936. Plant distribution as a guide to age determination. J. Wash. Acad. Sci. 26: 313–324.

Chaney, R.W. 1940. Tertiary forests and continental history. Bull. Geol. Soc. Amer. 51: 469–488.

Chaney, R.W. 1947. Tertiary centers and migration routes. Ecol. Monogr. 17: 139–148.

Chaney, R.W. 1964. Some observations on climatic relations of Tertiary floras bordering the North Pacific basin. In: Nairn, A.E.. (ed.), Problems in palaeoclimatology, pp. 40–43. Interscience Publ., London.

Chang, H. 1983a. Periglacial landforms in the eastern part of the main ridge of Mt. Chiri, South Korea. J. Korean Geogr. Soc. 27: 31–50 (in Korean).

Chang, H.S. 1983b. Techniques to measure the age. In: The National History Compilation Committee (ed.), Korean history Vol. 12 (Korean Archaeology), pp. 40–61. Seoul (in Korean).

Chang, K.C. 1968a. The-archaeology of ancient China. Yale Univ. Press, New Haven.

Chang, K.C. 1968b. Archaeology of ancient China. Science 162: 519–526.

Chang, K.C. 1970. The beginning of agriculture in the Far East. Antiquity 44: 175–185.

Chapin, F.S. & Shaver, G.R. 1985. Arctic. In: Chabot, B.F. & Mooney, H.A. (eds.), Physiological ecology of north plant communities. Chapman & Hall, London.

Chard, C.S. 1960a. Neolithic archaeology in North Korea. Asian Perspectives 4: 151–155.

Chard, C.S. 1960b. The Neolithic in northern Asia: A cultural area approach. Anthropologica 11: 241–246.

Chard, C.S. 1974. Northeast Asia in prehistory. Univ. Wisconsin Press, Madison, Wisconsin.

Charlesworth, J.K. 1957. The quaternary era. Vols. 1, 2, Edward Arnold, London.

Chejudo, 1985. Report of the academic survey of Hallasan (Mountain) Natural Preserve. Province of Cheju (in Korean).

Cherov, Yu.I. 1985. The living tundra. Cambridge Univ. Press, Cambridge (translated by Löve, D.).

Chiu, T.N. 1984. Major geological ages European and Chinese Nomenclature. In: Whyte, R.O. (ed.), The evolution of the East Asian environment, Vol. 1, p. 3. Univ. Hong Kong.

Choi, M.C. 1983. Comparison of Korean Palaeolithic culture in comparison with the Chinese. In: The National History Compilation Committee (ed.), Korean history Vol. 12 (Korean Archaeology), pp. 307–366. Seoul (in Korean).

Choi, M.C. 1986. The Palaeolithic culture in Korea. Yeomun Publ. Co., Taejon, Korea (in Korean).

Choi, S.R. 1982. A study on some problems related to radiocarbon dates in Korea. Jour. Kor. Archaeol. Stud. 13: 61–95 (in Korean).

Chung, C.H., Kim, J.Y. & Lee, Y.Y. 1984. Geology of the Soraksan district. In: Province of Kangwon (ed.), Scientific report on Mt. Sorak, pp. 429–457. Province of Kangwon, Chunchon (in Korean).

Chung, T.H. 1933, Flora of Eocheong Island. Rep. For. Exper. Stat. 5: 31–43 (in Japanese).

Chung, T.H. 1943. An atlas of Korean forest distribution. Kor. Soc. Nat. Hist. Seoul (in Japanese).

Chung, T.H. 1958. Flora of Mt Kaeryong. Theses Coll. Sungkyunkwan Univ. 3: 1–40 (in Korean).

Chung, T.H. 1965. Illustrated flora and fauna of Korea, Vol. 5. Plants, Ministry of Education, Seoul (in Korean).

Chung, T.H. & Lee, I.K. 1959. The flora of Mt. Sorak. Theses Coll. Kyunghee Univ. 2: 289–353 (in Korean).

Chung, T.H. & Lee, W.C. 1961. Flora of Chungbuk. Theses Coll. Sungkyunkwan Univ. 6: 231–289 (in Korean).

Chung, T.H. & Lee, W.C. 1962. Flora of Mt. Pukhan. Theses Coll. Sungkyunkwan Univ. 7: 373–396 (in Korean).

Chung, T.H. & Lee, W.C. 1963, Flora of Mt. Sorak. Theses Coll. Sungkyunkwan Univ. 8: 231–269 (in Korean).

Chung, T.H. & Lee, W.C. 1965. Horizontal and vertical distribution of the Korean woody plants. Theses Coll. Sungkyunkwan Univ. 10: 367–435 (in Korean).

Chung, T.H. & Lee, W.C. 1966. Vegetation of Is. Keomun. Theses Coll. Sungkyunkwan Univ. 11: 335–365 (in Korean).

Chung, T.H. & Paik, S.E. 1963. Flora of Puyeo. Kor. J. Bot. 6: 11–18 (in Korean).

Chung, Y.H. 1986. The history of Korean plant taxonomy. Academic Book Co., Seoul (in Korean).

Clapham, A.R., Tutin, T.G. & Warburg, E.F. 1981. Excursion Flora of the British Isles. Cambridge University Press, Cambridge.

CLIMAP Project Members 1976. The surface of the Ice-Age earth. Science, 191: 1131–1137.

CLIMAP Project Members 1981. Seasonal reconstructions of the Earth's surface at the last glacial maximum. Geol. Soc. Amer. Map and Chart Series, MC-36.

CLIMAP Project Members 1984. The last interglacial ocean. Quat. Res. 21: 123–224.

Collinson, A.S. 1980. Introduction to world vegetation. George Allen & Unwin, London.

Conolly, A.P. 1961. Some climatlc and edaphic indications from the late-glacial flora. Proc. Linn. Soc. Lond. 172: 56–62.

Conolly, A.P. & Dahl, G. 1970. Maximum summer temperature in relation to the modern and Quaternary distributions of certain arctic-montane species in the British Isles. In: Walker, D. & West, R.G. (eds.), Studies in the vegetational history of the British Isles, pp. 159–223. Cambridge University Press, Cambridge.

Cornwell, I. 1970. Ice ages: their nature and effects. John Baker Ltd., London.

Costin, A.B. 1981. Alpine and sub-alpine vegetation. In: Groves, R.H. (ed.), Australian vegetation, pp. 361–376. Cambridge Univ. Press, Cambridge.

Courtin, G.M. & Mayo, J.M. 1975. Arctic and alpine plant water relations. In: Vernberg, F.J. (ed.), Physiological adaptation to the environment, pp. 201–204. Intext Educational Publishers, New York.

Cox, C.B. 1974. Vertebrate palaeodistributional patterns and continental drift. J. Biogeogr. 1: 75–94.

Crawford, R.M.M. 1989. Studies in plant survival. Blackwell Scientific Publications, Oxford.

Cressey, G.B. 1926. The climate of the glacial period in eastern Asia. Proc. 3rd Pan Pacific Sci. Congr. 2: p. 1818.

Critchfield, W.B. & Little, E.L. Jr. 1966. Geographic distribution of the pines of the world. Misc. Publ. 991: U.S.D.A., Washington D.C.

Crovello, T.J. 1981. Quantitative biogeography: An overview. Taxon, 30: 563–575.

Cultural Property Preservation Bureau 1970. Cultural properties of Korea, Ministry of Culture and Information, Seoul (in Korean).

Dahl, E. 1946. On different types of unglaciated areas during the ice ages and their significance to phytogeography. New Phytol. 45: 225–242.

Dahl, E. 1951. On the relation between summer temperature and the distribution of alpine vascular plants in the lowlands of Fennoscandia. Oikos, 3: 22–52.

Dahl, E. 1953. Present day and late-glacial distribution of some alpine vascular plants in Scandinavia and their interpretation. In: Lousley, J.E. (ed.), The changing flora of Britain, pp. 77–83. T. Buncle & Co. Ltd., Arbroath, U.K.

Dahl, E. 1955. Biogeographic and geologic indicators of unglaciated areas in Scandinavia during the glacial ages. Bull. Geol. Soc. Amer. 66: 1499–1520.

Dahl, E. 1963. On the heat exchange of a wet vegetation surface and the ecology of *Koenigia islandica*. Oikos, 14: 190–211.

Dahl, E. 1964. Present-day distribution of plants and past climate. In: Hester, J.H. & Schoenwetter, J. (eds.), The reconstruction of past environment, No. 3, pp. 52–59. Fort Burgin Res. Centre.

Dahl, E. 1966. The heat exchange of plants and its importance to plant morphology and distribution. Blyttia, 24: 105–129 (in Norwegian).

Dahl, E. 1986. Zonation in arctic and alpine tundra and fellfield ecobiomes. In: Polunin, N. (ed.), Ecosystem theory and application, pp. 35–62. John Wiley & Sons, Chichester.

Dahl, E. 1987. The nunatak theory reconsidered. Ecol. Bull. (Copenhagen) 38: 77–94.

Dallimore, W. & Jackson, A.B. 1954. A handbook of Coniferae including Ginkgoaceae. Edward Arnold, London.

Daly, C. 1984. Snow distribution patterns in the alpine krummholz zone. Progress in Physical Geogr. 8: 157–175.

Damman, A.W.H. 1965. The distribution patterns of northern and southern elements in the flora of Newfoundland. Rhodora, 67: 363–392.

Damman, A.W.H. 1976. Plant distribution in Newfoundland especially in relation to summer temperatures, measured with the sucrose inversion method. Can. J. Bot. 54: 1561–1585.

Darlington, A. 1978. Mountains and moorlands. Hodder & Stoughton, London.

Daubenmire, R. 1954. Alpine timberlines in the Americas and their interpretation. Butler Univ. Bot. Stud. 11: 119–136.

Daubenmire, R. 1968. Plant communities. Harper & Row, New York.

Day, R.T. & Scott, P.J. 1964. The biology of *Diapensia lapponica* in Newfoundland. Can. Field Nat. 98: 425–439.

Den Ouden, P. & Boom, B.K. 1965, Manual of cultivated conifers. Martinus Nijhoff, The Hague, Netherlands.

Dickins, J.M. 1984. Evolution and climate in the Upper Palaeozoic. In: Brenchley, P. (ed.), Fossils and climate, pp. 317–327. John Wiley & Sons, Chichester.

Dmitrieva, E.V. 1978. *Pinus pumila* at the southern boundary of its spreading (Taxonomical-typological characterization). Bot. Zhur. 63: 1352–1358 (in Russian).

Dorf, E. 1942. Upper Cretaceous floras of the Rocky Mountain region. Carnegie Inst. Wash. Publ. 508, 168 pp.

Du Rietz, G.E. 1931. Life-forms of terrestrial flowering plants. Acta Phytogeogr. Suec. 3: 1–95.

Du Rietz, G.E. 1935. Glacial survival of plants in Scandinavia and the British Isles. Proc. Roy. Soc. Lond. Ser. B, 118: 226–229.

Du Rietz, G.E. 1940. Problems of bipolar plant distribution. Acta Phytogeogr. Suec, 13: 215–282.

Egler, F.E. 1962. Unifying concepts in vegetation study as applied to the Pacific Basin. Proc. 9th Pacific Sci. Congr. 4: 136–138.

Elkington, T.T. 1971. Biological flora of the British Isles, *Dryas octopetala*. J. Ecol. 59: 887–905.

Ellenberg, H. & Mueller-Dombois, D. 1965–66. Tentative physiognomic-ecological classification of plant formations of the earth. Separ. Aus. Ber. Geobot. Inst. 37: 21–55.

Ellenberg, H. & Muller-Dombois, D. 1967. A key to Raunkiaer plant life forms with revised subdivisions. Ber. Des. Geobot. Inst. Der. Eidg. Techn. Hochsch Sti. Rubel 37: 56–73.

Emerick, J.C. & Webber, P.J. 1982. The effects of augmented winter snow cover on the canopy structure of alpine vegetation. In: Halfpenny, J.C. (ed.), Ecological studies on the Colorado Alpine, pp. 63–72. Univ. Colorado, INSTAAR Occasional Paper 37, Boulder, Colorado.

Eshbaugh, W.H. 1968. Biosystematic and phytogeographic studies of the genus *Empetrum*. Yearbook Amer. Phil. Soc. Geol. & Biol. Sci. pp. 297–298.

Etherington, J.R. 1983. Plant physiological ecology. Edward Arnold, London.

Eyre, S.R. 1984. Vegetation and soils. Edward Arnold, London.

FAO, UN, 1976. Forest resources in the Asia and Far East region. FAO, UN, Rome.

Farmer, G. & Ford, M.J. 1984. The contribution of biological sciences to the reconstruction of past climates. In: Whyte, R.O. (ed.), The evolution of the East Asian environment, Vol. 1, pp. 188–198. Univ. Hong Kong.

Favarger, C. 1972. Endemism in the montana floras of Europe. In: Valentine, D.H. (ed.), Taxonomy, phytogeography and evolution, pp. 191–204. Academic Press, London.

Fernald, M.L. 1925. Persistence of plants in unglaciated areas of boreal America. Amer. Acad. Arts Sci. Mem. 15: 237–342.

Fitter, A.H. & Hay, R.K.M. 1981. Environmental physiology of plants. Academic Press, London.

Fitter, R., Fitter, A. & Blamey, M. 1985. The wild flowers of Britain and Northern Europe. Collins, London.

Flenley, J.R. 1984. Timescales in biogeography. In: Taylor, J.A. (ed.), Themes in biogeography, pp. 63–105. Croom Helm, London.

Flint, R.F. 1971. Glacial and quaternary geology. John Wiley & Sons, New York.

Flint, R.F. & Dorsey, H.G. 1945. Glaciation of Siberia. Bull. Geol. Soc. Amer. 56: 89–106.

Florin, R. 1963. The distribution of conifer and taxad genera in time and space. Acta Horti Bergiani Band 20 (4): 121–312.

Ford, M.J. 1982. The changing climate: response of the natural fauna and flora. George Allen & Unwin, London.

Fosberg, F.R. 1958. On the possibility of a rational general classification of humid tropical vegetation. Proc. Unesco Humid Tropics Symposium, Tjiaai, Indonesia, pp. 34–59.

Fosberg, F.G. 1961. A classification of vegetation for general purpose. Trop. Ecol. 2: 1–28.

Foster, R. 1982. Rock garden and alpine plants. David & Charles, Newton Abbot.

Franklin, J.F., Maeda, T., Ohsumi, Y., Matsui, M. Yagi, H. & Hawk, G.H. 1979. Subalpine coniferous forests of central Honshu, Japan. Ecol. Monogr. 49: 311–334.

Frenzel, B. 1973a. Climatic fluctuations of the Ice Age (Trans.

Nairn, A.E.M.). The Press of Case Western Reserve Univ. Cleveland.

Frenzel, B. 1973b. Some remarks on the Pleistocene vegetation. Eiszeitalter U. Gegenwart 23/24, 281–292.

Garrels, R.M. & Mackenzie, F.T. 1971. Evolution of sedimentary rocks. Norton & Co. New York.

Gauch, H.G. Jr. 1982. Multivariate analysis in community ecology. Cambridge Univ. Press, Cambridge.

Gauslaa, Y. 1984. Heat resistance and energy budget in different Scandinavian plants. Holarctic Ecol. 7: 1–78.

Gauslaa, Y. 1985. Climatic limitations on the distribution of alpine plants. A historical review. Blyttia, 43: 75–86 (in Norwegian).

Gellert, J.F. 1976. Warm- und Kaltzeiliche vegetationszonen und Klimagürtel am Ostrand Asiens in der Insulinde und Australien. Peterm. Geogr. Mitteil, 120: 295–302 (in German).

Gerasimov, I.P. & Markov, K.K. 1939. The glacial period in the territory of USSR. USSR Akad. Sci. Inst. Geogr. Tr. 33: 443–462 (in Russian).

Giterman, R.E. 1984. Kolyma lowland vegetation in the Pleistocene cold epochs and the problem of Polar Beringia landscapes. In: Kontrimavichus, V.L. (ed.), Beringia in the Cenozoic Era, pp. 214-220. Oxonian Press, New Delhi.

Giterman, R.E. & Golubeva, L.V. 1965. The history of development of the vegetation of Eastern Siberia in the Anthropogene. In: Osnovaye Problemy Izucheniya Chetvertichnogo Perioda, pp. 365–375. Nauka, Moscow (in Russian).

Gjaerevoll, O. 1963. Survival of plants on nunataks in Norway during the Pleistocene glaciation. In: Löve, A. and Löve, D. (eds.), North Atlantic biota and their history, pp. 261–283. Pergamon Press, London.

Gjaerevoll, O. & Bringer, K.G. 1965. Plant cover of the alpine regions. Acta Phytogeogr. Sue. 50, 257–268.

Godwin, H. 1966. Introductory address. In: World climate from 8000 to B.C. Roy. Met. Soc. Lond. pp. 3–14.

Goldsmith, F.B. & Harrison, C.M. 1976, Description and analysis of vegetation. In: Chapman, S.B. (ed.), Methods in plant ecology, pp. 85–155. Blackwell, London.

Golte, W. 1976. Ecological and phylogenetic bases of the distribution of conifers on the earth. Intl. Geogr. '76, Vol. 4, pp. 17–19, Moskva (in Russian).

Golubeva, L.V. 1964. Types of Pleistocene vegetation in eastern Siberia. Dokl. Akad. Nauk USSR 155: 74–77 (in Russian).

Good, R. 1927. The genus *Empetrum*. J. Linn. Soc. Bot. 47: 489–523.

Good, R. 1947. The geography of the flowering plants. Longman, London.

Gorchakovsky, P.C. & Shiyatov, S.G. 1978. The upper forest limit in the mountains of the boreal zone of the USSR. Arct. & Alp. Res. 10 (2), 349–363.

Goudie, A. 1979. Environmental change. Oxford Univ. Press, Oxford.

Goudie, A. 1985. The nature of the environment. Basil Blackwell, Oxford.

Grace, J. 1977. Plant response to wind. Academic Press, London.

Grace, J. 1981. Some effects of wind on plants. In: Grace, J., Ford, E.C. & Harvis, P.G. (eds.), Plants and their atmospheric environments, pp. 31–56. Blackwell Sci. Publ. Oxford.

Grace, J. 1987. Climatic tolerance and the distribution of plants. New Phytol. 106 (suppl.): 113–130.

Grant, M.C. & Mitton, J.B. 1977. Genetic differentiation among growth forms of Engelmann spruce and subalpine fir at tree line. Arct. Alp. Res. 9: 259–263.

Green, F.H.W. 1983. Soil temperature and the tree line: A note. Scot. Geogr. Mag. 99: 44–47.

Grey-Wilson, C. & Blamey, M. 1986. The alpine flowers of Britain and Europe. Collins, London.

Grichuk, V.P. 1984. Late Pleistocene vegetation history. In: Velichko, A.A., Wright, H.E. Jr. & Barnosky, C.W. (eds.), Late Quaternary environments of the Soviet Union, pp. 155–178. Longman, London.

Griffin, J.B. 1960. Some prehistoric connections between Siberia and America. Science 131: 801–812.

Grigg, D.B. 1984. The agricultural systems of the world. Cambridge Geogr. Ser. No. 5. Cambridge Univ. Press, Cambridge.

Griggs, R.G. 1930. A contribution to the arctic-alpine problem. Ecol. 11: 607–609.

Grime, J.P. 1979. Plant strategies and vegetation processes. John Wiley & Sons, London.

Grosset, G.E. 1959. A study of the ecology of *Pinus pumila* (the mechanism of active bending down to the ground on the onset of frost). Mosk. Obsch. Ispy. Prito. Okd. Biol. Biul. 2: 65–96 (in Russian).

Grubb, P.J. 1985. Plant populations and vegetation in relation to habitat, disturbance and competition: problem of generalization. In: White, J. (ed.), The population structure of vegetation, pp. 595–621. Dr. W. Junk, Dordrecht.

Haber, E. 1986. Flora of the circumpolar Arctic. In: Sage, B. (ed.), The Arctic and its wildlife, pp. 59–71. Croom Helm, London.

Hadac, E. 1963. On the history and age of some arctic plant species. In: Löve, A. and Löve, D. (eds.), North Atlantic biota and their history, pp. 207–219. Pergamon Press, Oxford.

Hagerup, O. 1914. Caprifoliaceae. Meed. Gronl. 37: 153–164.

Hagman, M., Feilberg, L., Lagerström, T. & Sanda, J.E. 1978. The Nordic Arboretum expedition to South Korea 1976. Helsinki.

Hämet-Ahti, L. 1979. The dangers of using the timberline as the zero line in comparative studies on altitudinal vegetation zones. Phytocoenologia 6: 49–54.

Hämet-Ahti, L., Ahti, T. & Koponen, T. 1974. A scheme of vegetation zones for Japan and adjacent regions. Ann. Bot. Fenn. 11: 588.

Hanson, H.C. 1958. Principles concerned in the formation and classification of communities. Bot. Rev. 24: 65–125.

Hare, F.K. 1954. The boreal conifer zone. Geogr. Studies 1: 4–18.

Heath, D. & Williams, D.R. 1979. Life at high altitude. Edward Arnold, London.

Hedberg, O. 1972. On the delimitation and subdivision of the high mountain region of Eurasian high mountains. In: Troll, C. (ed.), Geoecology of the high-mountain region of Eurasia, pp. 107–109. Franz Steiner Verlag, Wiesbaden.

Hegi, G. 1930. Alpine flowers. Blackie & Sons Ltd. London.
Heuberger, H. 1974. Alpine Quaternary glaciation. In: Ives, J.D. and Barry, R.G. (eds.), Arctic and alpine environments, pp. 319–338. Methuen, London.
Heusser, L.E. & Morley, J.J. 1985. Pollen and radiocarbon records from deep-sea core RC 14–103. Quat. Res. 24: 60–72.
Heywood, V.H. 1985. Flowering plants of the world. Croom Helm, London.
Higgins, P.D. & Spomer, G. 1976. Soil temperature effects on root respiration and the ecology of alpine and subalpine plants. Bot. Gaz. 137: 110–120.
Hill, M.O. 1979a. TWINSPAN - A FORTRAN program for arranging multivariate data in an ordered two-way table by classification of the individuals and attributes. Cornell Univ. Ithaca, N.Y.
Hill, M.O. 1979b. DECORANA - A FORTRAN program for detrended correspondence analysis and reciprocal averaging. Cornell Univ. Ithaca, N.Y.
Hill, M.O. & Gauch, H.G. 1980. Detrended correspondence analysis, an improved ordination technique. Vegetatio 42: 47–58.
Hoffman, R.S. & Taber, R.D. 1967. Origin and history of Holarctic tundra ecosystems, with special reference to their vertebrate faunas. In: Arctic and alpine environments, pp. 145–170. Indiana Univ. Press, Bloomington.
Holland, P.G. 1981. Pleistocene refuge areas and the revegetation of Nova Scotia, Canada. Progr. Physic. Geogr. 5: 535–562.
Holmes, D. 1978. Principles of physical geology. Van Nostrand Reinhold (UK) Co., Wokingham.
Holroyd, E.W. III, 1970. Prevailing winds on Whiteface Mountain as indicated by flag trees. For. Sci. 10: 222–229.
Holtmeier, F.K. 1973. Geoecological aspects of timberlines in northern and central Europe. Arct. Alp. Res. 5 (2): A45–A54.
Holtmeier, F.K. 1980. Infiuence of wind on tree-physiognomy at the upper timberline in the Colorado Front Range. In: Benecke, U. & Morris, M.R. (eds.), Mountain environments and subalpine tree growth, pp. 247–261. For. Res. Inst. N.Z. For. Ser. Tech. Paper No. 70, Wellington.
Honda, S. 1922. Forest zones of Japan. Tokyo (in Japanese).
Hong, W.S. 1959. Investigation report on plant communities on Mt. Sorak II. Theses Coll. Catholic Medical Coll. 3: 396–409 (in Korean).
Howe, G.M. 1986. The Soviet Union: A geographical survey. Longman, New York.
Hu, H.H. 1934. Distribution of taxads and conifers in China. Proc. 5th Pacific Sci. Congr. Canada 1933, Vol. 3, pp. 3373–3388.
Huh, K.S. & Lee, I.K. 1981. Ecological study on the flora of Wi Island. Kor. J. Ecol. 4: 68–79 (in Korean).
Hultén, E. 1927. Flora of Kamchatka and the adjacent islands I. Kungl. Svenska Vet. Handl. Tred. Ser. Band 5 (1), Almqvist & Wiksell, Stockholm.
Hultén, E. 1928. On the American component in the flora of eastern Siberia. Svenskbot. Tidskrift 22: 220–229.
Hultén, E. 1937. Outline of the history of Arctic and boreal biota during the Quaternary period. Bokf. Akt. Thule, Stockholm.
Hultén, E. 1950. Atlas of the distribution of vascular plants in NW Europe. Gen. Lito. Anst. For. Stockholm (in Swedish).
Hultén, E. 1955. The isolatlon of the Scandinavian mountain flora. Acta Soc. Pro. Fauna et Flora Fenn. 72 (8): 1–22.
Hultén, E. 1958. The amphi-atlantic plants and their phytogeographical connections. Sv. Vetensk. Akad. Handl. 7: 1–340.
Hultén, E. 1959. Studies in the genus *Dryas*. Svensk Bot. Tidskr. 53: 507–542.
Hultén, E. 1962. The circumpolar plant I. Kungl. Svenska Veten. Handl. Fjarde Serien Band 8 (5): 1–275.
Hultén, E. 1963. The distribution conditions of the flora of Beringia. In: Gressitt, J.L. (ed.), Pacific basin biogeography, pp. 7–22. 10th Pacific Sci. Congr. Honolulu, Hawaii, 1961, Bishop Mus. Press.
Hultén, E. 1968. Flora of Alaska and neighbouring territories. Stanford Univ. Press, Stanford, U.S.A.
Hultén, E. 1970. The circumpolar plants. II. Dicotyledons. Kungl. Svenska Vetenskap Handl. Fjarde Serien Band 13 (1): 1–463.
Hustrich, I. 1953. The boreal limits of conifers. Arctic 6: 149–162.
Huxley, A. 1967. Mountain floras in colour. Blandford Press, London.
Huzioka, K. 1951. Notes on some Tertiary plants from Korea II. Trans. Proc. Palaeontol. Soc. Japan N.S. 3: 57–74.
Huzioka, K. 1972. The Tertiary floras of Korea. J. Min. Coll. Akita Univ. Ser. A, Vol. 5, 1–83.
Hwang, K.S. 1979. Flora of Is. Seonyoo (in Korean).
Hyun, M.S. 1956. Korea. In: Haden-Guest, S., Wright, J.K. & Teclaff, E.M. (eds.), A world geography of forest resources, pp. 561–572. The Ronald Press Co., New York.
Igarashi, Y. & Kumano, S. 1981. Vegetational changes during the last Glacial Age in Hokkaido. Quat. Res. Jap. 20: 129–141 (in Japanese).
Imbrie, J. & Imbrie, K.P. 1979. Ice Age. Harvard Univ. Press, Cambridge, Mass.
Ingram, M.J., Farmer, G. & Wigley, T.M.L. 1981a. Past climates and their impact on man: A review. In: Wigley, T.M.L., Ingram, M.J. & Farmer, G. (eds.), Climate and history, pp. 3–50. Cambridge Univ. Press, Cambridge.
Ingram, M.J., Underhill, D.J. & Farmer, G. 1981b. The use of documentary sources for the study of past climate. In: Wigley, T.M.L., Ingram, M.T. & Farmer, G. (eds.), Climate and history, pp. 180–213. Cambridge Univ. Press, Cambridge.
Ingwersen, W. 1976. Ingwersen's manual of alpine plants. Will Ingwersen & Dunnsprint Ltd., Eastbourne, U.K.
Ishizuka, K. 1974. Mountain vegetation. In: Numata, M. (ed.), The flora and vegetation of Japan, pp. 173–210. Kodansa Ltd., Tokyo.
Ito, K. & Nishikawa, T. 1976. Alpine communities of the northern Taisetsu Mountain Ranges (I): Air and soil temperatures, soil acidity and a cyclical change of the Vaccinio – Pinetum pumilae. Rep. Taisetsuzan Inst. Sci. 11: 1–18.
Ito, K. & Nishikawa, T. 1977. Alpine communities of the northern Taisetsu Mountain Range (2). Rep. Taisetsuzan Inst. Sci. Asahikawa Coll. Hokkaido Univ. Educ. 12: 1–26.

Ives, J.D. 1978. Remarks on the stability of timberline. In: Troll, C. & Lauer, W. (eds.), Geoecological relations between the southern temperate zone and the tropical mountains, pp. 313–317. Franz Stein Verlag, Wiesbaden.

Ives, J.D. 1980. Remarks on the stability of timberline. In: Ives, J.D. (ed.), Geoecology of the Colorado Front Range: A Study of Alpine and Subalpine Environments, pp. 351–358. Westview Press, Boulder, U.S.A.

Iwata, S. 1986. Vertical zonation of patterned ground in the Japanese high mountains. Geogr. Rep. Tokyo Metropolitan Univ. 21: 209–221.

Jahn, G. 1977. Destructive effects made by mankind on natural landscape and forests. In: Miyawaki, A. & Tuxen, R. (eds.), Vegetation science and environmental protection, pp. 257–263. Maruzen, Tokyo.

Jahn, A. 1970. Soil movements under the influence of freezing. In: Unesco (ed.), Ecology of the subarctic region, pp. 119–123. Unesco, Paris.

Jardine, N. 1972. Computational methods in the study of plant distributions. In: Valentine, D.H. (ed.), Taxonomy, phytogeography and evolution, pp. 381–393. Academic Press, London.

Jinno, N. & Suzuki, H. 1973. A revised system of the subalpine coniferous forest of Japan. Hikkobia 6: 229–237.

Jo, W.R. 1980a. Sea. In: Korean Institute of Geography (ed.), Korean regional geography, pp. 252–257. Seoul (in Korean).

Jo, W.R. 1980b. Holocene sea-level changes on the east coast of the Korean peninsula. Geogr. Rev. Japan 53: 317–326 (in Japanese).

Jo, W.R. 1980c. Geomorphology. In: Korean Institute of Geography (ed.), Korean regional geography, pp. 160–165. Seoul (in Korean).

Johnson, P.L. & Billings, W.D. 1962. The alpine vegetation of the Beartooth Plateau in relation to cryopodogenic processes and patterns. Ecol. Monogr. 32: 105–135.

Jonasson, S. 1981. Plant communities and species distribution of alpine *Betula nana* heaths in northmost Sweden. Vegetatio 44: 51–64.

Jones, H.G. 1983. Plants and microclimates. Cambridge University Press, Cambridge.

Kaizuka, S. 1965. Geomorphological changes. In: Sugihara, S. (ed.), The archaeology of Japan, vol. 1, pp. 325–46. Kawade Publ. Co., Tokyo (in Japanese).

Kang, S.J. 1987. High moor of Mt. Daeam. In: Cho, R.S. *et al.* (eds.), Nature in the D.M.Z. Korea, pp. 169–201. Kangwon Univ. Press, Chunchon (in Korean).

Kang, S.J. & Iizumi, S. 1981. A historical review of the shifting cultivation and on some studies related to the burned field ecosystem in the Korean peninsula. Ecol. Rev. Jap. 19: 237–252.

Kangnung District Meteorological Service 1986. Climatic table of Kangwon by month and year. Kangnung District Meterological Service, Kangnung (in Korean).

Kano, T. 1937. Glacial topography of northeastern Korea. Geogr. Rev. Jap. 13: 1126–1145 (in Japanese).

Kappan, L. 1981. Ecological significance of resistance to high temperature. In: Lange, O.L. (ed.), Physiological plant ecology I: Response to the physical environment, pp. 439–474. Springer-Verlag, New York.

Karlstrom, T.N.V. 1966. Quaternary glacial record of the North Pacific region and world-wide climatic changes. In: Blumenstock, D.I. (ed.), Pleistocene and post-Pleistocene climatic variations in the Pacific Area, pp. 158–182. 10th Pacific Sci. Congr. Honolulu, Hawaii, 1961, Bishop Mus. Press.

Kawano, S. 1971. Studies on the alpine flora of Hokkaido, Japan. J. Coll. Lib. & Arts, Toyama Univ. 4: 13–96.

Kim, C.H., Lee, K.S. & Park, K.H. 1982. Ecological study on the flora of Bijin Island. Kor. J. Ecol. 5: 164–175 (in Korean).

Kim, D.J. 1970. Uber die Strukturbodenerscheinungen auf dem Halla-San. Naksan Chiri 1: 3–10 (in Korean).

Kim, D.S. 1972. Report on the woody plants of Mt. Samak. Theses Coll. Kangwon Univ. 6: 159–174 (in Korean).

Kim, I.T. & Lee, I.K. 1980. Ecological study on the flora of Muie Island. Kor. J. Ecol. 3: 21–30 (in Korean).

Kim, J.H., Jo, D.S., Jo, K.J. & Min, K.M. 1984. Ecological study on the evergreen broadleaved forest of Is. Chishim. Kor. J. Bot. 27: 51–60 (in Korean).

Kim, J.M. 1973. The ecology of Korean plants. Chonpakwahaksa, Seoul (in Korean).

Kim, J.M. 1980. Environmental changes and agricultural origin in Korea. Kor. J. Ecol. 3: 40–51 (in Korean).

Kim, K.W. 1981. Primitive-ancient art of Korea. Chungumsa, Seoul (in Korean).

Kim, T.W. & Kim, S.S. 1978. A study on the distribution of vascular plants at six islands near Is. Keojae. Rep. KACN 14: 35–58 (in Korean).

Kim, W. & Yang, I.S. 1973. Flora of Mt. Keumo. Kor. J. Pl. Tax. 5: 33–42 (in Korean).

Kim, Y.O. 1983. Past climatic environment in Korea: 51 B.C. (A.D. 921), Tasks and approaches in geography, pp. 231–273. Kyohaksa, Seoul (in Korean).

Kim, Y.O. 1984a. Climatic environment of Koryo Dynasty. J. Kor. Cul. Res. Inst. 44: 113–135 (in Korean).

Kim, Y.O. 1984b. The Little Ice Age in Korea. Geography – Education 14: 1–16 (in Korean).

Kim, Y.O. 1985. A climate and culture of Korea. Ewha Woman's Univ. Press, Seoul (in Korean).

Kim, Y.S., Ko, S.C. & Choi, B.H. 1981. Distribution atlas of plants of Korea (IV). Atlas of Pinaceae in Korea. Kor. J. Pl. Tax. 11: 53–75 (in Korean).

Kim, Y.S., Park, I.D. & Sim, J.G. 1976. Flora of Mt. Mugap. Kor. J. Pl. Tax. 8: 69–81 (in Korean).

Kind, N.V. 1967. Radiocarbon chronology in Siberia. In: Hopkins, D.M. (ed.), The Bering land bridge, pp. 172–192. Stanford Univ. Press, California.

Kira, T. 1977. A climatologlcal interpretation of Japanese vegetation zone. In: Miyawaki, A. & Tuxen, R. (eds.), Vegetation sciences and Environmental protection, pp. 21–30. Maruzen, Tokyo.

Kirkpatrick, J.B. 1982. Phytogeographical analysis of Tasmanian alpine flora. J. Biogeogr. 9: 255–271.

Kitagawa, M. 1939. Lineamenta Florae Manchuriae. Report Inst. Sci. Res. Manchukuo, Vol. III, pp. 1–487.

Kitagawa, M. 1979. Neo-Lineamenta Florae Manchuriae, Florae et Vegetatio Mundi Band 10. J. Cramer, Germany.

Klein, R.G. 1971. The Pleistocene prehistory of Siberia. Quat. Res. 1: 133–161.

Knystautas, A. 1987. The natural history of the USSR. Century, London.

Kobayashi, K. 1967. Phytosociological studies on *Pinus pumila* scrubs of the Daisetsu and the Hidaka ranges in Hokkaido, Japan. Jap. J. Ecol. 17: 189–198 (in Japanese).

Kobayashi, K. 1971. Phytosociological studies on the scrub of dwarf pine (*Pinus pumila*) in Japan. J. Sci. Hiroshima Univ. Ser. B Div. 2 14: 1–52 (in Japanese).

Koizumi, T. 1979. Periglacial processes and alpine plant communities on the high mountains in Japan, in relation to lithology. Jap. J. Ecol. 29: 71–81 (in Japanese).

Kojima, S. 1979. Biogeoclimatic zones of Hokkaido Island, Japan. J. Coll. Lib. Arts, Toyama Univ, Japan 1: 97–141 (in Japanese).

Kong, W.S. 1985. A phytogeographical study on the distribution of bamboos in the Korean peninsula. Kor. J. Ecol. 8: 89–98 (in Korean).

Kong, W.S. & Chung, S.R. 1984. A climatological study on the distribution of tea plant in Korea. Chirihakyonku 9: 583–594 (in Korean).

Kong, W.S. & Chung, S.R. 1985. An analysis of physical geography on the location of tea cultivation in Boseong, Korea. Jour. Regional Devel. Korea 13: 77–88 (in Korean).

Kong, W.S. & Watts, D. 1987. Biogeographic division and environment in Korea, II. Pacific Sci. Ass. 16th Congr. Abstr. Vol. I, p. 131.

Kong, Z.C., Tu, N.C., Shi, Y.C. & Tao, J.R. 1976. Vegetational and climatic changes in Peking during the past one hundred million years. Acta Phytotax. Sinica 14: 86–89 (in Chinese).

Korean Commission for Conservation of Nature and Natural Resources 1971. Nature in Korea. Ministry of Culture and Information, Seoul (in Korean).

Korean Institute of Geography 1980. Korean regional geography. Seoul (in Korean).

Korean Ministry of Culture and Information 1970. Cultural properties of Korea. Seoul (in Korean).

Korean Ministry of Education 1969. Natural conservation area Mt. Sorak. Korean Ministry of Education, Seoul (in Korean).

Korean Ministry of Home Affairs 1972. Protected trees of Korea. Seoul (in Korean).

Korean Ministry of Home Affairs 1979. A white paper on the conservation of nature. Samwha Publ. Co., Seoul (in Korean).

Korean National Park Authority 1984. National park Mt. Sorak. Korean National Park Authority, Seoul (in Korean).

Korean Overseas Information Service 1978. A handbook of Korea. Ministry of Culture and Information, Seoul.

Kornas, J. 1972. Corresponding taxa and their ecological background in the forests of temperate Eurasia and North America. In: Valentine, D.H. (ed.), Taxonomy, phytogeography and evolution, pp. 37–59. Academic Press, London.

Kotlyarov, I.I. 1978. Ecology of *Pinus pumila* by the Sea of Okhatsk. Sov. J. Ecol. 9: 470–473 (in Russian).

Krajina, V.J. 1960. Can we find a common platform for the different schools of forest type classification. Silva Fennica 105: 50–55.

Krebs, C.J. 1985. Ecology: The experimental analysis of distribution and abundance. Harper & Row Publ., London.

Krog, J. 1955. Notes on the temperature measurements indicative of special organisation in arctic and subarctic plants for utilisation of radiated heat from the sun. Physiol. Plant. 8: 836–839.

Kryshtofovich, A.N. 1929. Evolution of the Teritiary flora in Asia. New Phytol. 28: 303–312.

Kryshtofovich, A.N. 1935. A final link between the Tertiary flora of Asia and Europe. New Phytol. 34: 339– 344.

Küchler, A.W. 1949. A physiognomic classification of vegetation. A.A.A.G. 39: 201–209.

Küchler, A.W. 1962. Vegetation mapping in the Pacific region. Proc. Ninth Pacific Sci. Congr. 1957, Vol. 4, pp. 83–85.

Küchler, A.W. & Takenouchi, M. 1948. A new vegetation map of Manchuria. Ecol. 29: 514.

Kung, S.C. 1976. The Pleistocene spruce – fir floras of China. Lecture at Canberra A.N.U., 24th Nov. 1976 (Mimeographed material).

Kurentsova, G.E. 1984. Importance of the Beringia Land Bridge for the formation of flora and vegetation of the Sikhote–Alin Range. In: Kontrimavichus, V.L. (ed.), Beringia in the Cenozoic Era, pp. 228–234. Oxonian Press PVT Ltd., New Delhi.

Kown, S.S. 1980. Solifluction phenomenon on Mt. Keumjong, southeastern Korea. Chirihak Yonku 5: 295–306 (in Korean).

Kwon, S.S. 1987. A study on Quaternary periglacial cryogenic structures of granite regolith, Korea. Ph.D. Thesis, Seoul National Univ., Seoul, Korea (in Korean).

Kwon, T.W. & Shin, Y.H. 1977. An estimation of population during the Yi Dynasty. Dong – A culture 14: 289–300 (in Korean).

Laegaard, S. 1980. Alpine life-forms and snowbed vegetation. Geogr. Tids. 8: 13–15.

Lamb, H.H. 1970. Our changing climate. In: Perring, F. (ed.), The flora of a changing Britain, pp.11–24. Bot. Soc. Brit. Isles, Middlesex Conf. Rep.

Lamb, H.H. 1982. Climate, history and the modern world. Methuen, London.

Lance, G.N. & Williams, W.T. 1968. Note on new information – statistical classificatory program. Comput. J. 11: 195.

Landolt, E. & Aeschimann, D. 1986. Notre flore alpine. Editions du Club Alpin Suisse, Zurich (in French).

Larcher, W. 1980. Physiological plant ecology. Springer-Verlag, Berlin.

Larsen, J.A. 1980. The boreal ecosystem. Academic Press, New York.

Lautensach, H. 1941. Die Natürlichen Vegetationstufen in Korea. Petermanns Geographisch 87: plate 5 (in German).

Lautensach, H. 1945. Korea. Koehler Verlag, Leipzig (in German).

Lebedev, D.V. 1968. Union of Soviet Socialist Republics. In: Küchler, A.W. (ed.), Vegetation maps of the Union of Soviet Socialist Republics, Asia and Australia, Intl. Biblio. Vegetation Maps, Vol. 3, pp. 1–93. Univ. Kansas, Lawrence.

Lee, B.H., Lee, Y.N., Kim, J.H. & To, S.H. 1957. Flora of

Dukjeok Island. Theses Coll. Chung-ang Univ. (Pharm.) 1: 37–50 (in Korean).
Lee, C.R. & Ahn, H.S. 1963. Nomina plantarumkoreanum. Hyanghaksa, Seoul (in Latin & Korean).
Lee, D.B. 1955. Flora of Mt. Keumkang. 30th Ann. Theses Coll. Chungang Univ. pp. 353–381 (in Korean).
Lee, D.B. 1957. Flora of Cheju Island. Theses Coll. Korea Univ. (Coll. Lib. & Arts) 2: 339–412 (in Korean).
Lee, D.B. 1959. Flora of Mt. Sokri. Theses Coll. Korea Univ. (Coll. Lib. & Arts) 4: 202–253 (in Korean).
Lee, D.B. 1968. Flora of Kwangnung. Theses Coll. Chungang Univ. 13: 47–65 (in Korean).
Lee, D.B. 1970. A history of biology in Korea. Korea Univ. Press, Seoul (in Korean).
Lee, D.B. & Joo, S.W. 1958. Flora of Ulreung Island. Theses Coll. Korea Univ. (Coll. Lib. & Arts) 3: 223–295 (in Korean).
Lee, D.B. & Kim, Y.C. 1961. A historical review of some plants of American origin in Korea. Kor. J. Bot. 4: 25–30 (in Korean).
Lee, E.B. & Chang, Y.S. 1984. Distribution of deciduous broadleaved plants on Mt. Sorak. In: Province of Kangwon (ed.), Scientific report on Mt. Sorak, pp. 52–64. Province of Chunchon (in Korean).
Lee, H.J. 1981. Vegetation of Sangil Island. Kor. J. Ecol. 4: 80–92 (in Korean).
Lee, H.J., Choi, S.C. & Lee, Y.H. 1982. Taxonomical and ecological study on the vegetation of Seungbong Island. Kor. J. Ecol. 5: 105–116 (in Korean).
Lee, H.J. & Lee, W.H. 1962. List of Mt. Myungji flora. Kor. J. Bot. 5: 9–16 (in Korean).
Lee, H.Y. 1987. Palaeontology of Korea. Minumsa, Seoul (in Korean).
Lee, I.K. 1964. Flora of Mt. Myungji. Theses Coll. Kyunghee Univ.3: 245–267 (in Korean).
Lee, I.K. 1980. Flora of Bogil Island. Theses Coll. Konkuk Univ. 24: 41–62 (in Korean).
Lee, I.K. 1981. Flora of Wooie Island. Theses Coll. Konkuk Univ. 25: 33–53 (in Korean).
Lee, I.K. & Lee, C.B. 1959. Flora of Hong Island. Theses Coll. Kyunghee Univ. 3: 58–74 (in Korean).
Lee, I.K., Hwang, K.S. & Song, C.S. 1980. On the distribution and conservation state for the evergreen broadleaved trees in island of south-western seashore. Rep. Nat. Conservation 2: 13–33 (in Korean).
Lee, I.K. & Kim, I.T. 1980. Vegetation of Kogunsan Island. Rep. KACN 18: 87–101 (in Korean).
Lee, I.K., Kim, W., Lee, H.J. & Yoon, H.S. 1973. Flora of Dolsan Island. Kor. J. Pl. Tax. 3: 23–32 (in Korean).
Lee, I.K., Lee, H.J. & Kim, I.T. 1980. Flora of Chungsan Island. Theses Coll. Konkuk Univ. (Sci.) 6: 45–56 (in Korean).
Lee, I.K., Moon, H.J. & Lee, Y.H. 1982. Flora of Island Poong. Kor. J. Ecol. 5: 154–163 (in Korean).
Lee, J.S. 1962. Flora of Mt. Ipam. Theses Coll. Chonnam Univ. 7: 181–212 (in Korean).
Lee, K.N. & Chae, J.K. 1981. Studies on the distribution of woody plants in Bogil Island. Rural Develop. Rev. Chonnam Univ. 16:39–42.
Lee, S.W. 1982. Marine climate. In: Kim, K.S. *et al.* (eds.), The climate in Korea, pp. 121–133. Iljisa, Seoul (in Korean).
Lee, T.B. 1980a. Conservation of threatened plants in Korea. Bull. Kwanak Arbor. 3: 190–196 (in Korean).
Lee, T.B. 1980b. Rare and endangered species in the area of Mt. Sorak. Bull. Kwanak Arb. 197–201 (in Korean).
Lee, T.B. 1982a. Illustrated flora of Korea. Hyangmunsa (in Korean).
Lee, T.B. 1982b. Endemic plants and their distribution in Korea. J. Natl. Acad. Sci. R.O.K. Ser. 21: 71–113 (in Korean).
Lee, T.B. 1984. Endemic and rare plants of Mt. Sorak. Seoul Natl. Univ. Coll. Agr. Res. 9: 1–6 (in Korean).
Lee, T.B. & Yu, J.D. 1984. Endemic and rare plants. In: Province of Kangwon (ed.), Scientific report on Mt. Sorak, pp. 169–191. Province of Kangwon, Chunchon (in Korean).
Lee, W.C. 1969. On the natural monument of Korea. J. Kor. Plant Tax. 1: 27–35 (in Korean).
Lee, W.C. 1976. Vegetation of Mt. Chiak. Theses Coll. Kangwon Univ. 10: 259–275 (in Korean).
Lee, W.C. 1977. Flora of Mt. Kari. Theses Coll. Kangwon Univ. 11: 227–241 (in Korean).
Lee, W.C. & Chon, S.K. 1975. Vegetation of Mt. Daeryong. Theses Coll. Kangwon Univ. 9: 309–324 (in Korean).
Lee, W.C. & Chung, H.B. 1976. Flora of Mt. Samak and Juk Island. Kor. J. Pl. Tax. 7: 1–20 (in Korean).
Lee, W.C., Hong, M.P. & Park, C.J. 1980. Vegetation of Lake Songji. Kor. J. Pl. Tax. 10: 105–111 (in Korean).
Lee, W.C. & Paik, W.K. 1984. Flora of Mt. Daesung. Kor. J. Pl. Tax. 14: 109–132 (in Korean).
Lee, W.C. & Yim, Y.J. 1975. Vegetation of Juk Island. Kor. J. Pl. Tax. 6: 9–15 (in Korean).
Lee, W.C. & Yim, Y.J. 1978. Studies on the distribution of vascular plants in the Korean peninsula. Kor. J. Pl. Tax. 8: Appendix 1–32 (in Korean).
Lee, W.H. 1975. Plant resources in Mt. Mugap. Theses Coll. Seoul City Coll. 57–72 (in Korean).
Lee, Y.C. 1975. Radiocarbon dating and chronological problems in Korean prehistory. Kor. Hist. Rev. 68: 53–92 (in Korean).
Lee, Y.N. 1957. Flora of Chinwoo Island. Daedongdang 1–9 (in Korean).
Lee, Y.N. 1976. Illustrated flora and fauna of Korea, Vol. 18 (Flowering Plants). Korean Ministry of Education (in Korean).
Lee, Y.N. & Oh, Y.J. 1970. Flora of limestone Todam. J.K.R.I.B.L. 5: 101–115 (in Korean).
Lee, Y.N. & Oh, Y.J. 1982. An investigation of monocotyledons on Mts. Kaebang, Sokaebang and Kachil. Rep. K.A.C.N. 20: 43–61 (in Korean).
Levitt, J. 1966. Winter hardiness in plants. In: Merryman, H.T. (ed.), Cryobiology, pp. 495–563. Academic Press.
Levitt, J. 1980. Responses of plants to environmental stresses. Academic Press, New York.
Li, H.L. 1952. Floristic relationships between eastern Asia and eastern North America. Trans. Amer. Phil. Soc. 42: 371–429.
Lim, J.W., Lee, H.J., Kim, C.H. & Kim, S.H. 1982. Flora of Cheju Island. Kor. J. Ecol. 5: 187–203 (in Korean).
Lindroth, C.H. 1970. Survival of animals and plants on ice-

free refugia during the Pleistocene glaciations. Endeavour 29: 129–134.

Lindsey, C.C. 1981. Arctic refugia and the evolution of arctic biota. In: Scudder, G.E.E. and Reveal, J.C. (eds.), Evolution today, pp. 7-10. Carnegie-Mellon Univ., Pittsburgh.

Liu, T.S. 1971. A monograph of the genus Abies. Dept. Forestry, Natl. Taiwan Univ., Taipei.

Löve, A. 1963. Conclusion. In: Löve, A. and Löve, D. (eds.), North Atlantic biota and their history, pp. 391–397. Pergamon Press, London.

Löve, A. 1967. The evolutionary significance of disjunctions. Taxon 16: 324–334.

Löve, A. & Löve, D. 1967. Continental drift and the origin of the arctic-alpine flora. Rev. Roum. Biol. Ser. Bot. 12: 163–167.

Löve, A. & Löve, D. 1974. Origin and evolution of the arctic and alpine floras. In: Ives, J.D. and Barry, R.G. (eds.), Arctic and alpine environments, pp. 571–603. Methuen, London.

Löve, A. & Löve, D. 1975a. Cryophytes, polyploidy and continental drift. Phytocoenologia 2: 54–65.

Löve, A. & Löve, D. 1975b. Cryotaxonomical atlas of the arctic flora. J. Cramer, Leutershausen, Germany.

Löve, D. 1970. Subarctic and subalpine: where and what?. Arctic & Alpine Res. 2: 63–73.

Lozhkin, A.V. 1984. Late Pleistocene and Holeocene vegetation in western Bering Land. In: Kontrimavichus, V.L. (ed.), Beringia in the Cenozoic Era, pp. 88–95. Oxonian Press, New Delhi.

Maarel, E. van der 1976. On the establishment of plant community boundaries. Bet. Deut Bet. Ges. Bol. 89: 415–443.

Maarel, E. van der 1979. Multivariate methods in phytosociology, with reference to the Netherlands. In: Werger, M.J.A. (ed.), The study of vegetation, pp. 161–225. Dr. W. Junk, The Hague.

MacMahon, J.A. & Andersen, D.C. 1982. Subalpine forests: a world prospective with emphasis on western North America. Progress in Physical Geogr. 6: 368–425.

Maekawa, F. 1974. Origin and characteristics of Japan's flora. In: Numata, M. (ed.), The flora and vegetation of Japan, pp. 33–86. Kodansa Ltd., Tokyo.

Maekawa, F. & Shidei, T. 1974. Geographical background to Japan's flora and vegetation. In: Numata, N. (ed.), The flora and vegetation of Japan, pp. 1–31. Kodansa Ltd., Tokyo.

Mani, M.S. 1962. Introduction to high altitude entomology. Methuen, London.

Mani, M.S. 1973. Fundamentals of high altitude biology. Oxford & IBM, New Delhi.

Mani, M.S. 1978. Ecology and phytogeography of high altitude plants of the Northwest Himalaya. Chapman & Hall, London.

Mani, M.S. & Giddings, L.E. 1980. Ecology of highlands. Dr. Junk Publ., The Hague.

Marr, J.W. 1977. The development and movement of tree islands near the upper limit limit of tree growth in the southern Rocky Mountains. Ecol. 58: 1159–1164.

Matthews, J.R. 1937. Geographical relationship of the British Flora. J. Ecol. 25: 1–90.

McCown, B.H. 1975. Physiological responses of root systems to stress conditions. In: Vernberg, F.J. (ed.), Physiological adaptation to the environment, pp. 225–237. Intext Educational Publ., New York.

McCune, S. 1956. Korea's Heritage: A regional and social geography. Tokyo.

McLean, R.C. & Ivimey-Cook, W.R. 1967. Textbook of theoretical botany III. Longman, London.

Melzack, R.N. & Watts, D. 1982a. Cold hardiness in the yew (*Taxus baccata*) in Britain. J. Biogeogr. 9: 231–241.

Melzack, R.N. & Watts, D. 1982b. Variations in seed weight, germination, and seedling vigour in the yew (*Taxus baccata*) in England. J. Biogeogr. 9: 55–63.

Meusel, H., Jäger, E. & Mörchen, G. 1977. The study of growth-forms of higher plants in relation to application of biocides. In: Miyanwaki, A. & Tuxen, R. (eds.), Vegetation science and environmental protection, pp. 71–76. Maruzen, Tokyo.

Mezhennyi, A.A. 1979. The distribution of *Pinus pumila*. For. Abstr. 1979, p. 146.

Minato, M. 1972. Late quaternary geology in northern Japan. Intl. Geol. Congr. Canada Sect. 12: 63–71.

Mirov, N.T. 1967. The genus Pinus. The Ronald Press, New York.

Mityga, H.G. & Lanphear, F.O. 1971. Factors influencing the coldhardinessof *Taxus cuspidata* roots. J. Amer. Hort. Sci. 96: 83–86.

Moggi, G. 1985. The Macdonald encyclopedia of alpine flowers. MacDonald, London.

Molozhnikov, V.V. 1969. The distribution of *Pinus pumila* in the Barguzin Range (NE of Lake Baikal), USSR. Bot. Zhu. 54: 902–909 (in Russian).

Monk, C.D. 1966. An ecological significance of evergreenness. Ecol. 47: 504–505.

Mooney, H.A. & Billings, W.D. 1961. Comparative physiological ecology of arctic and alpine populations of *Oxyria dignya*. Ecol. Monogr. 31: 1–29.

Mooney, H.A. & Billings, W.D. 1965. Effect of altitude on carbohydrate content of mountain plants. Ecol. 46: 750–751.

Morton, J.K. 1972. Phytogeography of the west African mountains. In: Valentine, D.H. (ed.), Taxonomy, phytogeography and evolution, pp. 221–236. Academic Press, London.

Mt. Sorak National Park Authority 1987. Mt. Sorak national park. Sokcho, Korea (in Korean).

Mueller-Dombois, D. & Ellenberg, 1974. Aims and methods of vegetation ecology. John Wiley & Sons, New York.

Müller, P. 1986. Biogeography. Harper & Row Publ., New York.

Murata, G. 1977. Phytogeographical consideration on the flora and vegetation of Japan. Acta Phytotax. Geobot. 28: 65–83.

Murray, D.F. 1981. The role of arctic refugia in the evolution of the arctic vascular flora. In: Scudder, G.G.E. and Reveal, J.L. (eds.), Evolution today, pp. 11–20. Proc. Second Intl. Congr. Syst. Evol. Biol. Hunt. Bot. Documentation, Carnegie-Mellon Univ., Pittsburg.

Nakai, T. 1915–1939. Flora Sylvatica Koreana (in Japanese).

Nakai, T. 1919. Flora of Ulreung Island. The Government of Chosen (Korea) (in Japanese).

Nakai, T. 1926. The floras of Tsusima and Quelpaert as related to those of Japan and Korea. Proc. 3rd Pan-Pacific Congress, Tokyo, pp. 893–911.
Nakai, T. 1935. East Asian flora. Tokyo (in Japanese).
Nakai, T. 1952. A synoptic sketch of Korean flora. Bull. Natl. Sci. Mus., Tokyo.
Nelson, S.M. 1982. The effects of rice agriculture on prehistoric Korea. J. Asian Studies 41: 531–543.
Nikolskaya, V.V. & Fedorova, R.V. 1973. Determination of the long-term natural trends in the development of the vegetation on Sakhalin and Iturup (Kurile Island). Biol. Nauki 2: 69–77 (in Russian).
Nilsson, T. 1982. The Pleistocene: geology and life in the Quaternary Ice Age. D. Reidel Publ. Co., Dordrecht.
Noda, M. 1971. Flora of the N.E. Province (Manchuria) of China. Kazama Bookshop Co., Tokyo.
Noguchi, Y. 1979. Deformation of trees in Hawaii and its relation to wind. J. Ecol. 67: 611–628.
Noh, E.R. 1983. Studies on the growth range and optimum site determination of the tree species using climatological factors in Korea. J. Kor. For. Soc. 62: 1–18 (in Korean).
Numata, M. 1972. Ecological interpretations of vegetation zonation of high mountains particularly in Japan and Taiwan. Erdwissensch Forsch 4: 288–299.
Numata, M., Miyayaki, A. & Itow, P. 1972. Natural and semi-natural vegetation in Japan. Blumea 10: 435–496.
Oberbauer, S. & Miller, P.C. 1981. Some aspects of plant water relations in Alaskan arctic tundra species. Arct. Alp. Res. 13: 205–318.
Obruchev, S.V. 1933. The Indigirka expedition of 1926. Geol. Sketch Vol. 2, United Geol. Prosp. Serv. USSR Tr. Vol. 229: 134 (in Russian).
Obruchev, S.V. 1939, Past glaciation and Quaternary history of Chukchee Land. USSR Akad. Sci. Bull. Ser. Geol. Geophysi. No. 2: 129–146 (in Russian).
Oh, C.Y. 1976. A pollen analysis in the peat sediments from Pyungtaek county, Korea. Journ. Kor. Archaeol. Stud. 1: 125–134 (in Korean).
Oh, S.Y. 1971a. Vegetation of Mt. Biseul. Theses Coll. Andong Teachers Coll. 4: 419–467 (in Korean).
Oh, S.Y. 1976b. Vegetation of Mt. Choijung. Kor. J. Pl. Tax. 3: 42–56 (in Korean).
Oh, S.Y. 1972. Vegetation of Mt. Palgong. Theses Coll. Andong Teachers Coll. 5: 313–351 (in Korean).
Oh, S.Y. 1973. Vegetation of Mt. Kaya. Theses Coll. Andong Teachers Coll. 6: 241–280 (in Korean).
Oh, S.Y. 1977. A floristic study of vascular plants in Korea. Nature and Life (Kyungpook Univ.) 7: 13–39 (in Korean).
Oh, S.Y. 1983. Flora of Yeoseo Island. Kor. J. Ecol. 6: 70–77 (in Korean).
Ohsawa, M. 1984. Differentiation of vegetation zones and species strategies in the subalpine region of Mt. Fuji. Vegetatio 57: 15–52.
Ohwi, J. 1965. Flora of Japan. Smithsonian Institution, Washington D.C.
Oka, S. 1972. Deformation of trees on Mt. Fuji. Geogr. Rep. Tokyo Metropol. Univ. 6: 15–29.
Okamoto, Y. 1972. Piedmont glaciation in the taiga forests of ice ages in Japan and northern Italy similar to those now present in southern Alaska. Intl. Geol. Congr. Canada Sect. 12: 175–186.
Okitsu, S. & Ito, K. 1984. Vegetation dynamics of the Siberian dwarf pine (*Pinus pumila*) in the Taisetsu mountain range, Hokkaido, Japan. Vegetatio 59: 105–113.
Okumura, S. 1974. Forest of Korean Peninsula. Doi Ringaku Shinkokai, Tokyo (in Japanese).
Orlóci, L. 1975. Multivariate analysis in vegetation. Dr. W. Junk, The Hague.
Osgood, C. 1951. The Koreans and their culture. The Ronald Press Co., New York.
Packer, J.G. 1974. Differentiation and dispersal in alpine floras. Artic & Alpine Res. 6: 117–128.
Park, B.K. 1966. The grassland types of Korea. Kor. J. Bot. 9: 39–45 (in Korean).
Park, B.K. 1971. Vegetation of Ulreung (Dagelet) Island. J.K.R.I.B.L. 8: 53–54 (In Korean).
Park, B.K. 1984. The flora and vegetation of Mt. Sorak. In: Province of Kangwon (ed.), Scientific report on Mt. Sorak, pp. 48–49. Province of Kangwon, Chunchon (in Korean).
Park, H.H. 1983. Fauna and flora. In: The national history compilation committee (ed.), Korean history Vol. 12 (Korean Archaeology), pp. 91–186. Seoul (in Korean).
Park, J.D. 1985. Ecological studies on the *Pinus pumila* association in Mt. Seolag. M.Sc. Dissertation, Korea University, Seoul (in Korean).
Park, M.K. & Hong, W.S. 1959. Investigation report on plant communities of Mt. Sorak I. Kor. J. Bot. 2: 1–21 (in Korean).
Park, M.K. & Lee, E.B. 1970. Flora of Sohuksan Island. Rep. K.A.C.N. 1: 21–23 (in Korean).
Park, S.E. & Kim, H.E. 1972. A study on the plant distribution in Mt. Wolak. Theses Coll. Kyungbuk Univ. 17: 77–88 (in Korean).
Park, S.H. 1970. Flora of Mt. Moodung. Kor. J. Pl. Tax. 2: 23–39 (in Korean).
Park, Y.A. 1977. Submergence of the Yellow Sea coast of Korea and stratigraphy of the Sinpyeongcheong marsh, Kimjae, Korea. J. Kor. Archaeol. Stud. 3: 81–91 (in Korean).
Parkhurst, D.F. & Loucks, O.L. 1972. Optimal leaf size in relation to environment. J. Ecol. 60: 505–537.
Pears, N. 1985. Basic biogeography. Longman, London.
Pearson, R. 1977. Palaeoenvironment and human settlement in Japan and Korea. Science 197: 1239–1246.
Pennington, W. 1974. The history of British vegetation. The English Universities Press, London.
Pianka, E.R. 1966. Latitudinal gradients in species diversity: A review of concepts. Amer. Nat. 100: 33–46.
Pielou, E.C. 1979. Biogeography. John Wiley & Sons, New York.
Polunin, N. 1951. The real Arctic: suggestions for its delimitation, subdivision and characterisation. J. Ecol. 39: 308–315.
Polunin, N. 1955. Aspects of arctic botany. Amer. Sci. 43: 307–322.
Polunin, N. 1959. Circumpolar Arctic Flora. Oxford Univ. Press, Oxford.
Polunin, N. 1960. Introduction to plant geography. Longman, London.

Polunin, O. 1969. Flowers of Europe. Oxford Univ. Press, Oxford.
Polunin, O. & Walters, M. 1985. A Guide to the vegetation of Britain and Europe. Oxford Univ. Press, Oxford.
Porsild, A.E. 1951. Plant life in the Arctic. Can. Geogr. Jour. 42: 120–145.
Powers, W.R. 1973. Palaeolithic man in northeast Asia. Arctic Anthropology 10: 1–106.
Precht, H., Christophersen, J., Hensel, H. & Larcher, W. 1973. Temperature and life. Springer-Verlag, Berlin.
Price, L.W. 1981. Mountain and man. Univ. California Press, Los Angeles.
Prime, C.T. 1978. Plant life. Readers Union Group of Books Clubs, Devon.
Proctor, M.C.F. 1967. The distribution of British liverworts: A statistical analysis. J. Ecol. 55: 119–135.
Province of Kangwon 1984. Scientific report on Mt. Sorak. Province of Kangwon, Chunchon (in Korean).
Pruitt, W.O. 1978. Boreal ecology. Edward Arnold, London.
Putman, P.C. 1948. Power from the wind. Van Nostrand, New York.
Qin, S.W., Kang, B., Lee, P.W., Ha, Y.A., Wang, M.H. & Wang, P.F. 1981. A preliminary research on the evolution of the natural environment of northeast China Sea since late glacial period. Acta Geogr. Sinica 36: 315–327 (in Chinese).
Quaternary Flora Research Group 1974. The geohistorical succession of the vegetation and the climatic change during the Wurm glacial age in Japan. Quat. Res. Jap. 12: 161–175.
Randall, R.E. 1978. Theories and techniques in vegetation analysis. Oxford Univ. Press, Oxford.
Rapp, A. 1970. Some geomorphological processes in cold climates. In: Unesco (ed.), Ecology of the Subarctic Region, pp. 105–114. Unesco, Paris.
Raup, H.M. 1953. Some botanical problems of the Arctic and subarctic regions. Arctic 6: 68–74.
Raven, J. & Walters, M. 1984. Mountain flowers. Collins, London.
Reader's Digest 1985. The high kingdom. The Reader's Digest Ass. Ltd., London.
Richardson, S.D. 1966. Forestry in Communist China. Johns Hopkins Press, Baltimore, U.S.A.
Richardson, S.G. & Salisbury, F.B. 1977. Plant responses to the light penetrating snow. Ecol. 58: 1152–1158.
Riley, D. & Young, A. 1987. World vegetation. Cambridge Univ. Press, Cambridge.
Rind, D. & Peteet, D. 1985. Terrestrial conditions at the last glacial maximum and CLIMAP sea-surface temperature estimate: Are they consistent?. Quat. Rev. 24: 1–22.
Ritchie, J.C. 1955. Biological flora of the British Isles: *Vaccinium vitis*. J. Ecol. 43: 701–708.
Rousseau, J. 1953. The value of botany as indicator of unglaciated areas. Memoirs du Jardin Botanique de Montreal 40: 1–8.
Ruddiman, W.F. & Duplessy, 1985. Conference on the last deglaciation: timing and mechanism. Quat. Res. 23: 1–17.
Russell, R.S. 1948. The effect of arctic and high mountain climates on the carbohydrate content of *Oxyria digyna*. J. Ecol. 36: 91–95.
Rydberg, P. A. 1913. Phytogeographical notes on the Rocky Mountain region I. Torrey Bot. Club Bull. 40: 677–686.
Rydberg, P.A. 1914. Phytogeographical notes on the Rocky Mountain region V. Torrey Bot. Club Bull. 41: 89–103.
Sage,B. 1986. The Arctic and its wildlife. Croom Helm, London.
Sakai, A. & Kurahashi, A. 1975. Freezing resistance of conifers in Japan with special reference to their distribution. Jap. J. Ecol. 25: 192–200.
Sakai, A. & Otsuka, K. 1970. Freezing resistance of alpine plants. Ecol. 55: 665–671.
Sakai, J. 1981. Late Pleistocene climatic changes in central Japan. J. Fac. Sci. Shinshu Univ. 16: 1–64.
Sample, L.L. 1974. Tongsamdong: A contribution to Korean Neolithic culture history. Arctic Anthropology 11: 1–125.
Sano, Y., Matano, T. & Ujihara, A. 1977. Growth of *Pinus pumila* and climatic fluctuation of Japan. Nature 266: 159–161.
Sasa, Y. & Tanaka, K. 1938. Glacial topography in the Kanbo Massif, Korea. Jour. Fac. Sci. Hokkaido Univ. 4: 193–212 (in Japanese).
Saville, D.B.D. 1972. Arctic adaptations in plants. Canada Dept. Agr. Monogr. No. 6, pp. 1–81, Ottawa.
Scarre, C. 1988. Past worlds. The Times atlas of archaeology. Times Books Ltd., London.
Schröter, L. 1900. Taschenflora des Alpen-Wanderens. Albert Rausten, Zurich (in German).
Sekiguti, T. 1951. Studies on local climatology (III). Papers in Met. & Geophysics 2: 168–179.
Selander, S. 1950. Floristic phytogeography of southwestern Lule Lappmark. Acta Phytogeogr. Sue. 27: 1–200.
Seoul National University 1980. The history of horticultural development in Korea. Seoul.
Shackleton, N.J. 1984. Oxygen isotope evidence for Cenozoic climatic change. In: Brenchley, P. (ed.), Fossils and climate, pp. 27–34. John Wiley & Sons, Chichester.
Sharp, R.P. 1956. Glaciers in the Arctic. Arctic 9: 78–117.
Shetler, S. 1963. Plants in the arctic-alpine environment. Ann. Reg. Board Regents of the Smithsonian Institute Publ. 4530: 473–497.
Shidei, T. 1963. Productivity of *Pinus pumila* community growing in alpine zone of Tateyama-Range. Journ. Jap. For. Sci. 45: 169–173.
Shidei, T. 1974. Forest vegetation zones. In: Numata, M. (ed.), The Flora and vegetation of Japan, pp. 87–124. Kodansa Ltd., Tokyo.
Shikama, T. 1962. Quaternary land connections of Japanese Islands with continent from the viewpoint of Palaeomamalogy. Quat. Rev. Jap. 2: 146–153.
Shimizu, T. 1982. The new alpine flora of Japan in colour, Vol. I. Hoikusha Publ. Co. Ltd., Osaka, Japan (in Japanese).
Shimizu, T. 1983. The new alpine flora of Japan in colour, Vol.II. Hoikusha Publ. Co. Ltd., Osaka, Japan (in Japanese).
Sibrava, V., Bowen, D.Q. & Richmond, G.M. (eds.) 1986. Quaternary glaciations in the Northern Hemisphere. Quaternary Science Review, Vol. 5. Pergamon Press, Oxford.
Slavik, B. 1977. A colour guide to familiar mountain flowers. Octopus Books Ltd., London.

Sohn, P.K. 1978. The Early Palaeolithic industries of Sokchangni, Korea. In: Ikawa-Smith, F. (ed.), Early Palaeolithic in South and East Asia, pp. 233–245. Mouton Publ., The Hague.

Sohn, P.K. 1984. The palaeoenvironment of middle and upper Pleistocene Korea. In: Whyte, R.P. (ed.), The evolution of the East Asian environment, Vol. 4, pp. 877–893. Univ. Hong Kong.

Song, J.S. & Nakanishi, S. 1984. Phytosociological study of the subalpine forests of Mt. Halla of Cheju Island, Korea. Jap. J. Ecol. 35: 317–328.

Specht, R.L. 1979. Heathlands and related shrublands of the world. In: Specht, R.L. (ed.), Heathlands and related shrublands, pp. 1–18. Elsevier, Amsterdam.

Speirs, D.C. 1979. Arctic and alpine ecology. Quat. Bull. Alpine Gdn. Soc. 47: 31–36.

Spomer, G.G. 1962. Physiological ecology of alpine plants. Ph.D. Dissertation, Colorado State Univ., Port Collins, Colorado.

Spomer, G.G. 1964. Physiological ecology studies of alpine cushion plants. Physiol. Plant. 17: 717–724.

Stebbins, G.L. 1972. Ecological distribution of centers of adaptative radiation in angiosperms. In: Valentine, D.H. (ed.), Taxonomy, phytogeography and evolution, pp. 7–34. Academic Press, London.

Stoutjesdijk, P. 1970. A note on vegetation temperatures above the timberline in southern Norway. Acta Bot. Neerl. 19: 918–928.

Street, H.E. & Öpik, H. 1984. The physiology of flowering plants. Edward Arnold, London.

Stewart, W.S. & Bannister, P. 1973. Seasonal changes in carbohydrate content of three *Vaccinium* species with particular reference to *V. uliginosum* and its distribution in the British Isles. Flora 162: 134–155.

Suslov, S.P. 1961. Physical geography of Asiatic Russia. W.H. Freeman & Co., San Francisco and London.

Sutcliffe, J. 1986. Plants and water. Edward Arnold, London.

Suzuki, H. 1962. Southern limit of periglacial landform at low level and the climatic classification of the Latest Ice Age in Japan. Geogr. Rev. Jap. 35: 67–76.

Swan, L.W. 1967. Alpine and aeolian regions of the world. In: Wright, H.E. Jr., & Osburn, W.H. (eds.), Arctic and alpine environments, pp. 29–54. Indiana Univ. Press, Bloomington.

Takahashi, M. 1944. Studies on the system of plant-ecology based on field investigations made in Northern East Asia. J. Fac. Sci. Imperial Univ. Tokyo, Sect. III Bot. Vol. 5 (5): 427–649.

Takhtajan, A. 1986. Floristic regions of the world. Univ. California Press, Berkeley.

Tanai, T. 1972. Tertiary history of vegetation in Japan. In: Graham, A. (ed.), Floristics and palaeofloristics of Asia and Eastern North America, pp. 235–285. Elsevier Publ. Co., Amsterdam.

Tatewaki, M. 1935. On the distribution of *Pinus pumila* in northern Japan. Ecol. Rev. 1: 23–36.

Terri, J.A. 1969. The phytogeography of subalpine black spruce in New England. Rhodora 71: 1–6.

Tikhomirov, B.A. 1946, The dwarf pine (*Pinus pumila*) associations. Nat. Hist. Flora & Vegetation USSR 2: 491–537 (in Russian).

Tikhomirov, B.A. 1966. Vascular plants of the Siberian Northeast Far East. Acad. Sci. USSR, Moskva (Translated by Philips, L., Israel Program for Scientific Translations, Jerusalem, 1969).

Tivy, J. 1982. Biogeography. Longman, London.

Tolmachev, A.I. 1965. The roles of migration and of autochthonous development in the formation of the high mountain flora. In: Sukachev, V.N. (ed.), Studies on the flora and vegetation of high-mountain areas, pp. 9–12. Acad. Sci. USSR, Israel Program for Scientific Translations, Jerusalem.

Tralau, H. 1961. A real development of the European arctic-montane plants. Bot. Not. 114: 213–238.

Tralau, H. 1963. The recent and fossil distribution of some boreal and arctic-montane plants in Europe. Arkv. Bot. 5 (3): 533–582.

Tralau, H. 1973. Some quaternary plants. In: Hallam, A. (ed.), Atlas of palaeogeography, pp. 499–503. Elsevier Sci. Publ., Amsterdam.

Tranquillini, W. 1963. Climate and water relations of plants in the subalpine region. In: Rutter, A.J. & Whitehead, F.H. (eds.), The water relations of plants, pp. 153–167. Blackwell Sci. Publ., Oxford.

Tranquillini, W. 1964. The physiology of plants at high altitudes. Ann. Rev. Plant Physiol. 15: 345–362.

Tranquillini, W. 1976a. Water relations and alpine timberline. Ecol. Stud. 19: 473–493.

Tranquillini, W. 1976b. Water relations and alpine timberline. In: Lange, O.L., Kappen, L. & Schulze, E.D. (eds.), Water and plant life, pp. 473–491. Springer-Verlag, Berlin.

Tranquillini, W. 1979. Physiological ecology of the alpine timberline. Springer-Verlag, Berlin.

Tranquillini, W. 1980. Winter dessiccation as the cause for alpine timberline. In: Benecke, U. & Morris, M.R. (eds.), Mountain environments and subalpine tree growth, pp. 263–267. For. Res. Inst. N.Z. For. Ser. Tech. Paper No. 70, Wellington.

Tranquillini, W. 1982. Frost-drought and its ecological significance. In: Lange, O.L., Nobel, P.S., Osmond, C.B. & Ziegler, H. (eds.), Physiological plant ecology II, pp. 379–400. Springer-Verlag, New York.

Troll, C. 1968. The cordilleras of the tropical Americas: Aspects of climatic, phytogeographical and agrarian ecology. In: Troll, C. (ed.), Geoecology of the mountainous regions of the tropical Americas, pp. 1–556. Ferd. Dummers Verlag, Bonn.

Troll, C. 1972. Geoecology and the world-wide differentiation of high mountain ecosystems. In: Troll, C. (ed.), Geoecology of the high-mountain regions of Eurasia, pp. 1–16. Franz Steiner Verlag GMBH, Wiesbaden.

Troll, C. 1973. High mountain belts between the polar caps and the equator: their definition and lower limit. Arct. Alp. Res. 5 (3): part 2, A19–A27.

Tseplyaev, V.P. 1965. The forests of the USSR. Israel Program for Scientific Translations, Jerusalem (translated by Gourevitch, A.).

Tsukada, M. 1983a. Vegetation and climate during the last glacial maximum in Japan. Quat. Res. 19: 212–235.

Tsukada, M. 1983b. Late Quaternary spruce decline and rise in Japan and Sakhalin. Bot. Mag. Tokyo 96: 127–133.

Tsukada, M. 1985. Map of vegetation during the Last Glacial maximum in Japan. Quat. Res. 23: 369–381.

Turner, J. 1970. Post-Neolithic disturbance of British vegetation. In: Walker, D. & West, R.G. (eds.), Studies in the vegetation history of the British Isles, pp. 97–116. Cambridge Univ. Press, Cambridge.

Udvardy, M.D.F. 1975. A classification of the biogeographical provinces of the world. IUCN Occasional Paper No. 18, Morges, Switzerland.

Ueno, J. 1979. *Pinus* and pollen analysis. II. In: Horie, S. (ed.), Paleolimnology of Lake Biwa and the Japanese Pleistocene, Vol. 7, 348–357.

Uyeki, H. 1911. On the forest zones in Korea. Bull. Kor. Agr. Sci. 6 (in Japanese).

Uyeki, H. 1926. Korean timber trees Ginkgoales and Coniferae. Forestry Experiment Station, Seoul, Korea (in Japanese).

Uyeki, H. 1933. On the forest zones of Korea. Acta Phytotax. Geobot. 2: 73–85 (in Japanese).

Uyeki, H. 1941. Northern distribution limit of Korean evergreen broadleaved trees, Acta Phytotax. Geobot. 10: 89–93.

Van Steenis, C.G.G.J. 1962. The mountain flora of the Malaysian tropics. Endeavor, 21: 183–193.

Vaskovsky, A.D. 1964. A brief outline of the vegetation, climate and chronology of the Quaternary period in the upper reaches of the Kolyma and Indigirka Rivers and the northern coast of the Sea of Okhetsk. In: Michael, H.N. (ed.), The archaeology and geomorphology of Northern Asia, pp. 464–572. Univ. Toronto Press.

Velichko, A.A. 1984a. Late Pleistocene spatial palaeoclimatic reconstructions. In: Velichko, A.A., Wright, H.E. Jr. & Barnosky, G.W. (eds.), Late quaternary environments of the Soviet Union, pp. 261–285. Longman, London.

Velichko, A.A. 1984b. Introduction. In: Velichko, A.A., Wright, H.E. Jr. & Barnosky, C.W. (eds.), Later quaternary environments of the Soviet Union, pp. 23–27. Longman, London.

Wade, J.E. & Hewson, E.W. 1979. Trees as a local climatic wind indicator. J. Applied Meteorol. 18: 1182–1187.

Walter, H. 1963. Climate diagram as a measure to comprehend the various climatic types for ecological and agricultural purposes. In: Rutter, A.J. & Whitehead, F.H. (eds.), The water relation of plants, pp. 3–9. Blackwell Sci. Publ., London.

Walton, D.W.H. 1982, Instruments for measuring biological microclimates for terrestrial habitats in polar and high alpine regions: a review. Arct. Alp. Res. 14: 275–286.

Walton, K.C. 1984, Snowdonia's arctic-alpine plants. Wildlife in Snowdonia 5: Inst. Terrestial Ecology, NERC.

Wang, C.W. 1961, The forests of China with a survey of grassland and desert vegetation. Maria Moors Cabot Foundation Publ. Ser. No. 5, Harvard Univ., Cambridge, Mass.

Wang, P.X. 1984. Progerss in Late Cenozoic palaeoclimatology of China. In: Whyte, R.O. (ed.), The evolution of the East Asian environment, Vol. I, pp. 165–187. Univ. Hong Kong.

Wardle,. P. 1965. A comparison of alpine timberlines in New Zealand and North America, N.Z. J. Bot. 3: 113–135.

Wardle, P. 1968. Engelmann spruce (*Picea engelmanii*) at its upper limits of the Front Ridge, Colorado. Ecol. 49: 483–495.

Wardle, P. 1971. An explanation for alpine timberline. N.Z. J. Bot. 9: 371–402.

Wardle, P. 1974. Alpine timberlines. In: Ives, J.D. & Barry, R.G. (eds.), Arctic and alpine environments, pp. 371–402. Methuen, London.

Wardle, P. 1977. Japanese timberlines and some geographic comparisons. Arct. Alp. Res. 9: 249–258.

Wardle, P. 1981a. Is the alpine timberline set by physiological tolerance, reproductive capacity or biological interaction? Proc. Ecol. Soc. Aust. 11: 53–66.

Wardle, P. 1981b. Winter dessication of conifer needles simulated by artificial freezing. Arct. Alp. Res. 13: 419–423.

Waring, R.H. & Franklin, J.F. 1979. Evergreen coniferous forests of the Pacific Northwest. Science, 204: 1380–1386.

Warren Wilson, J. 1957. Observations on the temperatures of arctic plants and their environment. J. Ecol. 45: 499–531.

Warren Wilson, J. 1959. Notes on wind and its effects in arctic-alpine vegetation. J. Ecol. 47: 415–427.

Watts, D. 1966. Man's influence on the vegetation of Barbados 1627 to 1800. Univ. Hull Occasional Papers in Geography No. 4, Hull.

Watts, D. 1971. Principles of biogeography. McGraw-Hill, London.

Watts, D. 1984. The spatial dimensions in biogeography. In: Taylor, J.A. (ed.), Themes in biogeography, pp. 25–62. Croom Helm, London.

Watts, D. 1987. The West Indies: patterns of development, culture and environment change since 1492. Cambridge Univ. Press, Cambridge.

Watts, D. & Kong, W.S. 1987. Biogeographic division and environment in Korea I. Pacific Sci. Ass. 16th Congr. Abstr. Vol. 1, p. 243.

Webb, D.A. 1965. Some difficulties in the establishment of phytogeographical divisions. Rev. Roum. Biol. Ser. Bot. 10: 33–41.

Wehrmeister, R.R. & Bonde, E.K. 1977. Comparative aspects of growth and reproductive biology in Arctic and Alpine populations of *Saxifraga cernua*. Arc. Alp. Res. 9: 401–406.

Wendland, W.M. & Bryson, R.A. 1974. Dating climatic episodes of the Holocene. Quat. Res. 4: 9–24.

Werger, M.J.A. & Springer, J.T.A. 1982. Comparison of floristic and structural classification of vegetation. Vegetatio 50: 175–183.

Westhof, V. 1967. Problems and use of structure in the classification of vegetation, Acta Bot. Neer. 15: 485–511.

Whitmore, T.C. 1984. A vegetation map of Malesia at scale 1:5 million. J. Biogeogr. 11: 461–471.

Whittaker, R.H. 1962. Classification of natural communities. Bot. Rev. 28: 1–239.

Whittaker, R.H. 1965. Dominance and diversity in land communities. Science 147: 250–260.

Whittaker, R.H. 1978. Approaches to classifying vegetation. In: Whittaker, R.H. (ed.), Classification of plant communities, pp. 1–31. Dr. W. Junk, The Hague.

Wielgolaski, F.E. 1986. Production-ecology and functioning of polar and alpine ecosystems. In: Polunin, N. (ed.), Ecosystem theory and application, pp. 12–34. John Wiley & Sons, Chichester.

Wielgolaski, F.E. & Kärenlampi, 1975. Plant phenology of Fennoscandia tundra areas. In: Wielgolaski, F.E. (ed.), Fennoscandia Tundra Ecosystems pp. 94–102. Springer-Verlag, Berlin.

Wilson, A. 1956. The altitudinal range of British plants. T. Buncle & Co., Arboath, U.K.

Wilson, E.H. 1916. The conifers and taxads of Japan. Publ. Arnold Arbo. No. 8, Cambridge Univ. Press.

Wilson, E.H. 1918. The vegetation of Korea. Trans. Roy. Asiatic Soc. 9: 1–16.

Wilson, E.H. 1919. A phytogeographical sketch of the ligneous flora of Korea. J. Arnold Arb, 1: 32–43.

Wilson, E.H. 1923. The Rhododendrons of Northeastern Asia exclusive of those belonging to the subgenus *Anthodendron*. J. Arnold Arbor. 4: 33–56.

Wissmann, H.V. 1956. On the role of nature and man in changing face of the dry belt of Asia. In: Thomas, W.L. Jr. (ed.), Man's role in the changing face of the earth, pp. 278–303. Univ. Chicago Press, Chicago.

Wolfe, J.A. 1978. A palaeobotanical interpretation of Tertiary climates in the Northern Hemisphere. Amer. Sci. 66: 694–703.

Wolfe, J.A. 1980. Tertiary climates and floristic relationships at high latitudes in the Northern Hemisphere. Palaeogeogr., Palaeoclimatol., Palaeoecol. 30: 313–323.

The World Resources Institute and The International Institute for Environment and Development 1986. World resources 1986. Basic Books, New York.

Wright, H.E. Jr. & Barnosky, C.W. 1984. Introduction to the English Edition. In: Velichko, A.A., Wright, H.E. Jr. & Barnosky, C.W. (eds.), Late quaternary environments of the Soviet Union, pp. xiii–xxii. Longman, London.

Wu, C.L. 1956, The taxonomic revision and phytogeographical study of Chinese pines. Acta Phytotax. Sin. 5: 161–164 (in Chinese).

Xu, R. 1984. Changes of the vegetation in China since the Late Tertiary. In: Whyte, R.O. (ed.), The evolution of the East Asian environments, Vol. II, pp. 426–432. Univ. Hong Kong.

Xu, Z.B., Dai, H. & Li, X. 1987. Rational management of broadleaved *Pinus koraiensis* forest and improvement of woodland productivity in north-east China. In: Yang, H.X., Wang, Z., Jeffers, J.N.R. and Ward, P.A. (eds.), The temperate forest ecosystem, pp. 59–67. Institute of Terrestrial Ecology, Cumbria, U.K.

Yamanaka, M. 1976. Vegetational history since the Late Pleistocene in northeast Japan. Ecol. Rev. 19: 1–36.

Yang, I.S. 1958. Flora of Chin Island. Theses Coll. Kyungbuk Univ. 2: 323–349 (in Korean).

Yang, I.S. 1961. Flora of Kyungbuk Province. Theses Coll. Kyungbuk Univ. (Sci.) 5: 1–65 (in Korean).

Yang, I.S. 1969. Flora of Keojae Island. Theses Coll. Kyungbuk Univ. (Sci.) 13: 63–81 (in Korean).

Yang, I.S. & Kim, W. 1970. Flora of Namhae Island. Kor. J. Pl. Tax. 2, 1–10 (in Korean).

Yang, I.S. & Kim, W. 1971. The distribution of evergreen tree in Wan Island. Kor. J. Pl. Tax. 3: 29–32 (in Korean).

Yang, I.S. & Kim, W. 1972. Conspectus relation between the distribution of evergreen broadleaved trees and the climatic factor in southern area of Korea. Kor. J. Pl. Tax. 4: 11–18 (in Korean).

Yasuda, Y. 1978. Prehistoric environment in Japan: palynological approach. Sci. Rep. Tohoku Univ. 17th Ser. Geogr. 28: pp. 117–281.

Yasuda, Y., Kim, J.M., Lee, S.T. & Yim, Y.J. 1980. Environmental changes and agricultural origin in Korea. Jap. Min. Educ. Overseas Res. Rep. pp. 1–19 (in Japanese).

Yasuda, Y. & Narita, K. 1981. Calculated thermal index maps for the reconstruction of past vegetation since the last glacial age in Japan. Geogr. Rev. Jap. 54: 369–381.

Yi, K.B. 1984. A new history of Korea (Trans. by Wagner, E.W.). Ilchogak, Seoul.

Yim, K.B. 1968. Principles of silviculture. Hyangmunsa, Seoul (in Korean).

Yim, Y.J. 1977. Distribution of forest vegetation and climate in the Korean peninsula III. Distribution of tree species along the thermal gradient. Jap. J. Ecol. 27: 177–189.

Yim, Y.J. & Baik, S.D. 1985. The vegetation of Mt. Sorak. Chungang Univ. Press, Seoul (in Korean).

Yim, Y.J. & Jeon, E.S. 1980. Distribution of naturalised plants in the Korean peninsula. Kor. J. Bot. 23: 69–83 (in Korean).

Yim, Y.J. & Kira, T. 1975. Distribution of forest vegetation and climate in the Korean peninsula I. Distribution of some indices of thermal climate. Jap. J. Ecol. 25: 77–88.

Yim, Y.J. & Kira, T. 1977. Distribution of forest vegetation and climate in the Korean vegetation IV. Zonal distribution of forest vegetation in relation to thermal climate. Jap. J. Ecol. 27: 269–278.

Yoo, E.K. 1970. Recent distribution of Globigerina pachyderma and interpretation of Upper Cenozoic climatic changes. J. Geol. Soc. Kor. 6: 119–127 (in Korean).

Yoon, H.S. 1982. Flora of Soan Island. Ko. J. Ecol. 5: 211–221 (in Korean).

Yoshida, N. 1986. Thermal characteristics of the ground on a wind-blown slope in the alpine zone of Mt. Kisokomagatake. Geogr. Rep. Tokyo Metrol. Univ. 21: 201–208.

Yoshing, M.M. 1964, Some local characteristics of the winds as revealed by wind-shaped trees in the Rhone Valley in Switzerland. Erdkunde 18: 28–39.

Yoshino, M.M. 1966a. Some aspects of air temperature climate of the high mountains in Japan. Jap. Progr. Climat. Nov. 271–27.

Yoshino, M.M. 1966b. Wind-shaped trees as indicators of micro and local climatic wind situation. Bioclimat. 2, 997–1005.

Yoshino, M.M. 1973. Wind-shaped trees in the subalpine zone of Japan. Arct. Alp. Res. 5: (2), A115–A126.

Yoshino, M.M. 1975. Climate in a small area. Univ. Tokyo Press, Tokyo.

Yoshino, M.M. 1978. Altitudinal vegetation belts of Japan with special reference to climatic conditions. Arct. Alp. Res. 10, 449–456.

Young, C.C. 1950. The Plio-Pleistocene boundary in China.

Intl. Geol. Congr. Great Britain 1948, Rep. 9, pp. 115–125.

Yurtsev, B.A. 1972. Phytogeography of Northeastern Asia and the problems of Transveringian floristic interrelations. In: Graham, A. (ed.), Floristics and palaeofloristics of Asia and Eastern North America, pp. 19–54. Elsevier Publ., Amsterdam.

Zehinder, L.R. & Lanpher, P.O. 1966. The influence of temperature and light on the cold hardiness of *Taxus cuspidata*. Proc. Amer. Soc. Hort. Sci. 89: 706–713.

Zhang, L.S. 1984. Reconstruction of the climate of the Late Pleistocene Ice Age in East Asia. In: Whyte, R.O. (ed.), The evolution of the East Asian environment, Vol. 1, pp. 252–270. Univ. Hong Kong.

Zhao, S.Q., 1986. Physical geography of China. Science Press, Beijing.

Zimma, R.P. 1973. Upper forest boundary and the subalpine belt in the mountains of the southern USSR and adjacent countries. Arctic & Alpine Res. 5: (2), A29–A32.

Zon, R. & Sparhawk, W.N. 1923. Forest resources of the world. McGraw-Hill, New York.

Zwinger, A.H. & Willard, B.E. 1972. Land above the trees: a guide to American alpine tundra. Harper & Row, New York.

Index of Subject & Place Names

Adaptation 38, 97, 127, 139, 155, 164, 168, 172, 173, 179, 180
Afforestation 76
Alpine plants 22, 23, 29, 30, 38, 39, 42, 43, 82, 83–104, 105–139, 140–172, 173, 174, 175–180
Anthropogenic disturbance 56, 64, 67, 68, 69, 72, 159, 169, 170, 172, 175, 176
Artic-alpine plants 22, 23, 29, 30, 38, 39, 42, 43, 64, 65, 82, 83–104, 105–140, 162, 173, 174–180

Bamboos 11, 13, 14, 18, 22, 23, 24, 27, 28, 29, 30, 31, 34, 35, 39–44, 72, 73, 76, 173, 174
Bamboo, large 72, 73, 76
Bamboo, small 72, 73, 74
Barley 75
Boxtree 72, 73
Barnyard grass 75
Buckwheat 75

Charcoal 55, 56, 58, 60, 70
Chestnut 73, 77
China 4, 10, 16, 23, 31, 50, 53, 55, 61–65, 67, 87, 94, 98, 107, 109–116, 147, 174
Chungcheonbong 151, 152, 153, 154, 155, 156, 157, 160–162, 168, 171, 172, 181, 182
Climate 5–7, 67, 128–137, 141–143, 146, 167–168
  historical records 70, 71, 72–77
  lapse rates 130–133, 167–168
  on Mt Sorak 141–143, 146
  temperature 128–137, 138, 146, 158–161, 167, 168, 172
Climatic amelioration 120, 139, 150, 171, 173–176, 179, 181
Cold hardiness 130–137, 138
Cold tolerance 18, 139
Continental shelf 65
Citrus 72, 73, 74
Cotton 72, 73, 78
Cryogenic activity 61

Deforestation 68–70, 72, 75
Distributions
  alpine plants 98–103, 105–140
  arctic-alpine plants 92–98, 105–140
  continuous 92–98
  disjunctive 92–98, 99, 104, 166, 171, 175, 178, 181
  endemic 100–103
Dwarfing 126, 127

East Asia 53, 60, 61, 63, 64, 85, 87, 92–98, 99–103, 113, 122, 130, 132, 138, 141, 145, 147, 154, 158, 162, 164, 165, 169, 172–174, 176–178
East Sea (Sea of Japan) 1, 4, 5, 62, 75, 111, 142, 162, 176
Endemic species 11, 13, 20, 22, 23–25, 31, 35, 36, 37, 42, 89, 90, 91, 92, 93, 100–103, 105, 107, 111, 113, 121, 133–139, 143, 144, 146, 147, 149, 150, 170, 171, 174, 177–181
Endemism 13–15, 17, 43, 103
Environment 1–10, 128, 166–170
Environmental gradients 122
Environmental limitations 122
Evergreen broadleaved plants 10, 11–45, 50, 68, 69, 82, 88, 89, 103–140, 145, 146, 158, 173–175, 179, 180
Evergreen coniferous plants 11–45, 50, 56, 60, 82, 88, 89, 105–140, 146, 148–151, 173–176, 179, 180
Evergreenness 123
Evolution 97, 144, 176

Famine 70, 71, 72
Fauna 53–55, 56, 57, 58, 60, 141, 147
Floristic reports 11, 12, 31, 107
Foxtail millet 75

Ginseng 72, 73
Glaciation 60–64, 95, 138, 175
Greenhouse effect 122, 170, 172

Hemp 72, 73
Hokkaido 63, 64, 107, 109–117, 154, 158, 165
Holocene period 47, 49, 66–76, 82, 173, 174
  Upper Holocene 70–76
Homo sapiens sapiens 58, 59, 60

Islands
  Anma 16, 34
  Anmyon 16, 33
  Bangchuk 16, 33
  Bogil 16, 33
  Changsan 16, 33
  Cheju 1, 2, 10, 16, 28, 33 44, 61, 65, 72, 73, 88, 92, 93, 95, 98, 100–102, 107, 113, 117, 118, 121, 122, 128, 131, 138, 139, 147, 158, 174, 175, 178–180
  Chin 16, 33
  Chinwoo 17, 34
  Chisim 17, 34
  Chuja 16, 34
  Chungsan 17, 34

Daechung 16, 33
Daemo 17, 34
Docho 16, 34
Dok 16, 33
Dolsan 17, 34
Dukjeok 16, 33
Eochung 16, 33
Hatae 16, 34
Hong 17, 33
Hucksan 16, 33
Imja 16, 34
Juk 16, 33
Kal 17, 34
Keojae 16, 33
Keomoon 16, 33
Kimjae 16, 33
Kokeum 17, 34
Kuk 17, 34
Kwang 17, 34
Kwangha 16, 33
Mal 16, 33
Mokpo 16, 33
Mooie 16, 33
Namhae 17, 34
Nowha 16, 34
Oenaro 17, 34
Oeyeon 16, 33
Ong 16, 33
Paikryung 16, 33
Pian 16, 34
Pijin 16, 33
Pikeum 16, 34
Poong 16, 33
Pyung 17, 34
Pyunsan 16, 34
Sangil 17, 34
Seonyoo 16, 33
Seunghong 16, 33
Shinji 17, 34
Shinsi 16, 34
Soan 16, 34
Sohucksan 16, 34
Songjiho 16, 33
Sonjuk 16, 33
Ulreung 1, 2, 10, 17, 74, 111, 131, 147
Wan 16, 33
Wi 16, 34
Wooie 16, 34
Yami 16, 34
Yeonwha 17, 34
Yeoseo 17, 34
Yungjong 16, 33
Yungkwang 16, 33

Japan 1, 10, 23, 31, 50, 53, 61, 62, 64, 65, 67, 69, 87, 92, 94, 107, 109–117, 132, 134, 154, 158, 164, 169, 174, 176
Jujube 72, 73 74

Kampfzone 157

Korea
adjacent oceanic waters 4, 5, 7, 64
archeology 64, 67, 69–70
biogeographic divisions 9, 13–43, 77–81, 173, 174
Cenozoic period 47, 70
geology 1, 2, 3, 67, 141
geomorphology 1
Mesozoic period 1, 47–49, 141
meteorological stations 130–132
palaeogeography 48
Palaeozoic period 46, 47
Pleistocene period 1, 31, 38, 42, 47, 49, 50, 53–70, 76–81, 84–87, 95, 97, 118, 138, 139, 148, 150, 168, 171, 173–179, 182
Quaternary period 1, 2, 53–76, 78–82, 85, 103, 141
Tertiary period 1, 2, 47, 49–53, 85
Korean pine 72, 73, 74
Korean Straits 1, 88, 176
Korean War 76
Krummholz 16, 64, 84, 98, 123, 141, 148, 154, 157, 158, 160, 161–163, 169, 171, 172, 181, 182

Lacquer 72, 73
Lake Younglang 64–66, 68, 78
Little Ice Age 72, 73
Loess 64

Machilus 72, 73
Maize 75
Manchuria 1, 10, 69, 107, 109–116, 164, 165
Mandarin orange 72–74
Mesozoic period – *see* Korea, Mesozoic period
Meteorological stations – *see* Korea, Meteorological stations
Migration of plants 85–87, 97, 104, 136, 139, 144, 171, 176, 178, 181
Millet 75, 79
Miocene flora 50–53, 84, 175
Mongolia 31, 55, 87, 94, 109, 111–113, 164, 165
Mountains
Biseul 16, 33, 109
Booso 16, 33
Changsoo 16, 32, 107, 108, 109, 131
Chayoo 14, 32, 107, 108, 109, 113, 116, 131
Chiak 16, 32, 107, 108, 109, 110, 131
Chilbo 13, 32, 107, 108, 109, 111, 113, 114, 131
Chiri 4, 6, 16, 29, 33, 62, 107, 108, 109, 110, 113, 115, 131, 141, 142, 143, 145, 147, 174, 180
Chogae 16, 33
Choijung 16, 33, 109
Chuae 14, 32, 107, 108, 109, 111, 112, 131
Chuheul 16, 32
Chuwang 16, 33, 107, 109
Daedun 16, 33, 108
Daeryong 16, 32, 107, 109
Daesung 16, 32, 107, 109
Dancheon 13, 32, 113
Dodamsambong 16, 33
Doduk 16, 33
Dukyoo 16, 32, 107, 108, 109, 110, 113, 131
Halla 1, 2, 4, 5, 24, 28, 33, 42, 61, 65, 88, 107, 108, 109,

110, 113, 117, 126, 131, 141, 142, 143, 147, 158, 174, 175, 180
Haram 14, 32, 107, 108, 109–111, 131, 164
Hoochi 13, 32, 107, 108, 109, 110, 113, 114, 116, 131
Hyangro 16, 33, 109
Ilwol 16, 32, 107, 108, 109, 131
Ipam 16, 33
Jung 14, 32, 108, 110, 131
Ka 16, 32
Kachil 16, 32
Kaeryong 16, 33, 107, 109, 131
Kaji 16, 32, 107, 108, 109, 113, 131
Kari 16, 32, 109, 110
Kaya 16, 32, 107, 109, 113, 131
Koebang 14, 32, 107, 109
Kapsan 13, 32
Kaya 107, 108
Keelju 13, 32
Keumjung 16, 33, 61
Keumkang 14, 32, 107, 108, 109–111, 113, 116, 131, 147, 164
Keumo 16, 33
Keumpae 13, 32, 107, 108, 109–111, 113, 114, 116, 131
Koowol 16, 32, 108, 109, 131
Kunja 16, 32, 147
Kwanap 16, 32
Kwangnung 16, 32
Kwanmo 1, 13, 32, 61, 63, 65, 108, 111, 114–116, 158, 164
Manduk 16, 33
Mantap 13, 16, 33, 107, 108, 109–114, 116, 131, 164
Moodeung 16, 33, 108, 113, 131
Moosan 13, 22, 32, 113, 114, 116
Myohyang 14, 32, 107, 108, 109–112, 131, 164
Myulak 16, 32, 107, 108, 131
Myungcheon 17, 34
Myungji 14, 32, 107, 109
Naejang 16, 33, 107, 108, 131
Nampodae 1, 61, 63, 65, 164
Nangrim 14, 32, 108, 109–113, 131, 164
Odae 14, 32, 107, 108, 109, 110, 116, 131, 147
Paikdoo 1, 2, 4, 13, 16, 17, 22, 31, 32, 37, 61, 63, 65, 88, 98, 101, 102, 107, 108, 109, 111, 112–116, 130, 158
Paikyang 16, 33, 107, 108, 131
Palgong 16, 32, 107–109, 131
Peenanduk 14, 32, 107, 108, 109, 111, 131
Pirae 14, 32, 107, 108, 109–113, 131, 164
Pohyun 16, 32, 110, 164
Pukhan 16, 32, 107
Pukpodae 1, 61, 63, 65, 164
Pungsan 17, 34, 113
Pyungkang 17, 34
Ro 13, 22, 32, 108, 109–114, 116, 131, 164
Samak 16, 32, 109, 147
Sasoo 14, 32 107, 108, 109–113, 131, 164
Seungjuk 14, 32, 107, 108, 109–111, 131, 164
Sobaik 16, 18, 32, 107–110
Sokri 16, 32, 107, 108, 109, 131, 147
Solryong 1, 61, 63, 65
Songin (Ulreung) 108, 111, 131
Songjiin 14, 32, 107, 108, 110, 131
Sooyang 16, 32, 108, 109, 131
Sorak 2, 4, 23, 32, 107, 108, 109–112, 116, 122, 131, 140, 141–172, 175, 180–182
Sori 16, 32, 107, 109, 131
Taebaik 16, 32, 107, 108, 109–111, 131
Taeji 14, 32, 107, 108, 109, 110, 131
Wanu 17, 34
Whaak 16, 32, 107, 108, 109, 111, 131
Wolak 16, 32, 147
Wolchul 16, 33
Woonmoon 16, 33
Youngmoon 16, 32, 107, 108, 109, 131
Mountain ranges 3, 4, 141, 173
Mulberry 72, 73, 74

Nunataks 87
Nutmeg 72, 73

Oats 75

Palaeoclimates 47–50, 52, 53, 60–65, 67, 68, 69, 70, 71, 72–78, 87
Palaeozoic – *see* Korea, Palaeozoic
Paper mulberry 72, 73, 75
Patterned ground 1, 61, 164
Periglacial activity 1, 61, 62, 64, 65, 82
Permafrost 86
Persimmon 72
Physiological drought 159
Plant migration – *see* Migration of plants
Plant physiognomy 25–30, 38, 41–43, 89, 90, 91, 105, 121–128, 149, 150, 163, 164, 169, 173
Pollen data 55, 56, 58–60, 64–66, 67, 68, 70, 72, 80, 81, 175, 176
Pleistocene – *see* Korea, Pleistocene period
Pomegranate 72, 73

Quaternary – *see* Korea, Quaternary period

Ramie 72, 73
Ranges of plants
altitudinal 30–34, 35, 36, 38, 39, 40, 41, 85, 105–120, 129–137, 139, 140, 149–151, 155, 156, 157, 164, 166, 171, 172
latitudinal 30–37, 105–120
northernmost limits 144, 147, 170, 171, 180
southernmost limits 95, 104, 120, 141, 143, 145–147, 163, 167, 169–171, 176, 180
Refugia 38, 85, 86, 87, 88, 95, 97, 103, 104, 138, 148, 173, 176, 179
Rice 67–70, 75

Sea-level changes 67, 69, 88, 176
Siberia 1, 55, 61–64, 87, 88, 97, 103, 107, 109–112, 115, 116, 132, 168
Snow-line 1, 61, 62, 86
Soil erosion 141
Solifluction 1, 61, 84
South Sea 4, 62
Soybean 72, 73, 74
Species numbers per biogeographic region 17–19, 22, 30, 147

Steppe 64, 65, 66, 82, 175
Subalpine zone 84, 133, 134, 137, 141, 148–157, 159, 160, 162, 164, 166, 167, 173, 174, 181, 182
Summit effect 161, 182

Taecheongbong 141, 142, 147, 149, 151, 152, 153, 154, 155, 156, 157, 159–162, 167–169, 171, 172, 181, 182
Taiwan 64, 94
Tea plant 72, 73, 74
Tertiary period – *see* Korea, Tertiary period
Tertiary vegetation 85
Timberline 9, 16, 31, 36, 37, 83, 98, 101, 123, 157, 158, 161
Tobacco 73
Top phenomenon 161, 162, 182
Tree line 84, 86, 141, 147, 148, 157–161, 171, 172, 181
Tundra 64, 65, 66, 82–84, 86, 88, 164, 175

Ultraviolet radiation 127, 128

Vegetation changes, historic time 72–76, 77, 78, 79

Weeds 76
West Sea (Yellow Sea) 1, 4, 5, 62, 64, 65, 69
Wheat 75

Index of Plant Names

Abeliophyllum sp. 11, 17
Abeliophyllum distichum 25
Abies sp 9, 34, 43, 44, 50, 56, 58, 60, 64, 65, 67, 68, 70, 72, 77, 80, 81, 82, 85, 123, 130, 161, 175, 176
Abies holophylla 20, 21, 23, 28, 35, 40, 41, 88, 98, 106, 107, 119, 121, 122, 124, 130, 132–138, 143, 149, 178
Abies koreana 13, 19, 20–24, 31, 35–37, 40–42, 88, 101, 105–107, 109, 113, 115–117, 119, 121, 122, 124, 126, 127, 130, 133–139, 158, 174, 179
Abies nephrolepis 19, 38–40, 98, 106, 107, 109–111, 119, 121, 122, 124, 129, 132–138, 143, 147–152, 154–156, 158, 160–163, 171, 172, 178, 181, 182
Abies sachalinensis 107
Abies veitchii 107
Abietaceae 89, 90, 105, 106, 138
Acanthopanax chiisanensis 146
Acer sp. 9, 24, 43, 44, 50–53, 162
Acer formosum var. coreanum 143
Acer ginnala 151, 154
Acer mono 143
Acer pseudo-sieboldianum 143, 148
Acer tschnoskii var. ruprides 148–151, 154–156, 172, 182
Aceraceae 50, 58, 60
Achillea paektussanense 100
Aconitum napiforme 148
Adenophora sp. 65
Adenophora divaricata 100
Adenophora glandiflora form. alba 144, 146
Adenophora lamarkii form. albiflora 100, 144, 146, 148, 155
Adenophora lillifolia 98
Adenophora triphylla var. canenscens 98
Aesculus sp. 50, 51, 53
Agrostis canina subsp. trinii 93, 95
Agrostis flaccida 100
Agrostis hiemalis 98
Alangium sp. 50–53
Aletris fauriae 146
Alliaceae 90–92
Allium schoenoprasum 93, 95
Allium senescens 98, 145
Allium taquetii 145, 148
Allium victorialis var. platyphyllum 146, 148
Alnus sp. 51–53, 58, 60, 64, 67, 68, 70
Alnus fruticosa var. mandshurica form. normalis 101, 145, 146, 148–151, 155, 156
Alnus hirsuta var. japonica 99
Alnus maximowiczii 98, 154, 158
Alnus tinctoria var. typica 99
Alnus vermicularis 99, 102
Alsinaceae 90–92
Amygdalaceae 89
Anaphalis sinica subsp morii 102
Andromeda sp. 77, 80
Andromeda polifolia 19, 26, 43, 105, 106, 113, 124, 125, 127, 138, 178
Andromeda polifolia form. acerosa 19, 26, 43, 99, 105, 106, 114, 124, 126, 127, 138
Androsace capitata 37
Androsace cortusaefolia 144, 146
Anemone dichotoma 93, 95
Anemone glabrata 146
Anemone koraiensis 144, 146
Anemone narcissiflora 98, 148, 149
Antennaria dioica 100
Apiaceae 90, 91
Aquifoliaceae 55, 56 58
Arabia lyrata var. kamschatica 100
Arabis serrata var. hallaiensis 102
Araceae 90, 91
Araliaceae 58, 60, 89, 90
Arceria nana 154
Arctous alpina 88
Arctous ruber 98, 145, 146, 148
Artemisia sp. 56, 59, 60, 65, 67, 68, 70, 175
Artemisia angustissima 102
Artemisia laciniata 100
Artemisia sinensis 100
Artemisia stelleriana 100
Artemisia triroba 100
Aruncus aethusifolius 102
Asperula lasiantha 146
Asperula platyglium var. alpina 100
Aspidaceae 90–92
Asteraceae 89–91
Astralagus adsurgens var. alpinus 102
Astralagus davuricus 98
Astralagus membranaceus 98
Astralagus membranaceus var. alpinus 102
Astralagus membranaceus var. mandshuricus 102
Astralagus setsureianus 102
Astralagus uliginosus 100
Athyrium crenatum var. glabrum 100

Athyrium melandepsis 100
Athyrium rupestre 98
Aucuba japonica var. typica 22, 28, 41, 42
Avena sativa 75

Benzoin obtusilobum 55, 56
Berberidaceae 89, 90
Berberis amurensis var. latifolia 98
Berberis amurensis var. quelpartensis 102
Berberis koreana 101, 146
Berberis poiretii var. angustifolia 100
Bergenia pacifica 100
Betula sp. 43, 44, 50–53, 55, 56, 58, 60, 64, 68, 158, 181
Betula chinensis 98, 149
Betula chinensis var. angusticarpa 99
Betula cyclophylla 99
Betula costata 98, 149, 150, 154, 155
Betula ermanii 36, 64, 148, 158
Betula ermanii var. acutifolia 98, 143, 148, 149
Betula ermanii var. ganjuensis 98
Betula ermanii var. genuina 98, 147, 150–152, 154–157, 160–163, 171, 172, 181, 182
Betula ermanii var. incisa 98
Betula ermanii var. parvifolia 101
Betula fusenensis 101
Betula gmelini 99
Betula microphylla var. coreana 101
Betula middlendorfii 156
Betula paisanensis 101
Betula platyphylla 39, 102
Betula platyphylla var. latifolia 98
Betula platyphylla var. mandshurica 99
Betula schmidtii 98
Betulaceae 50, 89, 90
Bifaria sp. 80
Bifaria japonica 19, 23, 40
Bilderdykia ochotensis 100
Bilderdykia pacifica 100
Bilderdykia vulgaris var. yezoensis 98
Biota sp. 80
Biota orientalis 19, 22, 23, 34, 35, 38–41
Biota orientalis var. stricta 19, 23, 40
Bistorta vulgaris 148
Bladhia crenata 20, 42
Bladhia crenata var. taquetii 19, 20, 28, 36, 40, 41
Bladhia crispa var. typica 20
Bladhia japonica var. typica 19, 20, 22, 23
Bladhia villosa var. typica 20, 40, 42
Boehmeria frutescens 72
Boraginaceae 92
Botrychiacae 92
Botrychium virginianum 95
Brasenia schreberi 95
Brassicaceae 90–92
Bromus japonicus 92, 93
Broussonetia kazinoki 72
Bryales 59, 60
Buplerum euphorbiodes 144, 146–148
Buxus sp. 35, 36, 80
Buxus koreana 13, 19, 22, 23, 27, 28, 31, 34, 38, 39, 72
Buxus koreana var. insularis 13, 20, 23, 28
Buxus microphylla 19, 20, 28, 40–42

Calamagrostis canadensis subsp langsdorfii 93, 95
Calamagrostis langsdorfii var. ramosa 102
Calocedrus sp. 51, 52
Caltha palustris var. sibirica 93, 95
Calypso bulbsa 93, 95
Camellia sp. 9, 44, 68, 79, 80
Camellia japonica 20, 23, 130
Campanulaceae 90, 91
Cannabis sativa 72
Caprifoliaceae 89, 90, 92, 105, 106, 138
Cardamine amaraeformis 146
Cardamine bellidifolia 100
Cardamine resedifolia var. morii 102
Cardamine vallida 101
Carex sp. 9
Carex capillaris 100
Carex fusanensis 146
Carex lasiolepis 145
Carex loliaceae 93, 95
Carex okamatoi 146, 147
Carex pseudo-curaica 93, 95
Carex rostrata 95
Carex rotundata 92, 93, 95
Carex subumbellatum var. koreana 100
Carpinus sp. 44, 50–53, 58, 60, 66–68, 70
Carpinus erosa 143
Carpinus laxiflora 143
Carpites sp. 52, 53
Castanea sp. 51–53, 55, 56, 58, 60, 68, 70
Castanopsis sp. 43, 44, 50, 52, 53, 68, 78, 81
Castonopsis cuspidata var. sieboldii 22
Castanopsis cuspidata var. thunbergii 22
Castanopsis latifolia 20
Celastrus sp. 50, 53
Celtis sp. 55, 56, 60, 66, 68
Cephalotaxus sp. 43, 80
Cephalotaxus koreana 19, 22, 23, 40, 41, 42
Cephalotaxus nana 19, 40
Cercidiphyllum sp. 50–53
Cercis sp. 52, 53
Chamaecyparis obtusa var. nepalensis 20, 22
Chamaedaphne calyculata 19, 26, 92, 95, 96, 106, 114, 124, 125, 127, 138, 177, 178
Chamaepericlinemum canadense 99
Chenopodiaceae 65, 67, 68, 92
Chenopodium sp. 59, 60
Chosenia sp. 44
Chosenia bracteosa 145
Chrysanthemum zawadskii 101, 148
Cinna latifolia 95
Cinnamomum sp. 43, 44, 51–53, 79, 81, 175
Cinnamomum camphora 20, 40, 41
Cinnamomum loueiri 20
Cirsium chanroenicum 146
Citrus sp. 44, 72
Citrus junos 72, 73
Citrus kinokuni 20

Citrus media var. sarcodactylus 20
Citrus natsudaidai 20
Citrus nobilis 20
Citrus sinensis 20
Citrus unshiu 20
Cladophebis sp. 48
Cladrastris sp. 50, 51, 53
Clematis chiisanensis 101
Clematis fusca 149
Clematis fusca var. coreana 144, 146
Clematis koreana 98, 148–150
Clematis mandshurica var. lancifolia 101
Clematis nobilis var. plea 101
Clematis ochotensis 100
Clematis subtriternata var. tenuifolia 101
Clematis trichotoma 146
Cnidium tachitoei 98, 148
Compositae 59, 60, 65, 68, 76
Comptonia sp. 51, 52
Convolvulaceae 76
Cornaceae 89
Cornus sp. 50, 51, 53
Cornus alba 99
Cornus controversa 143
Corylus sp. 50, 52, 67, 68, 70
Crassulaceae 90, 91
Crataegus maximowiczii 99, 144, 146
Crataegus pinnatifida form. betulifolia 101, 102
Crataegus pinnatifida var. partita 102
Crataegus pinnatifolia var. major 99
Cruciferae 76
Cryptocarya sp. 52, 53
Cryptogramma crispa 100
Cryptomeria japonica 20, 22
Cupressaceae 58, 60, 80, 81, 88, 90, 105, 106, 138
Cyclobalanopsis sp. 44, 51–53, 68, 78, 79, 81, 175
Cyclobalanopsis acuta 22, 28, 41, 42
Cyclobalanopsis gilva 20, 28
Cyclobalanopsis myrsinaefolia 2
Cyclobalanopsis stenophylla 20, 40
Cyclobalanopsis stenophylla var. latifolia 20
Cynodon dactylon 95
Cyperaceae 59, 60, 65, 67, 68, 90, 91, 92

Damnacanthus indicus 20, 28, 40, 42
Damnacanthus major 20, 23, 40, 42
Daphne kamtschatica 88
Daphniphyllum glaucescens 20, 40, 41
Dasiphora fruticosa var. coreana 99
Dasiphora fruticosa var. glabra 99
Deschampsia caespitosa 100
Deutzia coreana var. triradiata 146
Dianthus morii 102
Dianthus repens 100
Dianthus superbus var. speciosus 100
Diapensia sp. 36
Diapensia lapponica subsp obovata 5, 20, 23, 28, 40, 42, 43, 65, 88, 92, 95, 96, 104, 106, 107, 117, 119–122, 124, 125, 127, 129, 130, 132–136, 138, 139, 154, 174, 177–178
Diapensiaceae 89, 90, 92, 105, 106, 138
Dicalix lucida 20
Dicalix prunifolia 20, 40, 42
Digitaria ischaemum 92, 95
Diospyros sp. 50, 51, 53
Diospyros kaki 72
Dipsacaceae 90, 91
Distylium racemosum var. latifolium 13, 20, 23, 28
Draba incana var. multicaulis 100
Drosera sp. 59, 60
Drosera angelica 95
Droseraceae 92
Dryas sp. 88
Dryas octopetala var. asiatica 22, 92, 95, 97, 106, 113, 119–122, 124, 125, 127, 129, 132–134, 138, 158, 177, 179
Dryopteris sp. 52, 53
Dryopteris filix-mas 95

Echinochloa frusanentacea 98
Echinosophora sp. 11, 17
Echinosophora koraiensis 24, 25
Elaeagnaceae 58, 60
Elaeagnus glabra var. euglaba 20, 22, 23, 28, 41, 42
Elaeagnus macrophylla 19, 20, 22, 23, 28, 41, 42
Elaeagnus maritima 20
Elaeocarpus sp. 36
Elaeocarpus sylvestrus var. ellipticus 20, 28, 40, 41
Elatinaceae 92
Elatine triandra 92, 95
Eleutherococcus santicosus 98
Empetraceae 89, 90, 92, 105, 106, 138
Empetrum sp. 36, 77, 80, 130
Empetrum nigrum var. asiaticum 5
Empetrum nigrum var. japonicum 19, 20, 23, 26, 40, 42, 43, 65, 88, 92, 93, 95, 97, 104–107, 113, 114, 119, 121, 122, 124, 126, 127, 130, 133, 135, 136, 138, 139, 154, 174, 177–179
Engelhardia sp. 51, 52
Entada sp. 52, 53
Epipactis aphyllum 100
Equisitaceae 92
Equisitum sp. 52, 53, 55, 56
Equisitum hiemale 93, 95
Eranthis stellata 146, 147
Ericaceae 60, 78, 80, 89, 90, 92, 105, 106, 138
Eriogeron thunbergii subsp. glabratus 100
Eriophorum alpina 92, 93, 95
Eriophorum angustifolium 100
Eriophorum gracile 145
Eriophorum japonicum 98
Eriophorum mandshuricum 100
Eriophorum polystachis 98, 148
Eriophorum vaginatum 98
Erythronium japonicum 155
Euonymus sp. 43, 51–53, 79, 81
Euonymus chibai 20
Euphorbiaceae 90, 91
Euphrasia corealpina 102
Euphrasia mucronulata 102
Euphrasia multifolia 100
Eurya emarginata 20, 28, 36, 40, 42

Eurya japonica var. aurescens 20, 28
Eurya japonica var. integra 20, 23
Eurya japonica var. montana 19, 20, 22, 28, 40–42

Fabaceae 89–92
Fagaceae 58, 60, 89
Fagophyrum esculentum 75
Fagus sp. 50–53, 56, 58, 60, 158
Fatsia japonica 20, 28
Festuca sp. 9
Ficus sp. 50, 53, 79, 81, 174
Ficus erecta 22
Ficus erecta var. longependulata 20
Ficus nipponica 22, 28
Filipendula glaberrima 146
Forsythia ovata 144, 146
Fraxinus sp. 44, 50, 51, 53
Fraxinus rhychophyllus 143

Galarhoeus fauriei 102
Galeopsis bifida 92, 95
Galium aparine 95
Galium pusillum 102
Galium spuricum 92, 95
Gardenia jasminoides form. glandifolia 20
Gentiana algida 100
Gentiana axillariflora var. coreana 144, 146
Gentiana jamesii 98
Gentiana makinoii 98
Gentianaceae 90, 91
Geranaceae 90, 91
Geranium sp. 65
Geranium dahuricum 146, 148
Geranium eriostemon var. glabrescens 102
Geranium eriostemon var. hypoleueus 102
Geranium eriostemon var. megalanthum 100, 155
Geranium eriostemon var. orientale 100, 148
Geranium maximowizcii 100, 148
Geranium skikokianum var. quelpaertense 102
Ginkgo sp. 52
Ginkgoaceae 58
Glaux maritima 92, 95
Gleditsea sp. 51, 52
Gleditsea japonica var. koraiensis 146
Glycine max 72, 73
Glyptostrobus sp. 50–53
Gossypium nanking 72
Gramineae 59, 60, 65, 67, 68, 70, 76, 90–92
Grewia parviflora 147

Hackelia deflexa 95
Hammamelis sp. 52, 53
Hanabusaya sp. 11, 17, 90
Hanabusaya asiatica 24, 25, 90, 146, 148, 149
Hanabusaya asiatica form. alba 146
Hanabusaya asiatica form. albescens 144, 146
Hanabusaya asiatica form. intermedia 144, 146
Hanabusaya asiatica form. rubescens 146
Hanabusaya asiatica form. rubropurpurea 146
Hanabusaya latisepala 24, 90
Hedera sp. 35, 50, 79, 80, 81, 174
Hedera tobleri 20, 22, 23, 28, 40–42
Helleboraceae 90–92
Hemitrapa sp. 50–53
Herminium monorchis 98
Hieracium japonicum 98
Hieracium umbellatum 100
Hierochloe alpina 100
Homopterya nakaiana 102
Hordeum sativum 75
Hugeria japonica 100
Hydrangea sp. 50, 51, 53
Hymenophyllaceae 92
Hymenophyllum wrightii 95, 97
Hypear sp. 31, 80
Hypear tanakae 19, 23, 31, 38–40
Hypericaceae 90, 91
Hypericum vaniotii 98

Ilex sp. 43, 44, 50–53, 175
Ilex cornata var. typica 20, 22, 28
Ilex cornata var. microphylla 22
Ilex integra var. typica 22, 28, 41, 42
Imperata sp. 9
Iridaceae 59, 60
Iris uniflora var. caricina 146
Isopyrum leveileanum 146
Isopyrum raddeanum 146, 147

Juglandaceae 55, 56, 58, 60
Juglans sp. 43, 50–53, 58, 60, 66–68, 70
Juncaceae 90–92
Juncus brachyspathus 100
Juncus triglumis 100
Juniperus sp. 43, 44, 60, 78, 80–82, 85, 123, 158, 175
Juniperus chinensis var. sargentii 88
Juniperus communis var. montana 88, 92, 93, 95
Juniperus coreana var. rigida 13, 19, 20, 22, 23
Juniperus dahurica 59, 60
Juniperus sargentii 88
Juniperus sibirica 19, 22, 26, 31, 36, 38, 101, 105, 106, 113, 121, 124, 125, 130, 132, 133, 138, 179
Juniperus utilis 19, 31, 36, 38
Juniperus utilis var. longicarpa 13, 19, 23
Juniperus utilis var. modesta 13, 22, 23
Juniperus virginia 19

Kalopanax sp. 51–53
Kalopanax pictum 154, 155
Keteleeria sp. 51–53
Kobresia myosuroides 100
Koelreuteria sp. 52

Labiatae 89, 92
Lamium amplexicaule 92, 95
Larix sp. 43, 44, 50, 59, 60, 64, 65, 67, 68, 158
Larix gmelini 158
Larix olgensis 16, 39, 99, 102, 111
Larix olgensis var. koreana 101
Larix olgensis var. viridis 101

Lasiagrostis mongolica 100
Lastrea quelpartensis 98
Lathyrus maritimus 95
Lauraceae 58
Ledum sp. 31, 77, 80
Ledum palustre 88
Ledum palustre var. anguistrum 26, 102
Ledum palustre var. decumbens 19, 26, 36, 38, 92, 95, 96, 104, 106, 119, 121, 124, 125, 127, 132–135, 139, 173, 178
Ledum palustre var. diversipilosum 92, 95, 96, 104, 106, 114, 119, 121, 122, 124, 125, 127, 132–138, 154, 176, 178
Ledum palustre var. latifolium 19, 26, 99, 105, 106, 114, 124, 125, 127
Ledum palustre var. maximum 19, 22, 26, 31, 36, 38, 102, 105, 106, 113, 115, 122, 124, 125, 127, 138, 139, 173, 179
Ledum palustre var. minus 19, 26, 36, 38, 99, 105, 106, 115, 119, 121, 122, 124, 125, 127, 132–135, 138, 173, 178
Ledum palustre var. nipponicum 99
Ledum palustre var. procumbens 19, 26, 36, 38, 106, 173
Ledum palustre var. yezoense 19, 26, 99, 105, 106, 114, 124, 125, 127, 178
Leguminoseae 76
Lentibulariaceae 90–92
Leontice microrhyncha 146
Leontopodium coreanum 101, 144, 146
Leontopodium japonicum 98
Leontopodium leoiolepis 102, 146, 148
Lepidodendron sp. 47
Lespedeza bicolor var. japonica 143
Lespedeza thunbergii var. intermedia 146
Ligularia fischeri 98, 151, 152
Ligularia jamesii 102
Ligularis stenocephala 98
Ligustrum sp. 50, 58, 60, 79, 81
Ligustrum lucidum 20, 28, 40, 42
Ligustrum ovalifolium 22
Ligustrum quihoui var. latifolium 13, 20, 23
Liliaceae 60, 90, 91
Lilium cernuum 146, 147, 149
Lilium distichum 146
Lindera sp. 52, 53
Lindera obtusifolium 143
Linnaea borealis 19, 26, 92, 95, 96, 106, 116, 124, 125, 138, 139, 177, 178
Linopteris sp. 47
Liquidambar sp. 52, 53
Liriodendron sp. 51–53
Lloydia sp. 43
Loisekeuria procumbens 154
Lonicera caerula subsp edulis 98, 149, 150, 154, 155
Lonicera caerula var. emphyllocalyx 98
Lonicera caerula var. Iongibracteata 100
Lonicera caerula form. ovata 98
Lonicera caerula var. venulosa 100
Lonicera cerasoides 102
Lonicera chrysantha 98
Lonicera coreana 146
Lonicera monantha 102
Lonicera nigra var. barbinervis 102
Lonicera okamotoana 146
Lonicera ruprechtiana 100
Lonicera sachalinensis 98, 148–150
Lonicera sachalinensis var. barbinervis 100
Lonicera sachalinensis var. stenophylla 100, 102
Lonicera subhispidis 100
Lonicera tatarinowii var. leptantha 100
Luzula oligantha 98
Luzula rufescens 93, 95, 178
Luzula sudetica var. nipponica 100
Luzula wahlenbergii 95
Lycopodiaceae 58, 60, 90–92
Lycopodium alpinum 102
Lycopodium complanatum 92, 95
Lycopodium selgo 100
Lycopodium sitchense 100

Maackia sp. 43
Maackia fauraei 102
Machilus sp. 35, 43, 44, 50
Machilus japonica 20, 22, 23
Machilus thunbergii 22, 41, 42, 72
Machilus thunbergii form. obovata 20, 40, 41
Magnolia sp. 50, 51, 53
Magnoliaceae 58, 60
Malvaceae 76
Marsdenia tomentosa 20
Masakia sp. 35, 80
Masakia japonica 19, 22, 23, 28, 34, 40–42
Masakia japonica var. latifolia 19, 23, 28, 39–41
Masakia radicans 19, 20, 22, 23, 28, 40–42
Megaleranthis sp. 11, 17
Megaleranthis saniculifolia 24, 25, 146, 147
Melampyrum arcuatum 100
Melampyrum roseum var. alpinum 100
Melanthaceae 90, 91
Meliosma sp. 43
Menispernum dahuricum var. pilosum 145
Metasequoia sp. 50–53
Minuartia arctica var. major 102
Minuartia laricina 100
Miscanthus sp. 9
Moneses uniflora 92, 95
Monotrapa hypopitys subsp japonica 95, 97
Morus alba 72
Myrica sp. 36, 50, 52, 53, 68, 174
Myrica rubra 20, 28, 40, 41

Nandina domestica 20, 28
Nelumbo sp. 51–53
Neolitsea sp. 44
Neolitsea sericea 22, 41, 42
Neuropteris sp. 47
Nicotiana tabacum 73
Nilssonia sp. 48
Nymphaceae 58, 92

Odontoperis sp. 47
Oleaceae 60, 67, 68
Onchyopsis sp. 48
Ophioglossaceae 92

Ophioglossum vulgatum 95
Oplopanax elatum 146, 148–150, 154, 155
Orchidaceae 90–92
Orchis bracteata 100
Oryza sp. 67–70, 72
Osmanthus zentaroanus 20
Osmunda sp. 59, 60
Osmunda cinnamomea var. asiatica 95, 97
Osmundaceae 92
Ostrya sp. 50, 51, 53
Oxycoccus microcarpus 19, 26, 43, 92, 95, 106, 115, 124, 125, 127, 138, 139, 177, 178
Oxycoccus quadripetalus 19, 26, 43, 92, 93, 95, 106, 115, 124, 125, 127, 138, 139, 177, 178
Oxyria digyna 92, 95
Oxytropis sp. 59, 60
Oxytropis anertii 100, 101
Oxytropis koreana 100
Oxytropis stroloilacea 100

Paliurus sp. 52, 53
Panax schinseng 72
Panicum miliaceum 75
Papaver coreanum 102
Papaveraceae 90, 91
Papillionoidae 60
Parageum calithifolium var. nipponicum 100
Parrotia sp. 52, 53
Passania sp. 52, 53
Patrinia saniculaefolia 146
Pedicularis apodochila 100
Pedicularis glandiflora 100
Pedicularis lunaris 102
Pedicularis mandshurica 98, 146
Pedicularis nigrescens 102, 148
Pedicularis spicata 98
Pedicularis sudetica 100
Pedicularis verticillata 98, 146, 148
Pentactina sp. 11, 17
Pentactina rupicola 24, 102
Persicaria lyrata 102
Peucedanum coreanum 102
Phegopteris decursive-pinnata 147
Philadelphaceae 89
Philadelphus pekinensis 98
Phleum alpina 100
Phleum coreanum 102
Phoebe sp. 52, 53
Phragmites sp. 9
Phyllodoce caerula 19, 26, 31, 36, 38, 43, 92, 93, 95, 96, 104, 106, 116, 119, 121, 122, 124, 126, 127, 132–135, 138, 139, 158, 173, 176, 178
Picea sp. 9, 43, 44, 50–53, 59, 60, 64, 65, 67, 68, 77, 80–82, 123, 158, 161, 175, 176
Picea intercedens 13, 22, 101, 105, 106, 113, 124, 138, 139, 179
Picea jezoensis 19, 22, 29, 31, 38, 39, 88, 98, 102, 106, 107, 109, 111, 119, 121, 122, 124, 129, 130, 132–138, 158, 178
Picea koraensis 39, 101, 102, 105, 106, 111, 113, 118, 119, 121, 122, 124, 133–139, 179
Picea pungsanensis 13, 22, 23, 101, 105–107, 113, 124, 138, 139, 179
Picea schrenkiana 100, 106, 116, 124, 139, 178
Picea tonaiensis 13, 22, 101, 106, 113, 124, 138, 139, 179
Pinaceae 89, 90, 105, 106, 138
Pinguicula villosa 92, 95
Pinus sp. 9, 43, 44, 50–53, 55, 56, 58–60, 65, 67–70, 77, 78, 80–82, 123, 158, 175, 176
Pinus bungeana 19
Pinus densiflora 8, 9, 20, 130, 143
Pinus densiflora form. erecta 143, 146
Pinus densiflora var. globosa 19, 22
Pinus globosa 19
Pinus koraiensis 19, 20, 22, 28, 31, 35, 38–42, 59, 60, 72, 85, 87, 98, 106, 107, 109, 119, 121, 122, 124, 127, 130, 132–137, 143, 175, 178, 179
Pinus parvifolia 19
Pinus ponderosa 133
Pinus pumila 19, 26, 31, 36, 38, 39, 59, 60, 64, 80, 85, 87, 98, 105–107, 111, 119, 121, 122, 124, 129, 130, 132–138, 141, 143, 145–158, 160–172, 178, 181, 182
Pinus rigida 19, 20, 22
Pinus sibirica 59, 60, 85, 175
Pinus sylvestris 133
Pinus tabulaeformis 22, 99, 106, 107, 113, 124, 127, 138, 178
Pinus thunbergii 19, 22, 23, 35, 40–42, 130
Piper kadzura 20, 40, 41
Pittosporum sp. 44, 50
Pittosporum tobira 20, 22, 23, 41, 42
Platanthera maximowicziana 98
Platanthera sachaliensis 98
Platanus sp. 50, 52, 53
Platycarya sp. 43, 50, 51
Pleistoblastus simoni 22
Pleuropteropyrum ochreatum 100
Poa deschampsiodes 102
Poa glauca 92, 95
Poa kanboensis 100
Poa palustris 95
Poa trivialis 95
Polygonaceae 76, 90–92
Polypodiaceae 92
Polypodites sp. 50, 51
Polypodium vulgare subsp virginianum 95
Polystichum braunii 95
Poneorchis parciflora var. joo-iwokiana 100
Populus sp. 43, 50–53, 58, 60, 158
Populus maximowiczii 143
Populus maximowiczii var. barbinervis 99
Porana sp. 52
Potentilla bifurea var. glabrata 100
Potentilla dickinsii var. breviseta 155
Potentilla matsumurae 98, 148
Potentilla nivea 37
Potentilla nivea form. alpina 100
Potentilla nivea var. vulgaris 100
Primulaceae 90–92
Primula modesta var. faurinae 100
Primula modesta var. genuina 100
Prunus sp. 43, 51–53, 58, 60

Prunus buergeriana 100
Prunus densiflora 102
Prunus nakaii var. typica 149, 150, 154, 155
Prunus sachalinensis 88
Pseudolysimachion kiusiana var. diamantica 146
Pseudolysimachion rotundum var. coreanum 146
Pseudolysimachion rotundum var. subintegrum 146
Pseudosasa sp. 31, 55
Pseudosasa japonica 19, 38–41, 72
Pseudosasa japonica var. purpurascens 20
Pseudotsuga sp. 50–52, 59, 60, 175
Pterididaceae 60, 90, 91
Pteridium sp. 9, 50, 51
Pterocarya sp, 50–53
Pulsatilla nivalis 102
Punica granatum 72
Pyrola minor 95
Pyrola secunda 95
Pyrolaceae 90–92
Pyrus ussuriensis var. diamantica 144–146
Pyrus ussuriensis var. hakunensis 146

Quercus sp. 9, 43, 44, 50–53, 56, 58, 60, 65, 67–70
Quercus mongolica 8, 143, 144, 147–151, 154–157, 160–162, 171, 172, 182
Quercus mongolica var. mandshurica 100
Quercus mongolica var. tomentosa 148
Quercus urticefolia 143
Quercus serrata 143

Ranunculaceae 89–92
Ranunculus acris var. elatus 102
Ranunculus borealis 98
Ranunculus crucilobus var. chrysotrichus 102
Ranunculus hyperboreus 95
Raphiolepis umbellata 22
Rhamnaceae 89, 90
Rhamnus sp. 51, 52
Rhamnus koraiensis 146
Rhamnus taquetii 102
Rhodiola tachiroei 98
Rhododendron sp. 18, 36, 44, 50, 51, 53, 65, 78, 80, 82, 88, 126, 130, 154, 158, 162, 171, 175, 181
Rhododendron aureum 19, 26, 27, 31, 36, 38, 39, 43, 98, 105–107, 111, 119, 121, 122, 124, 125, 127, 132–138, I43, 145, 146, 148–151, 154–156, 173, 178
Rhododendron chrysanthum 154
Rhododendron dauricum 19, 20, 23, 27, 28, 31, 38–40, 42, 43, 98, 105–107, 110, 119, 121, 122, 124, 125, 127, 130, 132, 133, 135–138, 174, 178
Rhododendron fauriae var. roseum 19, 20, 23, 27, 28, 38, 39, 43
Rhododendron fauriae form. rufescens 19, 20, 22, 23 26, 27, 28, 31, 36, 38, 39, 43, 98, 105–107, 110, 119, 121, 122, 124, 125, 127, 132, 133, 135–138, 148, 150, 178
Rhododendron lapponicum 92, 95, 129
Rhododendron micranthum 19, 28
Rhododendron mucronulatum 88, 148, 149, 151, 155, 156
Rhododendron mucronulatum var. ciliatum 147–151, 156, 157, 160, 161, 172, 182
Rhododendron parvifolium 19, 26, 31, 36, 38, 92, 95, 104, 106, 116, 121, 125, 127, 132, 133, 135, 138, 139, 178
Rhododendron parvifolium var. albiflorum 100
Rhododendron parvifolium var. alpinum 99
Rhododendron redowskiana 154
Rhododendron schlippenbachii 143, 148–151, 155, 156
Rhododendron trinerve 100
Rhododendron tschonoskii 100
Rhus sp. 50, 51, 53
Rhus verniciflua 72
Ribes buriense 100
Ribes diacantha 100
Ribes distans var. typicum 98
Ribes glabellum 100
Ribes horridum 100
Ribes ussuriensis 100
Ribesiaceae 89, 90
Rodgersia podophylla var. viridis 149
Rosa sp. 50, 51, 53
Rosa koreana 98, 145, 146, 148–150
Rosa maretti 98, 145, 151, 155, 156
Rosa pimpinellifolia 98
Rosa suavis var. polyphylla 99
Rosa suavis var. rosea 145, 146
Rosaceae 89–92, 105, 106, 138
Rubiaceae 90–92
Rubus arcticus 92, 95
Rubus arcticus var. tentipetala 99
Rubus chamaemorus 92, 95
Rubus idaeus var. concolor 101
Rubus pedatus 154
Rumex maritimus 95
Ruppia maritima 95
Ruppiaceae 92

Sabina sp. 37, 80, 123
Sabina chinensis 19, 20, 38
Sabina chinensis var. horizontalis 13, 19, 22
Sabina davurica 99, 106, 113, 124, 127, 138, 179
Sabina pacifera 19, 20, 28
Sabina sargentii 19, 20, 22, 23, 28, 29, 31, 38, 39–42, 65, 92, 93, 105–107, 109, 119, 121, 122, 124–127, 130, 132–138, 151, 154–156, 178
Salicaceae 89, 90
Salix sp. 44, 50–53, 56, 58, 60, 68, 70
Salix berberifolia var. brayi 99
Salix berberifolia var. genuina 99
Salix bicarpa 101
Salix blinii 102
Salix floderusii var. glabra 98
Salix hallaiensis 98, 146, 148–151, 155
Salix hallaiensis var. longiflora 146, 148
Salix hultenii 98
Salix hultenii var. elongata 101
Salix integra 99
Salix kangensis 101
Salix maximowiczii 145, 146
Salix metaformosa 101
Salix myrtilloides var. mandschurica 98
Salix orthostemma 102

Salix pentandra var. intermedia 99
Salix pseudocrenata 100
Salix pseudolinearis 99
Salix roridaeformis 99
Salix rotundifolia var. lanata 102
Salix sericio-cinerea var. lanata 102
Salix sibirica var. brachypoda 99
Salix siuzevii 98
Salvia chanroenica 146
Salvinia sp. 50–52
Sambucus sp. 58, 60
Sambucus latipinna var. rhodoclotus 100
Sambucus velutina 102
Sanguisorba alpina 100
Sanguisorba argutipens 102
Sanguisorba hakusanensis 146, 148
Sanguisorba obtusa 100
Sanguisorba parcifolia 100
Sanguisorba parviflora 148
Sapindus sp. 50, 52, 53
Sapium japonicum 145
Sarcandra glabra 20, 28, 40, 41
Sasa sp. 81
Sasa coreana 13, 22, 23
Sasa quelpartensis 13, 20, 23
Sasakia ochnacea 20, 28, 40, 41
Sasamorpha sp. 9, 31, 35
Sasamorpha chiisanensis 13, 22, 23
Sasamorpha purpurascens var. borealis 19, 22, 38–42, 143
Sassafraga sp. 52, 53
Saussurea alpicola 37, 101
Saussurea diamantica var. longiflora 146
Saussurea mandshurica 98
Saussurea nutans 146
Saussurea rectinervis 102
Saussurea triangulata subsp. mandshurica 146
Saxifraga sp. 59, 60
Saxifraga cernua 100
Saxifraga furumii 102
Saxifraga laciniata 100
Saxifraga oblongiflora 146
Saxifraga octopetala var. albiflora 100
Saxifraga punctata 146
Saxifraga reniformis 146
Saxifragaceae 90, 91
Scabiosa japonica var. alpina 101
Scabiosa mansanensis 148
Scabiosa mansanensis form. pinnata 146
Scheuzeria palustris 92, 95
Scheuzeriaceae 90–92
Sciadopitys sp. 50–53
Scrophularia koraiensis 146
Scrophulariaceae 59, 90, 91
Scurrula yadoriki 20, 28, 40
Selaginella rupestris var. shakatanensis 98
Selaginella selaginoides 100
Selaginellaceae 90, 91
Senecio sp. 65
Senecio flanmeus 98
Senecio koreanus 146
Senecio nemorensis subsp. fuchsii 98
Serissa japonica 20
Setaria italica 75
Sieboldia sp. 65
Sieboldia coreana 101
Silene sp. 65
Silene fasciculata 102
Silene jenisseensis 146
Silene koreana 146
Silene tenuis 148
Sinoarundinaria sp. 31, 35, 81
Sinoarundinaria compressa 22
Sinoarundinaria nigra 20
Sinoarundinaria nigra var. henonis 19, 20, 22, 28, 40, 41
Sinoarundinaria pubescens 20
Sinoarundinaria reticulata 19, 20, 22, 38–41, 72
Sinomalus komarovii 101, 144
Sinomalus komarovii var. major 102
Sinomalus komarovii var. pilosa 102
Sinomalus sambucifolia 98
Smilax sp. 52, 60
Smilax china var. microphylla 58, 60
Solanaceae 76
Sorbaria stellipilla var. glanduosa 101
Sorbaria stellipilla var. typica 98, 149
Sorbus sp. 50, 51, 53
Sorbus amurensis 93, 95, 97
Sorbus amurensis var. rufea 98
Sorbus chinophylla 101
Sorbus commixta 93, 95, 97, 98, 149–151, 154–156
Sorbus matsumurae 154
Sorbus sambucifolia 154
Sphagnum sp. 59, 60
Spiraceae 89–91
Spiraea pubescens var. lasiocarpa 144, 146
Spiraea sylvestris 101
Spiraea ulmifolia 149
Stauntonia hexaphylla 20, 22, 28
Stellaria longiflora 93, 95
Stephania japonica 20, 40, 41
Styrax sp. 43, 52, 53
Styrax japonica 147
Styrax obassia 143
Suaeda maritima 92, 95
Symplocarpus buchensis 100
Syringa sp. 44
Syringa elatum 148
Syringa wolfii 148

Taxaceae 58, 60, 89, 90, 105, 106, 138
Taxus sp. 34, 37, 44, 56, 58, 60, 77, 80–82, 85, 123, 175
Taxus baccata 130
Taxus caespitosa 102, 105, 106, 116, 119, 124, 125, 133–135, 138, 139, 144, 148–151, 154–157, 162, 171, 179
Taxus cuspidata 19, 20, 22, 23, 28, 29, 31, 38–42, 88, 98, 106, 107, 119, 121, 122, 124, 129, 130, 132–138, 143, 148, 150, 178
Taxus cuspidata var. latifolia 19, 98, 105, 106, 111, 124, 138, 178
Ternstroemia sp. 50

Ternstroemia mokof 20, 28, 40, 41
Textoria sp. 43
Textoria morbifera 20, 23
Thalictrum coreanum 146
Thalictrum rochebrunianum var. glandisepalum 146
Thea sp. 35
Thea sinensis var. bohea 19, 22, 28, 40–42, 72
Therorhodion redawskianum 99
Thuja sp. 34, 43, 44, 55, 56, 59, 60, 78, 80–82, 85, 88, 123, 175
Thuja koraiensis 13, 19, 20, 22, 23, 28, 31, 34, 35, 38–40, 101, 105–107, 111, 112, 119, 121, 122, 124, 133–139, 143–152, 154–157, 161–163, 171, 172, 179, 181, 182
Thujopsis sp. 52
Thymus quinquecostatus var. ibukiensis 98
Tilia sp. 43, 44, 50–53, 56, 58, 60, 65, 68, 70
Tilia amurensis var. barbigera 143
Tilia megaphylla 144, 146
Tilia rufa 146
Tilia taquetii 146
Tofieldia fauriei 102
Tofieldia nutans var. rubescens 100
Torreya nucifera 20, 23, 35, 40, 41, 72
Trachelospermum asiaticum var. intermedium 19, 28, 40, 41
Trachelospermum asiaticum var. oblanceolatum 20, 28
Trachelospermum jasminoides var. pubescens 20, 28
Triadica japonica 147
Trientalis europaea var. arctica 98
Trientalis europaea var. eurasiatica 98
Triglochin maritimum var. asiaticum 93, 95, 97
Tripterygium regelii 149–151, 155, 156
Trisetum spicatum 100
Triticum aestivum 75
Trollius hondoensis 100
Trollius japonicus 100
Trollius macropetalus 100
Tsuga sp. 59, 60, 64, 175
Turritis glabra 92, 93
Typha sp. 59, 60
Typha latifolia 95
Typhaceae 65, 68, 92

Ulmaceae 89
Ulmus sp. 43, 50, 51, 53, 58, 60, 66–68, 70
Ulmus mandshurica 100
Umbelliferae 65, 68
Urtica sp. 59, 60
Utricularia intermedia 95

Vaccinium sp. 43, 44, 65, 77, 80, 148, 158
Vaccinium bracteatum 20, 23
Vaccinium koreanum 146, 148–151, 155, 156
Vaccinium ovalifolium 154
Vaccinium uliginosum 88, 95, 98, 129, 145–148, 151, 154, 155
Vaccinium uliginosum var. alpinum 92
Vaccinium uliginosum var. krushianum 148
Vaccinium vitis-idaea 130, 154, 178
Vaccinium vitis-idaea var. alpinum 93
Vaccinium vitis-idaea var. genuinum 19, 26, 43, 99, 100, 105, 106, 116, 124, 125, 127, 138
Vaccinium vitis-idaea var. minus 19, 20, 23, 27, 28, 31, 38, 39, 43, 92, 95, 106, 107, 111, 112, 119, 121, 122, 124, 125, 127, 132–138, 177, 178
Veratrum glandiflorum 155
Veratrum maximowiczii var. coreanum 100
Veratrum versicolor form. albidum 98
Veronica stelleri var. longistyla 100
Veronica subincanovelutina 101
Viburnum sp. 36, 50–53, 81
Viburnum awabuki 40, 42
Viburnum koreana 101
Viburnum wrightii var. stippellatum 101
Vicia venosa var. cuspidata 144, 146
Vicia venosissima 146
Viola biflora 100
Viola cressa 100
Viola koraiensis 102
Viola selkirkii 95, 148
Viola sylvestriformis 100
Viola xanthopetala 148
Violaceae 90–92
Viscum sp. 31, 35
Viscum coloratum 31
Viscum coloratum var. lutescens 19, 20, 27, 28, 38–40, 42
Viscum coloratum var. ruboroaurantiacum 20, 28
Vitex rotundifolia 20, 22, 28, 41, 42
Vitis sp. 52, 53

Waldstenia ternata 146
Weigelia subsessilis 146
Woodsia glabella 100
Woodsia ilvenis 100
Woodsia saitonana 100

Xylosma sp. 50
Xylosma apathis 20, 28, 40, 41

Zanthoxylum sp. 35, 43
Zanthoxlyum planispinum 20, 40, 41
Zea mays 75
Zelkova sp. 50–53
Zizyphus jujuba 72
Zostera marina 95
Zosteraceae 92
Zoysica sp. 9
Zygadenus sibirica 100

Geobotany

1. J.B. Hall and M.D. Swaine (eds.): *Distribution and Ecology of Vascular Plants in a Tropical Rain Forest.* Forest Vegetation in Ghana. 1981 ISBN 90-6193-681-0
2. W. Holzner and M. Numata (eds.): *Biology and Ecology of Weeds.* 1982 ISBN 90-6193-682-9
3. N.J.M. Gremmen: *The Vegetation of the Subantarctic Islands Marion and Prince Edward.* 1982 ISBN 90-6193-683-7
4. R.C. Buckley (ed.): *Ant-Plant Interactions in Australia.* 1982 ISBN 90-6193-684-5
5. W. Holzner, M.J.A. Werger and I. Ikusima (eds.): *Man's Impact on Vegetation.* 1983 ISBN 90-6193-685-3
6. P. Denny (ed.): *The Ecology and Management of African Wetland Vegetation.* 1985 ISBN 90-6193-509-1
7. C. Gómez-Campo (ed.): *Plant Conservation in the Mediterranean Area.* 1985 ISBN 90-6193-523-7
8. J.B. Faliński: *Ecological Studies in Białowieza Forest.* 1986 ISBN 90-6193-534-2
9. G.A. Ellenbroek: *Ecology and Productivity of an African Wetland System.* The Kafue Flats, Zambia. 1987 ISBN 90-6193-638-1
10. J. van Andel, J.P. Bakker and R.W. Snaydon (eds.): *Disturbance in Grasslands.* Causes, Effects and Processes. 1987 ISBN 90-6193-640-3
11. A.H.L. Huiskes, C.W.P.M. Blom and J. Rozema (eds.): *Vegetation Between Land and Sea.* Structure and Processes. 1987 ISBN 90-6193-649-7
12. G. Orshan (ed.): *Plant Pheno-morphological Studies in Mediterranean Type Ecosystems.* 1988 ISBN 90-6193-656-X
13. B. Dell, J.J. Havel and N. Malajczuk (eds.): *The Jarrah Forest.* A Complex Mediterranean Ecosystem. 1988 ISBN 90-6193-658-6
14. J.P. Bakker: *Nature Management by Grazing and Cutting.* 1989 ISBN 0-7923-0068-8
15. J. Osbornová, M. Kovářová, J. Lepš and K. Prach (eds.): *Succession in Abandoned Fields.* Studies in Central Bohemia, Czechoslovakia. 1990 ISBN 0-7923-0401-2
16. B. Gopal (ed.): *Ecology and Management of Aquatic Vegetation in the Indian Subcontinent.* 1990 ISBN 0-7923-0666-X
17. B.A. Roberts and J. Proctor (eds.): *The Ecology of Areas with Serpentinized Rocks.* A World View. 1991. ISBN 0-7923-0922-7
18. J.T.A. Verhoeven (ed.): *Fens and Bogs in the Netherlands.* Vegetation, History, Nutrient Dynamics and Conservation. 1992 ISBN 0-7923-1387-9
19. Woo-seok Kong and D. Watts: *The Plant Geography of Korea.* With an Emphasis on the Alpine Zones. 1993 ISBN 0-7923-2068-9
20. R. Aerts and G.W. Heil (eds.): *Heathlands.* Patterns and Processes in a Changing Environment. 1993 ISBN 0-7923-2094-8

KLUWER ACADEMIC PUBLISHERS – DORDRECHT / BOSTON / LONDON

MIX
Papier aus verantwortungsvollen Quellen
Paper from responsible sources
FSC® C105338

If you have any concerns about our products,
you can contact us on
ProductSafety@springernature.com

In case Publisher is established outside the EU,
the EU authorized representative is:
Springer Nature Customer Service Center GmbH
Europaplatz 3, 69115 Heidelberg, Germany

Printed by Libri Plureos GmbH
in Hamburg, Germany